Staphylococcus aureus

Akio Nakane • Krisana Asano
Editors

Staphylococcus aureus

Interplay between Bacteria and Hosts

Springer

Editors
Akio Nakane
Hirosaki University Graduate
School of Medicine
Hirosaki, Aomori, Japan

Krisana Asano
Department of Microbiology and
Immunology
Hirosaki University Graduate
School of Medicine
Hirosaki, Aomori, Japan

ISBN 978-981-99-9427-4 ISBN 978-981-99-9428-1 (eBook)
https://doi.org/10.1007/978-981-99-9428-1

This Springer imprint is published by the registered company Springer Nature Singapore Pte Ltd.
The registered company address is: 152 Beach Road, #21-01/04 Gateway East, Singapore 189721, Singapore

Paper in this product is recyclable.

Preface

Staphylococcus aureus is an opportunistic human pathogen that colonizes the mucosal surfaces. It is the causative agent of various serious acute and chronic infections. The anterior nares are the primary reservoirs of *S. aureus* in humans. Extranasal colonization of *S. aureus* also takes place in several locations, including the skin, rectum, axillae, vagina, pharynx, and gastrointestinal tract. *S. aureus* causes numerous infectious diseases including skin infections, surgical and trauma wound infections, urinary tract infections, gastrointestinal tract infections, pneumonia, osteomyelitis, endocarditis, thrombophlebitis, mastitis, meningitis, infections on indwelling medical devices, toxic shock syndrome, and septicemia. *S. aureus* produces various virulence factors such as capsule, coagulase, lipase, hyaluronidase, protein A, fibrinogen binding proteins, fibronectin binding proteins and secreted toxins such as enterotoxins, toxic shock syndrome toxin-1, Panton-Valentine leukocidin, hemolysins, and phenol-soluble modulins. Incidence of severe infectious diseases by multidrug-resistant *S. aureus* including methicillin-resistant *S. aureus* (MRSA) has been the serious problem in hospitals and community. Alternatively, a troublesome problem is chronic infection with *S. aureus*. Adaptation to intracellular persistence represents an important strategy of *S. aureus* to evade host defenses and contributes to chronic infection. It is possible that offense and defense responses are continuously executed between *S. aureus* and host to establish and maintain commensalism for *S. aureus*.

This book provides topics concerning mechanisms of *S. aureus* infection and unique strategies of interplays between the bacterium and hosts.

Hirosaki, Aomori, Japan — Akio Nakane
Hirosaki, Aomori, Japan — Krisana Asano

Contents

Chapter 1
Colonization and Persistence Strategies of *Staphylococcus aureus*

Karsten Becker

Abstract *S. aureus* is a prime example of a multitalented opportunistic pathogen that possesses a broad array of virulence factors, not infrequently redundant and multifunctional. Thus, the many lineages of the *S. aureus* complex are well-equipped for the colonization of various mammalian hosts, including humans, and, after breaching the barrier of the skin and the mucous membranes, for infectious processes. The capability to adhere on biotic and inserted abiotic surfaces, to form complex biofilms and to become internalized plays a key role in the pathogenesis of *S. aureus* infections. The spectrum of infections ranges from mild to severe, even life-threatening infections including those associated with inserted foreign bodies. Typically, *S. aureus* is known as an extracellular pathogen causing acute aggressive courses of infection. However, the pathogen is also able to persist intracellularly in some kinds of cells including epithelial cells after internalization by formation of an adapted phenotype designated "small colony-variant". The intracellular lifestyle leads to chronic infections with persistent and relapsing course of disease challenging the routine microbiological laboratory and the treating physician.

Keywords Biofilm · Colonization · Host cells · Internalization · Intracellular persistence · Invasion · Pathogenicity · Small colony variants · *Staphylococcus aureus* · Virulence factors

Abbreviations

agr	Accessory gene regulatory
Akt	Protein kinase B
Bap	Biofilm-associated protein
CF	Cystic fibrosis

K. Becker (✉)
Friedrich Loeffler-Institute of Medical Microbiology, University Medicine Greifswald, Greifswald, Germany
e-mail: karsten.becker@med.uni-greifswald.de

A. Nakane, K. Asano (eds.), *Staphylococcus aureus*,
https://doi.org/10.1007/978-981-99-9428-1_1

CLABSI	Central line-associated bloodstream infection
Clf	Clumping factor
CoNS	Coagulase-negative staphylococci
CRBSI	Central venous catheter-related bloodstream infection
CVC	Central venous catheter
CWA	Cell wall-anchored
Eap	Extracellular adherence protein
Ebh	Extracellular matrix-binding protein homologue
Ebp	Elastin-binding protein
ECM	Extracellular matrix
eDNA	Extracellular DNA
Efb	Extracellular fibrinogen binding protein
Emp	Extracellular matrix binding protein
EPS	Extracellular polymeric substance
FAK	Focal adhesion kinase
FnBP	Fibronectin-binding protein
GroP	Poly-glycerolphosphate
HemA	Glutamyl-tRNA reductase
HemG	Protoporphrinogen oxidase
Hla	Alpha-hemolysin
LPXTG	Leu-Pro-X [any amino acid]-Thr-Gly
LTA	Lipoteichoic acid
MenB	1,4-Dihydroxy-2-raphthoyl-CoA synthase
MRSA	Methicillin-resistant *Staphylococcus aureus*
MSCRAMM	Microbial surface component recognizing adhesive matrix molecule
NEAT	Near iron transporter
PI3K	Phosphatidylinositide 3-kinase
PIA	Polysaccharide intercellular adhesin
PNAG	Poly-N-acetylglucosamine
RboP	Poly-ribitolphosphate
SasG	*Staphylococcus aureus* surface protein G
SCV	Small colony variant
SdrD	Serine aspartate repeat-containing protein D
SEC	Secretory apparatus
SraP	Serine-rich adhesin for platelet
TA	Teichoic acid
ThyA	Thymidylate synthase
TMP-SMX	Trimethoprim-sulfamethoxazole
TNase	Thermonuclease
WTA	Wall teichoic acid

The clinically highly significant, coagulase-positive species *Staphylococcus aureus,* effectively described in 1884, meets the typical requirements for the definition of an opportunistic pathogen. This is evident from his ability to colonize mammalian hosts including humans, but also to cause mild to severe, even life-threatening infections, once it has overcome the natural barrier built by the skin and mucous membranes [1]. Its pathogenic capacity distinguishes this species from typical coagulase-negative staphylococci (CoNS), which are characterized as non-pathogenic to opportunists leading to rather mild courses of infection, mostly associated with the insertion of foreign bodies [2].

The recent taxonomic changes concerning a so-called "*S. aureus* complex" including (1) the delimitation of *S. argenteus* and *S. schweitzeri* [3] and (2) the description of *S. roterodami* and *S. singaporensis* [4, 5] as novel species within this complex is not addressed within this chapter. The ancestral *S. argenteus* lineage and *S. singaporensis* appear to be broadly similar to *S. aureus* in terms of pathogenicity and clinical infection features for the human host [5, 6]. For *S. roterdami*, studies on pathogenicity are lacking [4]. *S. schweitzeri* seems to be associated primarily with African wildlife [6] and the former *S. aureus* subspecies *anaerobius* [7], i.e. biovar. Anaerobius, is limited to sheep as host [8].

In terms of its relationship to the infected host, *S. aureus* can be considered as "Janus-faced" pathogen: On the one hand, this microorganism can act as a typical extracellular microorganism causing aggressive courses of acute infections, but on the other hand, *S. aureus* may exercise an intracellular lifestyle characterized by relapsing chronic infections. The intracellular persistence has been associated with the expression of the small colony-variant (SCV) phenotype [9, 10].

S. aureus harbors an extensive "toolbox" for adhesion, invasion and immune evasion processes (for evasion, please refer to respective literature, e.g [11, 12]. While the possession of respective virulence factors is in part strain-dependently and certain differences in virulence may occur, all isolates of *S. aureus* carry a core genome which basically renders this species a potential pathogen, thus, findings of *S. aureus* should never be ignored in the clinical setting.

1.1 *S. aureus* "Toolbox" for Adhesion and Invasion Processes

The attachment of bacterial species including staphylococci to biotic or abiotic surfaces can occur either directly or via so-called bridging molecules comprising extracellular matrix (ECM) and plasma proteins. The particular "success story" of *S. aureus* as colonizer of biological surfaces such as the skin and mucous membranes of the human host and many animal surfaces as well as colonizer of or at least as survivor on abiotic surfaces in the hospital environment and on foreign material inserted in patients is based on its equipment with a multitude of virulence factors. Besides exhibiting distinct roles, these factors are often multifunctional, e.g. combining enzymatic and adhesion activities, such as autolysins [13]. Some of the virulence factors must necessarily act together for optimal pathogenicity as shown e.g. for the secreted protein Eap (extracellular adherence protein) and the cell wall-anchored (CWA) protein SasG (*S. aureus* surface protein G) in biofilm formation [14]. Moreover, this pathogen is equipped with virulence factors showing redundant functions such as the secretable expanded repertoire adhesion molecules (SERAMs) [15].

Inevitably, multifunctionality and redundancy result in major research challenges in determining the real impact of a given virulence factor within the microbe's array of pathogenic capacities. Moreover, it aggravates e.g. the selection of candidates for vaccination strategies against *S. aureus*. Finally, in particular multifunctional factors, often referred to as "moonlighting" proteins [16], have repeatedly caused some confusion in nomenclature by re-naming in the course of determining their real functions and/or characterization their structure (see Table 1.1).

The exceptionally broad-ranged equipment renders *S. aureus* to a "multi-talented" microorganism allowing the pathogen to cope with a wide variety of environmental conditions by developing different lifestyles: (1) planktonic lifestyle in fluids, (2) sessile lifestyle on biotic and abiotic surfaces including formation of a

Table 1.1 Cell wall-anchored (CWA) proteins expressed by *S. aureus* (selection)

CWA[a]	CWA family	Ligand specificity	(Presumed) function	Reference
Biofilm-associated protein (Bap)	MSCRAMM	Gp96 chaperone	Intercellular adhesion, attachment to abiotic surfaces and epithelial cells, biofilm formation; internalization prevention	[232–234]
Bone sialoprotein-binding protein (Bbp)[b]	MSCRAMM	Bone sialoprotein, fibrinogen	Adhesion to ECM	[235–238]
Clumping factor A (ClfA)	MSCRAMM	Fibrinogen, von Willebrand factor-binding protein, annexinA2, complement factor I	Adhesion to immobilized fibrinogen; immune evasion by binding soluble fibrinogen and degradation of C3b	[93, 239–242]
Clumping factor B (ClfB)	MSCRAMM	Fibrinogen, cytokeratins 8 and 10, loricrin	Nasal colonization by adhesion to desquamated epithelial cells, platelet aggregation	[44, 243–247]
Extracellular adenosine synthase (AdsA)[c]	–	–	Innate immune evasion by promotion of survival in neutrophils via hindering the oxidative burst	[248–250]
Fibronectin-binding protein A (FnBPA)	MSCRAMM	Fibronectin, fibrinogen, elastin	Adhesion to ECM, activation of human platelets, invasion, biofilm promotion	[82, 251–256]
Fibronectin-binding protein B (FnBPB)	MSCRAMM	Fibronectin, elastin	Adhesion to ECM, activation of human platelets, invasion, biofilm promotion	[82, 253, 256–258]

Table 1.1 (continued)

CWA[a]	CWA family	Ligand specificity	(Presumed) function	Reference
Iron-regulated surface protein (IsdA)[d]	NEAT motif	Hemoglobin, hemin, human transferrin, fetuin, fibronectin, fibrinogen, cytokeratin 10, loricrin	Heme uptake, iron acquisition, adhesion to desquamated epithelial cells; lactoferrin resistance, protease inhibitor; abscess formation	[259–267]
Iron-regulated surface protein B (IsdB)	NEAT motif	Hemoglobin, hemin, platelet integrin $\alpha_{IIb}\beta_3$	Heme uptake, iron acquisition, promotion of adhesion to platelets, invasion (non-phagocytic cells); abscess formation	[264, 266, 268–272]
Iron-regulated surface protein C (IsdC)	NEAT motif	Hemin	Heme scavenging	[264, 267]
Iron-regulated surface protein H (IsdH)[e]	NEAT motif	Hemoglobin, haptoglobulin-hemoglobin complex, von Willebrand factor, vitronectin	Heme uptake, iron acquisition, acceleration of C3b degradation, adherence to endothelial cells	[268, 269, 273–276]
Protein A (SpA)	Three-helical bundle (domains B, E, and Z)	Immunoglobulin G (Fcγ region and Fab portion), IgM, platelet membrane protein gC1qR/p33, V_H3 type B cell receptor, von Willebrand factor, TNFR-1	Blocking opsonophagocytosis, killing of B cells (apoptotic collapse), adherence to platelets, cell aggregation, abscess formation	[266, 277–287]
S. aureus collagen adhesin (Cna)	MSCRAMM	Collagen type I, complement factor C1q, laminin	Tissue (collagen-rich) and basement membrane adhesion, prevention of complement activation (classical pathway)	[288–291]
S. aureus surface protein C (SasC)	G5–E repeat family	Unknown ligands	Cell aggregation and biofilm formation	[292]
S. aureus surface protein G (SasG)[f]	G5–E repeat family	Unknown ligands	Adhesion to desquamated epithelial cells and glycolipids; biofilm formation and thickness, intercellular adhesion, internalization prevention	[14, 248, 293–296]
S. aureus surface protein X (SasX)[g]	G5–E repeat family	Unknown ligands	Adhesion to host tissue (human nasal epithelial cells), cell aggregation and biofilm formation	[297]

(continued)

Table 1.1 (continued)

CWA[a]	CWA family	Ligand specificity	(Presumed) function	Reference
Serine-aspartate repeat-containing protein C (SdrC)	MSCRAMM	Hydrophobic surfaces, β-neurexin, SdrC homophilic interactions	Nasal colonization by adherence to desquamated epithelial cells, biofilm formation	[298–301]
Serine-aspartate repeat-containing protein D (SdrD)	MSCRAMM	Desmoglein-1, further yet unknown ligands	Nasal colonization by adherence to desquamated epithelial cells, abscess formation	[92, 266, 300, 302, 303]
Serine-aspartate repeat-containing protein E (SdrE)	MSCRAMM	Complement factor H	Platelet aggregation, complement evasion	[283, 300, 304, 305]
Serine-rich surface protein (SraP)[h]	Legume lectin-like, cadherin-like	Platelets, epithelial cells (sialylated receptors, trisaccharide ligand, N-acetylneuraminic acid)	Platelet binding; mediating intraspecies interaction and promoting bacterial aggregation	[306–308]

[a]Abbreviations: *CWA* cell wall-anchored, *ECM* extracellular matrix, *MSCRAMM* microbial surface component recognizing adhesive matrix molecules, *NEAT* near iron transporter
[b]Member of the staphylococcal Sdr family
[c]Originally described as SasH
[d]Originally described as Seg7, also described as FrpA and StbA
[e]Also referred to as staphylococcal haptoglobin receptor A (HarA)
[f]A SasG homologue in MRSA has been designated as plasmin-sensitive surface protein (Pls)
[g]Has been linked to the epidemic success of hospital-associated MRSA clones (ST239 and others), in particular in Asia [297]
[h]*Streptococcus gordonii* GspB homolog; surface-exposed serine-rich repeat glycoprotein (SRRP), originally described as SasA

biofilm environment and (3) intracellular lifestyle by formation of the SCV phenotype (Fig. 1.1).

From a pathogenic view, *S. aureus* is able to pursue the following strategies: (1) to colonize its host via adherence to various surfaces and to generate a biofilm, (2) to invade into the host environment and (3) to evade the immune response. If internalized by host cells, two strategies can be pursued by *S. aureus*: an aggressive course leading to heavy inflammation and host cell destruction manifesting as acute infectious disease entities versus a way of adaptation to a permanent intracellular

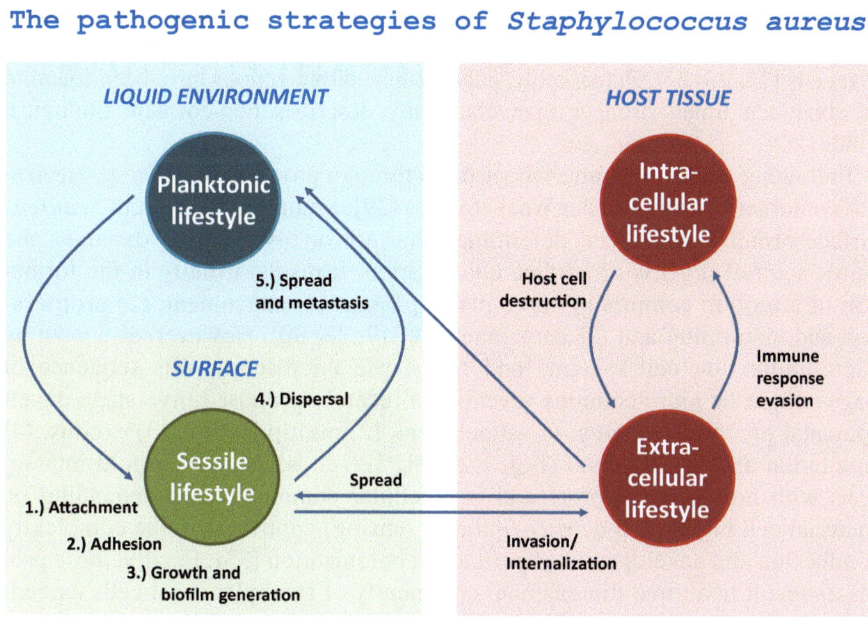

Fig. 1.1 The pathogenic strategies of *S. aureus* depending on the environmental conditions. Change (left) between a planktonic lifestyle in a fluid environment and the colonization of biotic and abiotic surfaces with biofilm formation (sessile lifestyle) both represented by an extracellular lifestyle; change (right) between the extracellular lifestyle displaying the "normal" wildtype phenotype and an intracellular lifestyle with the formation of the small-colony variant (SCV) phenotype

lifestyle leading to persistence within the host and a chronic course of infection, often with relapsing episodes of inflammation.

1.1.1 Adhesion and Biofilm Generation

The transition of a staphylococcal cell from a free-living planktonic microorganism to a sessile, surface-attached lifestyle are triggered by environmental factors, physicochemical forces, properties of the target surface, and the bacterial cell body and its surface structures including the production of adhesin molecules and small molecule signaling (Fig. 1.1) [17–19]. In particular, the high affinity of *S. aureus* to mammalian ECM and plasma proteins, such as fibrinogen, fibronectin and some collagens and cytokeratins, is one of the major underlying pathogenic principles what explains its success as colonizing and infectious agent [20–22]. The adherence to ECM and other proteins is an essential step in biofilm formation and invasion of host cells [21, 23–27]. Remarkably, adhesin-ligand interactions have been found to be about ten times stronger than classically describes non-covalent biological bonds [28].

Following weak attachment to surfaces through physical forces (e.g. electrostatic interactions and van der Waals forces) [29], a number of different *S. aureus* surface proteins have been determined that are involved in the dynamic and multi-factorial process of surface colonization. It results usually in the formation of a biofilm comprising three major phases: (1) attachment, (2) proliferation and maturation and (3) detachment [2, 19, 23, 30]. However, as shown by microfluidic flow-cell systems and time-lapse microscopy, this sequence of stages might be more complex revealing a recently proposed, five-stage developmental process including: (1) attachment, (2) multiplication, (3) exodus, (4) maturation and (5) dispersal (Fig. 1.2) [31, 32]. In addition, manifold interactions with host cell receptors and intercellular communication depending on bacterial cell population density (quorum sensing) contribute to the complexity of adhesion and development of permanent colonization [33]. Finally, these processes result in a three-dimensional community of staphylococcal cells embedded in an extracellular matrix that was originally called "slime" and is now often referred to as extracellular polymeric substance (EPS) [34]. The EPS of staphylococci consists mainly of polysaccharides, proteins and/or extracellular DNA (eDNA) [35–39].

The ability of *S. aureus* to colonize various host surfaces and, from an evolutionary perspective, more recently abiotic surfaces implanted in host tissues, has made it such successful colonizer. However, this necessitates to address the different conditions of the various host surfaces and environments such as the skin-like

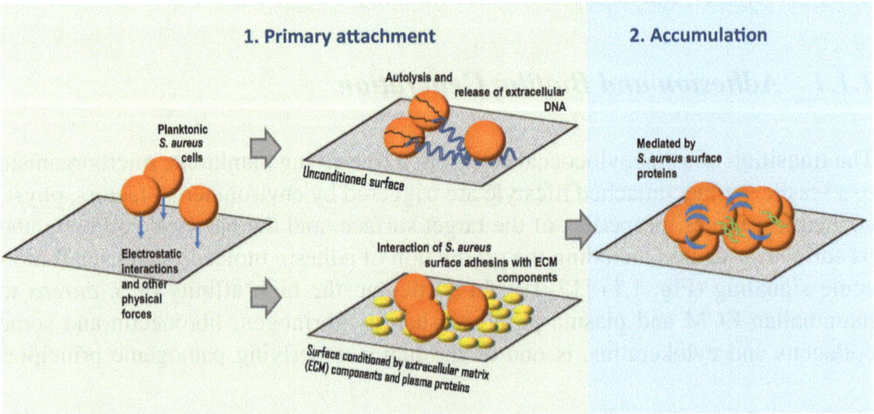

Fig. 1.2 Attachment (left) of planktonic *S. aureus* cells onto unconditioned and extracellular matrix (ECM)-conditioned surfaces and accumulation (right) of *S. aureus* cells mediated by the pathogen's surface proteins

squamous epithelium of the anterior nares containing glands and vibrissae as well as the respiratory epithelium of the deeper nasal cavity described as pseudostratified columnar ciliated epithelium [40]. Also, *S. aureus* is efficaciously able to cope with the specific conditions of the lung of cystic fibrosis (CF) patients where it is known as notorious colonizer and invasive agent [41, 42].

Staphylococci including *S. aureus* possess a broad armamentarium of factors mediating direct or indirect attachment to biotic and abiotic surfaces which comprise proteinaceous and non-proteinaceous adhesins. However, the CWA proteins are considered to be *S. aureus´* "main tool" for colonizing and acting as pathogen. These proteins foster attachment to surfaces that have been conditioned with host plasma proteins. They are covalently attached to the peptidoglycan, whereby anchorage is facilitated by characteristic sorting signals at their carboxyl termini [43, 44]. In addition to the sorting signal, which comprises a conserved Leu-Pro-X [any amino acid]-Thr-Gly (LPXTG) sortase cleavage motif, the open reading frames of the CWA proteins possess a secretory SEC (secretory apparatus)-dependent signal sequence at the amino terminus as well as two further subdomains (a stretch of positively charged residues and the hydrophobic domain) at the C terminus to retain the protein in the membrane during secretion as pre-condition for binding of the sortase and their acting as transpeptidase [21, 43, 45, 46].

Strain- and growth phase-dependent and affected by environmental conditions (e.g., access to iron), *S. aureus* expresses up to 20 or more different CWA virulence factors (see Table 1.1 and Chap. 2). CWA proteins have been classified into five main groups on the basis of structural motifs: (1) the microbial surface component recognizing adhesive matrix molecules (MSCRAMM) family as re-defined by Foster et al. [21, 47], (2) the near iron transporter (NEAT) motif family, (3) the three-helical bundle motif protein A, (4) the G5-E repeat family and (5) the legume-lectin-, cadherin-like domain protein [48]. Well-studied MSCRAMMs include e.g. the clumping factors (Clfs) A and B, the fibronectin-binding proteins (FnBPs) A and B and the biofilm-associated protein (Bap). As an example, the major MSCRAMM ClfA plays a key role in bloodstream infections by binding to platelets via the platelet-specific $\alpha_{IIb}\beta_3$ integrin and rapidly inducing thrombus formation. In the presence of fibrinogen, ClfA binds to endothelial cells through $\alpha_V\beta_3$ integrin, which results in endothelial cell apoptosis and increased permeability [49]. Integrins are transmembrane α/β heterodimers that mediate cellular adhesion and other processes requiring cell-matrix and cell-cell interactions [50].

Not all of the many proteinaceous adhesins of *S. aureus* are covalently attached to the cell wall (Table 1.2). Besides the autolysin/adhesins group, most of the studied molecules of those non-covalently-linked surface-associated proteins are members of the SERAM group [15]. They include e.g. the extracellular matrix binding protein (Emp), the extracellular fibrinogen binding protein (Efb), and the extracellular adherence protein (Eap, also known as Map and p70, respectively). The membrane-spanning proteins constitute a further group of non-covalently-linked adhesins including the extracellular matrix-binding protein homologue (Ebh), a giant 1.1 mDa fibronectin-binding protein described for *S. aureus* [51]. It was found that the secretion of full-length Ebh into the cross wall supports staphylococcal

Table 1.2 Non covalently-linked surface-associated and secreted proteins expressed by *S. aureus* (selection)

Protein	Ligand specificity	(Presumed) function	Reference
Autolysins/ adhesins of *S. aureus* (Aaa)[a]	Fibronectin, fibrinogen, vitronectin	Attachment to host factor-coated surfaces and host tissue; enzymatic activity as peptidoglycan hydrolase with involvement in bacterial cell-wall turnover, cell division and separation and antibiotic-induced lysis of bacteria	[13, 60, 309]
Atl	Fibronectin, fibrinogen, vitronectin, Hsc70	Attachment to abiotic surfaces; biofilm formation and thickness; internalization; release of cytoplasmic proteins; enzymatic activity as peptidoglycan hydrolase with involvement in cell separation and cell wall turnover and penicillin-induced lysis of the bacterial cells	[58, 59, 310–313]
Efb homologous protein (Ehp)[b]	C3 (C3d domain)	Inhibition of the alternative complement activation pathway	[314, 315]
Extracellular adherence protein (Eap)[c]	Fibronectin, fibrinogen, vitronectin, collagen (supramolecular structure-dependent), ICAM-1	Adhesion to abiotic and eukaryotic cell surfaces; biofilm formation; uptake enhancement by eukaryotic cells, inhibition of binding of neutrophils to the epithelium and their transmigration, decrease in phagocytic activity; abscess formation; angiostatic activity	[14, 15, 266, 316–323]
Extracellular fibrinogen-binding protein (Efb)	Fibrinogen, platelets	Promotion of platelet binding of fibrinogen; anti-thrombotic activity by inhibition of platelet activation; immunosuppressive role by blocking complement activation	[324–327]
Extracellular matrix protein-binding protein (Emp)	Fibronectin, fibrinogen, vitronectin, collagen	(low-iron-induced) biofilm formation; abscess formation	[266, 328–330]
Extracellular matrix-binding protein homologue (Ebh)	Fibronectin	Cellular adhesion; cell growth, envelope assembly; enabling complement resistance	[51, 52]
Elastin-binding protein of *S. aureus* (Ebps)	Elastin	Cell growth regulation	[53, 331, 332]

[a]Also known as N-acetylmuramyl-L-alanine amidase Sle1
[b]Also known as extracellular complement-binding protein (Ecb)
[c]Also known as MHC class II analogous protein (Map) and P70

replication by ensuring the stability of the envelope and protection against host defenses by complement resistance enabling [52]. The elastin-binding protein of *S. aureus* (Ebps) is another integral membrane protein [53]. As the name implies,

Ebps binds elastin, which is a major ECM component. Another class of non-covalently bound proteins is represented by the autolysin/adhesins, which are associated with the surface by hydrophobic or ionic interactions. Characteristically, they are multifunctional and exert both enzymatic (peptidoglycan hydrolases) and adhesive functions [54, 55]. These staphylococcal autolysins are mainly involved in cell-wall turnover, cell division, cell separation and daughter cell formation [56, 57]. The moonlighting *S. aureus* major autolysin Atl also mediates the attachment of the pathogen to polystyrene and subsequent biofilm formation and it is involved in *S. aureus* internalization [58, 59]. Aaa, the so-called autolysin/adhesin of *S. aureus,* represents a further member of this group, which displays bacteriolytic activity against staphylococcal cells and mediates adherence to various ECM proteins and, therefore, might promote the colonization of host ECM and tissue by *S. aureus* and [13, 60].

The adherence-mediating arsenal of *S. aureus* comprises also non-proteinaceous adhesins. These include complex teichoic acid (TA) glycopolymers, which, in addition to interactions with host or bacteriophage receptors, are involved in the protection against environmental stress and to prevent the passage of harmful molecules through the peptidoglycan layers [61]. Further crucial roles of them include the control of enzyme activities directing morphogenesis or cell division and of cation concentration in the cell envelope [61]. TAs represent up to 60% of the total cell wall mass and are composed of alternating phosphate and ribitol groups in the case of wall teichoic acids (WTAs) or glycerol groups in the case of the lipoteichoic acids (LTAs), which are substituted with D-alanine and N-acetylglucosamine [62, 63]. The highly charged cell wall polymers are of importance for the colonization of abiotic surfaces because it depends on the charge of the teichoic acids [62, 64]. While many CoNS including *S. epidermidis* produce poly-glycerolphosphate (GroP)-WTA, almost all *S. aureus* strains produce poly-ribitolphosphate (RboP)-WTA. In addition to the CWA proteins, the highly hydrophilic WTA is essential for nasal colonization. It facilitates the pathogen's efficient adherence to human nasal epithelial cells and provide resistance against hydrophobic antimicrobial fatty acids from human skin [65–67]. Moreover, this polymer is also involved in adherence to endothelial cells [68]. The RboP-WTA type has been recently shown as an essential factor for the pathogenicity of *S. aureus* since the *S. aureus*-type WTA renders *S. epidermidis* clones more virulent in invasive infections if transferred [69].

Another non-proteinaceous adhesins occurring also in many, but not all *S. aureus* strains, is the polysaccharide intercellular adhesin (PIA), which has been first described for *S. epidermidis* in the context of the slimy substance produced in biofilms [38, 70, 71]. The PIA exopolysaccharide is biosynthesized by the gene products of an operon composed of the regulatory *icaR* gene and the *icaADBC* genes resulting in a linear β-1,6-linked N-acetylglucosaminoglycan [38, 71–73]. The alternative designation as poly-N-acetylglucosamine (PNAG) is less accurate because 15–20% of the N-acetylglucosaminyl residues are non-N-acetylated [74, 75]. This partial deacetylation renders the polysaccharide positively charged, which governs its function as an intercellular adhesin by electrostatic attraction of the negatively charged teichoic acid to the staphylococcal cell surface [76]. In addition to its contribution to adhesion, PIA inhibits phagocytosis by shielding the bacterial surface from phagocyte attacks and by contributing to the biofilm complex [77]. However, possible *icaADBC*-independent mechanisms of biofilm development have been reported for clinical *S. aureus* isolates [78].

Staphylococcal cell lysis promoted by *cidA*-encoded murein hydrolase and followed by DNA release is a further critical factor of staphylococcal biofilm development and ubiquitous part of the biofilm architecture of staphylococci [79]. The tight control of cell lysis and eDNA release during biofilm development is mediated by opposing activities of the Cid-Lrg regulatory system whereas the staphylococcal thermonuclease (TNase) is involved in eDNA degradation to foster biofilm dispersal [80, 81].

1.1.2 Internalization by Host Cells

The capability to invade host cells is a common feature of *S. aureus* with strain-specific expression, but the subsequent fate could hardly be more different. The best known and most frequent course of infection is the one in which the pathogen acts in an aggressive manner which leads to the destruction of the invaded host cell. This way of infection is typically associated with the vast variety of acute infections caused by *S. aureus*. However, the internalization of *S. aureus* by host cells may also result in a persistent cellular infection by dynamic adaptation of the pathogen to the intracellular environment which results in a permanent intracellular lifestyle. From the clinical perspective, these kind of host-pathogen interaction leads to a chronic type of infection, often with a relapsing course. The phenotypic correlate on the pathogen side is the SCV phenotype (see below).

The most important internalization process is mediated by the FnBP-fibronectin-α5β1 integrin pathway, which is characterized by forming a fibronectin bridge between fibronectin-binding proteins on the staphylococcal cell and the α5β1 integrin on the host cell side [82–85]. Since one FnBP molecule is able to bind 6–9 fibronectin molecules, α5β1 integrins could become clustered on the cell surface [86]. This clustering may efficiently trigger the intracellular signaling cascade involving focal adhesion kinases (FAKs), activated Src protein-tyrosine kinase and phosphatidylinositide 3-kinase (PI3K)/protein kinase B (Akt) that are prerequisite for internalization [87–91]. Hereby, the actin cytoskeleton and the endocytosis machinery are mobilized by *S. aureus* to finally intrude into the host cell.

However, adhesion and internalization are also modified by other factors that may influence the internalization efficiency. Other mechanisms involved in adherence to and internalization into host cells are based on the activities of Atl and several MSCRAMMs such as the clumping factor A (ClfA), the bacterial serine aspartate repeat-containing protein D (SdrD) and the and serine-rich adhesin for platelets (SraP). These secondary, fibronectin-independent mechanisms include the interactions described to date such as Atl with Hsc70, SdrD with desmoglein 1, ClfA via fibrinogen bridges or directly with host cells and SraP via the salivary scavenger protein gp340 [59, 92, 93].

1.2 Colonization and Persistence: Epidemiological and Clinical Aspects

Of note, in contrast to defined *in vitro* study conditions, biofilm-associated infections in the clinical context are often of polymicrobial nature, in particular, if natural biotic surfaces are involved, such as the skin and mucous membranes. Various beneficial, but also antagonistic microbe-microbe interactions with other Gram-positive and –negative bacteria, yeasts and viruses have been described [40, 94–97]. These interactions, including co-evolutionary effects, may be beneficial to microbial colonization but may worsen the course and outcome of infection [41, 98–102]. Thus, the complexity of mixed biofilms is much higher. Subsequent infections, such as those of cystic fibrosis-associated and necrotizing pneumonia lungs or chronically infected wounds (e.g. diabetic foot), are more difficult to understand, can proceed in several phases with changing predominance of the pathogens involved and pose a real challenge for their treatment [103–105].

1.2.1 Colonization of Human and Animal Hosts: The Ancestral Evolutionary Niche

As a typical prerequisite for infection, *S. aureus* owns the ability to establish itself as commensal for humans and animals. For that, it is well equipped to colonize the skin as the natural surface of humans and other mammalian species (see above and Chap. 2). While the mere colonization by *S. aureus* does not fall into the category of diseases, it is an independent risk factor and source for subsequent infections once the skin barrier has been overcome, e.g. by injuries of the skin or respective medical interventions [106–110].

The human nasal cavity serves as the principal habitat for *S. aureus* [111]. Within the nasal vestibule, the skin-like squamous epithelium containing vibrissae and glands of the anterior nares undergoes a change into pseudostratified columnar ciliated epithelium, i.e. the typical respiratory epithelium, before entering the nasal fossae. In this respect, especially in literature concerning *S. aureus* colonization, the anterior nares are mostly mentioned and often equated to the nasal vestibule or even the complete nasal cavity. However, investigations using a surgical approach have shown that in fact the posterior vestibule is most frequently colonized if analyzed in detail [40]. Usually originating from the nasal colonization, *S. aureus* can be transferred to other areas of the skin [112], where the throat and the perineum, but also the intertriginous skin folds, the axillae, the intestine, and the vagina, are also found to be regularly colonized [113, 114].

Already in 1944, it was underlined that "the carriage of staphylococci appears to depend upon the person, not upon his environment" [115]. Mostly determined in hospital settings, nasal *S. aureus* carriage rates range from less than 10% to more than 50%, typically around 40% [40, 112, 116, 117]. Since certain studies do not support earlier findings, it is still a matter of debate whether and which factors may influence the *S. aureus* carriage rates including those that are screening method-dependent (e.g. molecular or culture-based, with or without broth enrichment, intensity of the swabbing procedure, swab material) and study population-related, such as age, geographic location, profession (healthcare worker), underlying disease of patients (e.g. long-term hemodialysis, insulin-dependent diabetes mellitus, chronic skin diseases, immunosuppression) and lifestyle-depending factors (e.g. intravenous drug abuse, nose picking) [118–122]. If several body sites, in particular nose, throat and perineum, were included and in-depth analyzed, even more than >50% of the patients are colonized by *S. aureus* [123]. Analyzing a large group of non-hospitalized volunteers recruited from a general population, an overall prevalence of nasal *S. aureus* carriage of 41% has been recently determined [117, 124].

Based on the analysis of several longitudinal studies performed in the 1940–1950s, three types of *S. aureus* nasal carriers were classically distinguished: (1) persistent carriers (16 to 33%, carrying one strain over a prolonged period), (2) intermittent and occasional carriers (15 to 70%, carrying different strains), and (3) persistent non-carriers (6 to 38%) [114]. More or less similar percentages were reported in subsequent studies. However, not before 2009, it was proposed by van Belkum et al. [125] that there are probably only two types of nasal carriers: persistent carriers and "others" because intermittent carriers and non-carriers show similar *S. aureus* nasal elimination kinetics. Recent in depth-analysis of a cohort of non-hospitalized volunteers (n = 1878), who were screened thrice in intervals of 6–8 months, revealed that among those, who constantly carried *S. aureus* (11%), clonal lineages changed in about 80% over time [117]. In this study, only 14.1% (1.5% of all participants) and 5.3% (0.6% of all participants) of the permanent carriers in fact persistently carried either isolates belonging to the same *spa* type or *spa*-CC, respectively [117].

Still not well understood is to what extent the co-colonizing bacterial microbiota (flora) and possibly specific bacteria-bacteria interactions may impact *S. aureus* carriage and skin diseases caused or aggravated by this species. Possible options include antagonistic interactions as suggested for *S. epidermidis* and *S. hominis* by the production of unique antimicrobial peptides known as lantibiotics [126]. Topical administration of those antimicrobial-producing strains to a limited number of individuals with atopic dermatitis decreased the *S. aureus* colonization [126]. As demonstrated by epidemiological studies, the presence of serine protease Esp-secreting cells of a subset of *S. epidermidis* in the nasal cavities of human volunteers is linked with the absence of *S. aureus* due to the inhibition of biofilm formation and destruction of pre-existing biofilms by degrading biofilm-associated proteins [127, 128]. However, applying an extensive culturomic approach, 95% of individuals found to be nasally colonized with *S. aureus* carried also *S. epidermidis* in their nasal cavity and, vice versa, 66% of nasal *S. epidermidis* carriers were also colonized by *S. aureus* [40]. A significant *S. aureus* antagonism has recently been shown for *S. lugdunensis* based on the expression of lugdunin, a thiazolidine-containing cyclic peptide antibiotic [129]. Of note, applying subinhibitory concentrations of this substance during continuous serial passaging, no development of *in vitro* lugdunin resistance was observed [129]. Moreover, in this study, the *S. aureus* nasal detection rate in the presence of *S. lugdunensis* was 5.9-fold ($P = 0.015$) lower. However, an in depth-analysis of the nasal microbiome revealed that if *S. lugdunensis* had been detected as nasal colonizer, also *S. aureus* was found in 66.7% of those cases [40]. Thus, 30% of the *S. aureus* carriers were also colonized by *S. lugdunensis* in this study [40]. Some investigations have indicated interactions also with non-staphylococcal species, such as a negative correlation with *Streptococcus pneumoniae* analyzing the nasopharynx, with corynebacterial species or *Finegoldia magna* [130–132]. However, other studies could not confirm these associations [40, 133]. Inter-individual host differences, strain-specific microbial features and different investigation strategies may explain these contradictory results.

Already in the 1930s, a interrelation between nasal *S. aureus* colonization and recurrent skin infections (furuncolosis) by this pathogen has been postulated by N. Danbolt in his PhD thesis [134]. Many decades later, it has been shown that nasal *S. aureus* carriage is in fact associated with an increased risk for health care-associated infections in defined patient groups including catheter exit-site infections in patients on continuous ambulatory peritoneal disease, wound infections after cardiac surgery and nosocomial *S. aureus* bacteremia [106–110]. This risk of nosocomial infection due to *S. aureus* is about three to six times the risk among non-carriers or those exhibiting only low-level colonization [109, 135]. Applying a genotyping approach, nasal colonization was verified as endogenous source of *S. aureus* bacteremia and other invasive infections [106]. In a large randomized, double-blind, multicenter study in the Netherlands, patients identified as *S. aureus* nasal carriers on admission to surgical and internal medicine departments were either treated with mupirocin nasal ointment and chlorhexidine soap or with placebo ointment and soap [136]. Particularly those patients with deep surgical site-infections benefited significantly from the eradication of nasal *S. aureus* carriage exhibiting a relative risk of infection (0.42%) reduction compared to the placebo group [136]. Thus, the authors concluded that rapid screening on admission followed by decolonizing treatment reduces can prevent hospital-acquired infections due to *S. aureus* [136].

In the case of the prevention of methicillin-resistant *S. aureus* (MRSA) strains, so-called "search and destroy" approaches have become standard procedure as laid down in many international, national or local guidelines and initiatives [137–140]. Moreover, the success of those strategies has been documented in a multitude of studies and meta-analyses, in particular if this approach has been part of a bundle of prevention measures and being part of concerted regional networks [141–146].

Quite inexplicably, the application of those strategies to prevent hospital-associated infections by *S. aureus* independently of being an MRSA or not is still less common. In a randomized, placebo-controlled trial enrolling surgical patients, those of the mupirocin–/chlorhexidine-treated group revealed not only a reduction of the incidence of nosocomial *S. aureus* infections but also a reduction of the 1-year mortality [147].

1.2.2 Colonization of Biomaterials: Conquest of an Evolutionary Novel, Man-Made Niche

The introduction of biomaterials in medicine to replace or restore the function of compromised tissues or organs led to a new medical entity, i.e. the foreign-body related infection (FBRI). The ability to generate biofilms on biotic and abiotic surfaces is of crucial importance for the colonization of indwelling medical devices and is a major virulence determinant of various staphylococcal species. Besides species belonging to the CoNS group, which are notoriously associated with FBRIs, *S. aureus* is one of the major pathogens involved in all kinds of the large range of FBRIs [148, 149]. Altogether, about 75% of orthopedic implant-associated infections are caused by staphylococci and about one third of all cases by *S. aureus* alone [150]. Central venous catheter (CVC)-related bloodstream infections (CRBSIs) and central line-associated bloodstream infection (CLABSI), respectively, represent a further FBRI entity typically caused by staphylococci [151]. In a retrospective, observational study analyzing home parenteral nutrition, the infecting agents of CRBSIs were determined with *S. aureus* as being responsible for 10% of all cases [152].

1.2.3 Invasion Into and Persistence in the Intracellular Environment as a Protected Niche

While *S. aureus* is well-described as extracellular pathogen, its intracellular lifestyle has been only scantily acknowledged in the past. Meanwhile, it is widely accepted that *S. aureus* can be internalized by a variety of non-professional phagocytes, such as epithelial, mesenchymal and endothelial cells.

The ability to persist intracellularly, in particular by formation of a variant subpopulation, provides *S. aureus* with a number of survival advantages that extends beyond mere increased resistance to antibiotics. The ability to switch into the SCV seems to be a general feature of *S. aureus* [153]. This phenomenon is based on a slow-growing subpopulation, which is defined by tiny colonies on solid media about 10 times smaller than the parent wildtype strain, hence, their recent designation as

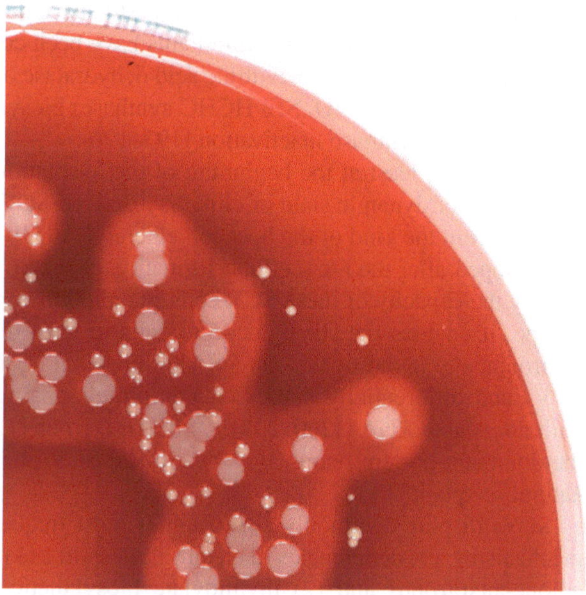

Fig. 1.3 *S. aureus* after 24 h of incubation on Columbia blood agar with 5% sheep blood. Picture section displays the typical colonial morphology of the "normal" *S. aureus* phenotype here with grayish-pigmented colonies surrounded by a broad hemolysis zone (arrays) and of the small-colony variant (SCV) phenotype characterized by tiny, non-hemolytic, and non-pigmented colonies (dashed arrays)

"small colony-variants", historically termed also as dwarf-colony or gonidial variants (Fig. 1.3) [154–156].

Although this specific phenotype has been recognized at least since the 1930s, connecting this phenotype to chronic and recurrent infections has only recently been appreciated [157]. Typical SCV-associated infections include persistent and relapsing cases of arthritis and osteomyelitis [157–162] and abscesses, (diabetic) ulcers and wound infections [163–169]. However, also other entities with chronic courses of disease have been reported including endocarditis [170, 171] and rhinosinusitis [172]. Moreover, SCV infections have been associated with the insertion of various foreign bodies [159, 162, 171, 173–175]. In children with CF lung disease, persistent colonization of the airways with *S. aureus* SCVs, typically thymidine-auxotrophic variants, is associated with worse CF respiratory outcomes [104, 176–178]. Also in the animal host, chronic *S. aureus* infections due to the SCV phenotype have been reported such as persistent bovine mastitis [179, 180].

The delayed growth of SCV cells in broth and on solid media, often overgrown by co-occurring normal phenotypes, various auxotrophies and rapid phenotype switch back into the wildtype complicates their investigation [9, 10, 181–184]. In particular, the instability of the vast majority of clinically derived SCVs contributes to the situation that the true extent of infections caused by this phenotype is unknown. Exposure to aminoglycosides and trimethoprim-sulfamethoxazole (TMP-SMX), serial subcultures of SCV phenotype-displaying colonies, host-generated chemical stresses and later adaptations to the intracellular lifestyle enable the selection of stable variant forms [185–189]. These are characterized by mutations in the MenB (1,4-dihydroxy-2-naphthoyl-CoA synthase), HemA (glutamyl-tRNA reductase) and HemG (protoporphyrinogen oxidase) encoding genes [190], however, also stable mutations in *thyA*, *relA*, *mgrA* and *rsbU* have been reported [188, 191, 192]. Also, several genetically defined mutants have been constructed, e.g. by interrupting hemin (HemB, 5-aminolevulinic acid dehydratase or porphobilinogen synthase) and menaquinone (MenD, SEPHCHC synthase) biosynthetic genes and as well as by thymidylate synthase inactivation [193–196].

However, at least at the beginning of the adaptation to the intracellular persistence, a rather dynamic state of clinically derived SCVs has been observed, which depends from the kind of the host cell entered [194, 197–199]. This dynamic mode is developed after the intracellular formation of the SCV phenotype and characterized by the capacity of these SCV cells to rapidly revert back to the wild-type phenotype. It enables the staphylococcal cells to immediately response to changing environmental conditions. Larger chromosomal rearrangements with unstable SCVs and differentially expressed non-protein-coding RNAs may be associated with non-stable RNAs [200–203]. Of interest, when comparing proteomic profiles, the dominant protein features of the clinical SCV phenotype were retained in the natural revertant [194].

SCVs display commonly auxotrophies most frequently based on two different mechanisms: (1) deficiencies in electron transport as known for hemin-and/or menadione-auxotrophs and (2) thymidylate biosynthetic-defective SCVs [190, 192, 193, 204–207]. Also, other auxotrophies including those for CO_2, fatty acids and thiamin have been described, however, for many SCVs the underlying auxotrophism is unclear [169, 208, 209]. Further typical SCV-associated characteristics comprise attenuated virulence, dropped membrane potential, decreased activation of hypoxia-inducible factors and reduced susceptibility towards some, but not all antibiotic agents and antimicrobial peptides [195, 210–219]. For the determination of the SCV auxotrophies and the phenotypic switch, refer to the literature [9, 181, 197].

In the past two decades, several studies yielded new insights into the mechanisms, metabolic and physiological adaptations and consequences of the bidirectional phenotype switch as prerequisite of the pathogen's dual lifestyle occurring as extracellular as well as intracellular microorganism [183, 194, 220–224]. Alpha-hemolysin (Hla) is the major cytotoxic agent released by *S. aureus*. Mutants defective in Hla production or wild-type *S. aureus* with reduced Hla production are able to persist intracellularly [225, 226]. Similarly, SCVs, which are characterized by low levels of Hla production have been associated with persistent infections [198, 227]. This toxin and some other virulence factors (e.g. TSST-1) are regulated via the *hld* product RNAIII by the accessory gene regulatory (*agr*) operon, which controls a quorum-sensing system [228]. Entering the host cell, the anaerobic intracellular environment reduces RNAIII and, thus, Hla expression and selects for a staphylococcal phenotype with exhibition of the SCV phenotype [184, 197, 229].

In the case of electron transport-defective SCVs, the metabolic consequences of the intracellular lifestyle are the down-regulation of citric acid cycle activity, in particular a reduced carbon flux through the tricarboxylic acid cycle (also known as the Krebs or citric acid cycle), and the down-regulation of proteins associated with the purine/pyrimidine and folate metabolism [194, 223]. This reduced metabolic status and alterations in global regulatory genes has been demonstrated in different types of SCVs, which may explain the consistent main features of SCV phenotypes of *S. aureus* regardless of the underlying mechanism of their formation.

If patients are treated with TMP-SMX for prolonged periods, thymidine-dependent SCVs may emerge [177, 227, 230, 231]. TMP-SMX inhibits the synthesis of tetrahydrofolic acid, a cofactor for the thymidylate synthase (ThyA). This enzyme plays a key role in DNA synthesis and is essentially for the conversion of thymidine from uracil. This kind of SCVs depends on an external supply of thymidine, which occurs in infected tissues as a result of cell degradation and pus formation.

1.3 Conclusion

The multitalented opportunistic pathogen *S. aureus* possesses a large, often redundant array of virulence factors, which are not uncommonly multifunctional. They form the basis for the colonization of humans and other mammalian hosts, including companion and farm animals. After accidental breaching a skin and the mucous membranes barrier the pathogen owns a large "tool box" to initiate infectious processes. At this, the pathogen's ability to adhere on biotic and inserted abiotic surfaces within the host complemented by the capability to form complex biofilms plays a key role in the pathogenesis of *S. aureus* infections, which may range from mild to severe, even life-threatening infections. *S. aureus* is notoriously known as an extracellular pathogen causing aggressive courses of infection. However, after internalization, the pathogen is able to persist intracellularly in epithelial and other cells by formation of the phenotype of the small colony-variant. At least at the beginning, the SCV generation during the course of the adaptation to an intracellular lifestyle is a dynamic process with the ability of the pathogen to switch back to the "normal" wildtype phenotype recognizable by the formation of unstable SCVs. External factors and evolutionary adaptation to continued intracellular persistence may lead to stable SCVs. Clinically speaking, the intracellular lifestyle leads to chronic infections with persistent and relapsing course of disease, which are difficult to diagnose and treat.

References

1. Tong SY, Davis JS, Eichenberger E, Holland TL, Fowler VG Jr (2015) *Staphylococcus aureus* infections: epidemiology, pathophysiology, clinical manifestations, and management. Clin Microbiol Rev 28(3):603–661. https://doi.org/10.1128/cmr.00134-14
2. Becker K, Heilmann C, Peters G (2014) Coagulase-negative staphylococci. Clin Microbiol Rev 27(4):870–926. https://doi.org/10.1128/cmr.00109-13
3. Tong SY, Schaumburg F, Ellington MJ, Corander J, Pichon B, Leendertz F et al (2015) Novel staphylococcal species that form part of a *Staphylococcus aureus*-related complex:

the non-pigmented *Staphylococcus argenteus* sp. nov. and the non-human primate-associated *Staphylococcus schweitzeri* sp. nov. Int J Syst Evol Microbiol 65(Pt 1):15–22. https://doi.org/10.1099/ijs.0.062752-0
4. Schutte AHJ, Strepis N, Zandijk WHA, Bexkens ML, Bode LGM, Klaassen CHW (2021) Characterization of *Staphylococcus roterodami* sp. nov., a new species within the *Staphylococcus aureus* complex isolated from a human foot infection. Int J Syst Evol Microbiol 71(9). https://doi.org/10.1099/ijsem.0.004996
5. Chew KL, Octavia S, Lai D, Lin RTP, Teo JWP (2021) *Staphylococcus singaporensis* sp. nov., a new member of the *Staphylococcus aureus* complex, isolated from human clinical specimens. Int J Syst Evol Microbiol 71(10). https://doi.org/10.1099/ijsem.0.005067
6. Becker K, Schaumburg F, Kearns A, Larsen AR, Lindsay JA, Skov RL et al (2019) Implications of identifying the recently defined members of the *Staphylococcus aureus* complex *S. argenteus* and *S. schweitzeri*: a position paper of members of the ESCMID Study Group for Staphylococci and Staphylococcal Diseases (ESGS). Clin Microbiol Infect 25(9):1064–1070. https://doi.org/10.1016/j.cmi.2019.02.028
7. Madhaiyan M, Wirth JS, Saravanan VS (2020) Phylogenomic analyses of the *Staphylococcaceae* family suggest the reclassification of five species within the genus *Staphylococcus* as heterotypic synonyms, the promotion of five subspecies to novel species, the taxonomic reassignment of five *Staphylococcus* species to *Mammaliicoccus* gen. Nov., and the formal assignment of *Nosocomiicoccus* to the family *Staphylococcaceae*. Int J Syst Evol Microbiol 70(11):5926–5936. https://doi.org/10.1099/ijsem.0.004498
8. de la Fuente R, Suarez G, Schleifer KH (1985) *Staphylococcus aureus* subsp. *anaerobius* subsp. nov., the causal agent of abscess disease of sheep. Int J Syst Bacteriol 35(1):99–102. https://doi.org/10.1099/00207713-35-1-99
9. Kahl BC, Becker K, Löffler B (2016) Clinical significance and pathogenesis of staphylococcal small colony variants in persistent infections. Clin Microbiol Rev 29(2):401–427. https://doi.org/10.1128/cmr.00069-15
10. Proctor RA, von Eiff C, Kahl BC, Becker K, McNamara P, Herrmann M et al (2006) Small colony variants: a pathogenic form of bacteria that facilitates persistent and recurrent infections. Nat Rev Microbiol 4(4):295–305. https://doi.org/10.1038/nrmicro1384
11. Kim HK, Thammavongsa V, Schneewind O, Missiakas D (2012) Recurrent infections and immune evasion strategies of *Staphylococcus aureus*. Curr Opin Microbiol 15(1):92–99. https://doi.org/10.1016/j.mib.2011.10.012
12. de Jong NWM, van Kessel KPM, van Strijp JAG (2019) Immune evasion by *Staphylococcus aureus*. Microbiol Spectr 7(2). https://doi.org/10.1128/microbiolspec.gpp3-0061-2019
13. Heilmann C, Hartleib J, Hussain MS, Peters G (2005) The multifunctional *Staphylococcus aureus* autolysin aaa mediates adherence to immobilized fibrinogen and fibronectin. Infect Immun 73(8):4793–4802. https://doi.org/10.1128/iai.73.8.4793-4802.2005
14. Yonemoto K, Chiba A, Sugimoto S, Sato C, Saito M, Kinjo Y et al (2019) Redundant and distinct roles of secreted protein Eap and cell wall-anchored protein SasG in biofilm formation and pathogenicity of *Staphylococcus aureus*. Infect Immun 87(4):e00894–e00818. https://doi.org/10.1128/iai.00894-18
15. Chavakis T, Wiechmann K, Preissner KT, Herrmann M (2005) *Staphylococcus aureus* interactions with the endothelium: the role of bacterial "secretable expanded repertoire adhesive molecules" (SERAM) in disturbing host defense systems. Thromb Haemost 94(2):278–285. https://doi.org/10.1160/th05-05-0306
16. Jeffery CJ (1999) Moonlighting proteins. Trends Biochem Sci 24(1):8–11. https://doi.org/10.1016/s0968-0004(98)01335-8
17. Berne C, Ellison CK, Ducret A, Brun YV (2018) Bacterial adhesion at the single-cell level. Nat Rev Microbiol 16(10):616–627. https://doi.org/10.1038/s41579-018-0057-5
18. Bjarnsholt T, Alhede M, Alhede M, Eickhardt-Sørensen SR, Moser C, Kühl M et al (2013) The in vivo biofilm. Trends Microbiol 21(9):466–474. https://doi.org/10.1016/j.tim.2013.06.002

19. Flemming HC, Wingender J, Szewzyk U, Steinberg P, Rice SA, Kjelleberg S (2016) Biofilms: an emergent form of bacterial life. Nat Rev Microbiol 14(9):563–575. https://doi.org/10.1038/nrmicro.2016.94
20. Patti JM, Höök M (1994) Microbial adhesins recognizing extracellular matrix macromolecules. Curr Opin Cell Biol 6(5):752–758. https://doi.org/10.1016/0955-0674(94)90104-x
21. Foster TJ, Geoghegan JA, Ganesh VK, Höök M (2014) Adhesion, invasion and evasion: the many functions of the surface proteins of *Staphylococcus aureus*. Nat Rev Microbiol 12(1):49–62. https://doi.org/10.1038/nrmicro3161
22. Heilmann C (2011) Adhesion mechanisms of staphylococci. Adv Exp Med Biol 715:105–123. https://doi.org/10.1007/978-94-007-0940-9_7
23. Otto M (2018) Staphylococcal biofilms. Microbiol Spectr 6(4). https://doi.org/10.1128/microbiolspec.GPP3-0023-2018
24. Speziale P, Pietrocola G, Foster TJ, Geoghegan JA (2014) Protein-based biofilm matrices in Staphylococci. Front Cell Infect Microbiol 4:171. https://doi.org/10.3389/fcimb.2014.00171
25. Lister JL, Horswill AR (2014) *Staphylococcus aureus* biofilms: recent developments in biofilm dispersal. Front Cell Infect Microbiol 4:178. https://doi.org/10.3389/fcimb.2014.00178
26. Archer NK, Mazaitis MJ, Costerton JW, Leid JG, Powers ME, Shirtliff ME (2011) *Staphylococcus aureus* biofilms: properties, regulation, and roles in human disease. Virulence 2(5):445–459. https://doi.org/10.4161/viru.2.5.17724
27. Moormeier DE, Bayles KW (2017) *Staphylococcus aureus* biofilm: a complex developmental organism. Mol Microbiol 104(3):365–376. https://doi.org/10.1111/mmi.13634
28. Dufrêne YF, Viljoen A (2020) Binding strength of gram-positive bacterial adhesins. Front Microbiol 11:1457. https://doi.org/10.3389/fmicb.2020.01457
29. Parreira P, Martins MCL (2021) The biophysics of bacterial infections: adhesion events in the light of force spectroscopy. Cell Surf 7:100048. https://doi.org/10.1016/j.tcsw.2021.100048
30. O'Toole G, Kaplan HB, Kolter R (2000) Biofilm formation as microbial development. Annu Rev Microbiol 54:49–79. https://doi.org/10.1146/annurev.micro.54.1.49
31. Moormeier DE, Endres JL, Mann EE, Sadykov MR, Horswill AR, Rice KC et al (2013) Use of microfluidic technology to analyze gene expression during *Staphylococcus aureus* biofilm formation reveals distinct physiological niches. Appl Environ Microbiol 79(11):3413–3424. https://doi.org/10.1128/AEM.00395-13
32. Moormeier DE, Bose JL, Horswill AR, Bayles KW (2014) Temporal and stochastic control of *Staphylococcus aureus* biofilm development. MBio 5(5):e01341–e01314. https://doi.org/10.1128/mBio.01341-14
33. Singh R, Ray P (2014) Quorum sensing-mediated regulation of staphylococcal virulence and antibiotic resistance. Future Microbiol 9(5):669–681. https://doi.org/10.2217/fmb.14.31
34. Ekstedt RD, Bernhard JM (1973) Preparation and characterization of a slime layer material produced by *Staphylococcus aureus*. Proc Soc Exp Biol Med 142(1):86–91. https://doi.org/10.3181/00379727-142-36964
35. Dengler V, Foulston L, DeFrancesco AS, Losick R (2015) An electrostatic net model for the role of extracellular DNA in biofilm formation by *Staphylococcus aureus*. J Bacteriol 197(24):3779–3787. https://doi.org/10.1128/JB.00726-15
36. Izano EA, Amarante MA, Kher WB, Kaplan JB (2008) Differential roles of poly-*N*-acetylglucosamine surface polysaccharide and extracellular DNA in *Staphylococcus aureus* and *Staphylococcus epidermidis* biofilms. Appl Environ Microbiol 74(2):470–476. https://doi.org/10.1128/AEM.02073-07
37. Foulston L, Elsholz AK, DeFrancesco AS, Losick R (2014) The extracellular matrix of *Staphylococcus aureus* biofilms comprises cytoplasmic proteins that associate with the cell surface in response to decreasing pH. MBio 5(5):e01667–e01614. https://doi.org/10.1128/mBio.01667-14
38. Mack D, Fischer W, Krokotsch A, Leopold K, Hartmann R, Egge H et al (1996) The intercellular adhesin involved in biofilm accumulation of *Staphylococcus epidermidis* is a linear β-1,6-

linked glucosaminoglycan: purification and structural analysis. J Bacteriol 178(1):175–183. https://doi.org/10.1128/jb.178.1.175-183.1996
39. Vinogradov E, Sadovskaya I, Li J, Jabbouri S (2006) Structural elucidation of the extracellular and cell-wall teichoic acids of *Staphylococcus aureus* MN8m, a biofilm forming strain. Carbohydr Res 341(6):738–743. https://doi.org/10.1016/j.carres.2006.01.012
40. Kaspar U, Kriegeskorte A, Schubert T, Peters G, Rudack C, Pieper DH et al (2016) The culturome of the human nose habitats reveals individual bacterial fingerprint patterns. Environ Microbiol 18(7):2130–2142. https://doi.org/10.1111/1462-2920.12891
41. Kahl BC, Duebbers A, Lubritz G, Haeberle J, Koch HG, Ritzerfeld B et al (2003) Population dynamics of persistent *Staphylococcus aureus* isolated from the airways of cystic fibrosis patients during a 6-year prospective study. J Clin Microbiol 41(9):4424–4427. https://doi.org/10.1128/JCM.41.9.4424-4427.2003
42. Goerke C, Wolz C (2010) Adaptation of *Staphylococcus aureus* to the cystic fibrosis lung. Int J Med Microbiol 300(8):520–525. https://doi.org/10.1016/j.ijmm.2010.08.003
43. Mazmanian SK, Liu G, Ton-That H, Schneewind O (1999) *Staphylococcus aureus* sortase, an enzyme that anchors surface proteins to the cell wall. Science 285(5428):760–763. https://doi.org/10.1126/science.285.5428.760
44. Ní Eidhin D, Perkins S, Francois P, Vaudaux P, Höök M, Foster TJ (1998) Clumping factor B (ClfB), a new surface-located fibrinogen-binding adhesin of *Staphylococcus aureus*. Mol Microbiol 30(2):245–257. https://doi.org/10.1046/j.1365-2958.1998.01050.x
45. Ton-That H, Liu G, Mazmanian SK, Faull KF, Schneewind O (1999) Purification and characterization of sortase, the transpeptidase that cleaves surface proteins of *Staphylococcus aureus* at the LPXTG motif. Proc Natl Acad Sci U S A 96(22):12424–12429. https://doi.org/10.1073/pnas.96.22.12424
46. Clarke SR, Foster SJ (2006) Surface adhesins of *Staphylococcus aureus*. Adv Microb Physiol 51:187–224. https://doi.org/10.1016/S0065-2911(06)51004-5
47. Foster TJ (2019) The MSCRAMM family of cell-wall-anchored surface proteins of gram-positive cocci. Trends Microbiol 27(11):927–941. https://doi.org/10.1016/j.tim.2019.06.007
48. Foster TJ (2019) Surface proteins of *Staphylococcus aureus*. Microbiol Spectr 7(4). https://doi.org/10.1128/microbiolspec.GPP3-0046-2018
49. McDonnell CJ, Garciarena CD, Watkin RL, McHale TM, McLoughlin A, Claes J et al (2016) Inhibition of major integrin $\alpha_V\beta_3$ reduces *Staphylococcus aureus* attachment to sheared human endothelial cells. J Thromb Haemost 14(12):2536–2547. https://doi.org/10.1111/jth.13501
50. Bennett JS (2005) Structure and function of the platelet integrin $\alpha_{IIb}\beta_3$. J Clin Invest 115(12):3363–3369. https://doi.org/10.1172/JCI26989
51. Clarke SR, Harris LG, Richards RG, Foster SJ (2002) Analysis of Ebh, a 1.1-megadalton cell wall-associated fibronectin-binding protein of *Staphylococcus aureus*. Infect Immun 70(12):6680–6687. https://doi.org/10.1128/IAI.70.12.6680-6687.2002
52. Cheng AG, Missiakas D, Schneewind O (2014) The giant protein Ebh is a determinant of *Staphylococcus aureus* cell size and complement resistance. J Bacteriol 196(5):971–981. https://doi.org/10.1128/JB.01366-13
53. Downer R, Roche F, Park PW, Mecham RP, Foster TJ (2002) The elastin-binding protein of *Staphylococcus aureus* (EbpS) is expressed at the cell surface as an integral membrane protein and not as a cell wall-associated protein. J Biol Chem 277(1):243–250. https://doi.org/10.1074/jbc.M107621200
54. Heilmann C, Hussain M, Peters G, Götz F (1997) Evidence for autolysin-mediated primary attachment of *Staphylococcus epidermidis* to a polystyrene surface. Mol Microbiol 24(5):1013–1024. https://doi.org/10.1046/j.1365-2958.1997.4101774.x
55. Hussain M, Steinbacher T, Peters G, Heilmann C, Becker K (2015) The adhesive properties of the *Staphylococcus lugdunensis* multifunctional autolysin AtlL and its role in biofilm formation and internalization. Int J Med Microbiol 305(1):129–139. https://doi.org/10.1016/j.ijmm.2014.11.010

56. Foster SJ (1995) Molecular characterization and functional analysis of the major autolysin of *Staphylococcus aureus* 8325/4. J Bacteriol 177(19):5723–5725. https://doi.org/10.1128/jb.177.19.5723-5725.1995
57. Oshida T, Sugai M, Komatsuzawa H, Hong YM, Suginaka H, Tomasz A (1995) A *Staphylococcus aureus* autolysin that has an *N*-acetylmuramoyl-L-alanine amidase domain and an endo-β-*N*-acetylglucosaminidase domain: cloning, sequence analysis, and characterization. Proc Natl Acad Sci U S A 92(1):285–289. https://doi.org/10.1073/pnas.92.1.285
58. Biswas R, Voggu L, Simon UK, Hentschel P, Thumm G, Götz F (2006) Activity of the major staphylococcal autolysin Atl. FEMS Microbiol Lett 259(2):260–268. https://doi.org/10.1111/j.1574-6968.2006.00281.x
59. Hirschhausen N, Schlesier T, Schmidt MA, Götz F, Peters G, Heilmann C (2010) A novel staphylococcal internalization mechanism involves the major autolysin Atl and heat shock cognate protein Hsc70 as host cell receptor. Cell Microbiol 12(12):1746–1764. https://doi.org/10.1111/j.1462-5822.2010.01506.x
60. Hirschhausen N, Schlesier T, Peters G, Heilmann C (2012) Characterization of the modular design of the autolysin/adhesin Aaa from *Staphylococcus aureus*. PloS One 7(6):e40353. https://doi.org/10.1371/journal.pone.0040353
61. Xia G, Kohler T, Peschel A (2010) The wall teichoic acid and lipoteichoic acid polymers of *Staphylococcus aureus*. Int J Med Microbiol 300(2–3):148–154. https://doi.org/10.1016/j.ijmm.2009.10.001
62. Weidenmaier C, Peschel A (2008) Teichoic acids and related cell-wall glycopolymers in gram-positive physiology and host interactions. Nat Rev Microbiol 6(4):276–287. https://doi.org/10.1038/nrmicro1861
63. van Dalen R, Peschel A, van Sorge NM (2020) Wall teichoic acid in *Staphylococcus aureus* host interaction. Trends Microbiol 28(12):985–998. https://doi.org/10.1016/j.tim.2020.05.017
64. Neuhaus FC, Baddiley J (2003) A continuum of anionic charge: structures and functions of D-alanyl-teichoic acids in gram-positive bacteria. Microbiol Mol Biol Rev 67(4):686–723. https://doi.org/10.1128/MMBR.67.4.686-723.2003
65. Weidenmaier C, Kokai-Kun JF, Kristian SA, Chanturiya T, Kalbacher H, Gross M et al (2004) Role of teichoic acids in *Staphylococcus aureus* nasal colonization, a major risk factor in nosocomial infections. Nat Med 10(3):243–245. https://doi.org/10.1038/nm991
66. Kohler T, Weidenmaier C, Peschel A (2009) Wall teichoic acid protects *Staphylococcus aureus* against antimicrobial fatty acids from human skin. J Bacteriol 191(13):4482–4484. https://doi.org/10.1128/JB.00221-09
67. Aly R, Shinefield HR, Litz C, Maibach HI (1980) Role of teichoic acid in the binding of *Staphylococcus aureus* to nasal epithelial cells. J Infect Dis 141(4):463–465. https://doi.org/10.1093/infdis/141.4.463
68. Weidenmaier C, Peschel A, Xiong YQ, Kristian SA, Dietz K, Yeaman MR et al (2005) Lack of wall teichoic acids in *Staphylococcus aureus* leads to reduced interactions with endothelial cells and to attenuated virulence in a rabbit model of endocarditis. J Infect Dis 191(10):1771–1777. https://doi.org/10.1086/429692
69. Du X, Larsen J, Li M, Walter A, Slavetinsky C, Both A et al (2021) *Staphylococcus epidermidis* clones express *Staphylococcus aureus*-type wall teichoic acid to shift from a commensal to pathogen lifestyle. Nat Microbiol 6(6):757–768. https://doi.org/10.1038/s41564-021-00913-z
70. Mack D, Siemssen N, Laufs R (1992) Parallel induction by glucose of adherence and a polysaccharide antigen specific for plastic-adherent *Staphylococcus epidermidis*: evidence for functional relation to intercellular adhesion. Infect Immun 60(5):2048–2057. https://doi.org/10.1128/iai.60.5.2048-2057.1992
71. Cramton SE, Gerke C, Schnell NF, Nichols WW, Götz F (1999) The intercellular adhesion (*Ica*) locus is present in *Staphylococcus aureus* and is required for biofilm formation. Infect Immun 67(10):5427–5433. https://doi.org/10.1128/IAI.67.10.5427-5433.1999

72. Heilmann C, Schweitzer O, Gerke C, Vanittanakom N, Mack D, Götz F (1996) Molecular basis of intercellular adhesion in the biofilm-forming *Staphylococcus epidermidis*. Mol Microbiol 20(5):1083–1091. https://doi.org/10.1111/j.1365-2958.1996.tb02548.x
73. Heilmann C, Götz F (1998) Further characterization of *Staphylococcus epidermidis* transposon mutants deficient in primary attachment or intercellular adhesion. Zentralbl Bakteriol 287(1–2):69–83. https://doi.org/10.1016/s0934-8840(98)80149-7
74. Götz F (2002) *Staphylococcus* and biofilms. Mol Microbiol 43(6):1367–1378. https://doi.org/10.1046/j.1365-2958.2002.02827.x
75. Maira-Litrán T, Kropec A, Abeygunawardana C, Joyce J, Mark G III, Goldmann DA et al (2002) Immunochemical properties of the staphylococcal poly-*N*-acetylglucosamine surface polysaccharide. Infect Immun 70(8):4433–4440. https://doi.org/10.1128/IAI.70.8.4433-4440.2002
76. Vuong C, Kocianova S, Voyich JM, Yao Y, Fischer ER, DeLeo FR et al (2004) A crucial role for exopolysaccharide modification in bacterial biofilm formation, immune evasion, and virulence. J Biol Chem 279(52):54881–54886. https://doi.org/10.1074/jbc.M411374200
77. Vuong C, Voyich JM, Fischer ER, Braughton KR, Whitney AR, DeLeo FR et al (2004) Polysaccharide intercellular adhesin (PIA) protects *Staphylococcus epidermidis* against major components of the human innate immune system. Cell Microbiol 6(3):269–275. https://doi.org/10.1046/j.1462-5822.2004.00367.x
78. Fitzpatrick F, Humphreys H, O'Gara JP (2005) Evidence for *icaADBC*-independent biofilm development mechanism in methicillin-resistant *Staphylococcus aureus* clinical isolates. J Clin Microbiol 43(4):1973–1976. https://doi.org/10.1128/JCM.43.4.1973-1976.2005
79. Rice KC, Mann EE, Endres JL, Weiss EC, Cassat JE, Smeltzer MS et al (2007) The *cidA* murein hydrolase regulator contributes to DNA release and biofilm development in *Staphylococcus aureus*. Proc Natl Acad Sci U S A 104(19):8113–8118. https://doi.org/10.1073/pnas.0610226104
80. Mann EE, Rice KC, Boles BR, Endres JL, Ranjit D, Chandramohan L et al (2009) Modulation of eDNA release and degradation affects *Staphylococcus aureus* biofilm maturation. PloS One 4(6):e5822. https://doi.org/10.1371/journal.pone.0005822
81. Beltrame CO, Côrtes MF, Bonelli RR, de Almeida Côrrea AB, Botelho AM, Américo MA et al (2015) Inactivation of the autolysis-related genes *lrgB* and *yycI* in *Staphylococcus aureus* increases cell lysis-dependent eDNA release and enhances biofilm development *in vitro* and *in vivo*. PloS One 10(9):e0138924. https://doi.org/10.1371/journal.pone.0138924
82. Sinha B, Francois PP, Nüße O, Foti M, Hartford OM, Vaudaux P et al (1999) Fibronectin-binding protein acts as *Staphylococcus aureus* invasin via fibronectin bridging to integrin $\alpha_5\beta_1$. Cell Microbiol 1(2):101–117. https://doi.org/10.1046/j.1462-5822.1999.00011.x
83. Grundmeier M, Hussain M, Becker P, Heilmann C, Peters G, Sinha B (2004) Truncation of fibronectin-binding proteins in *Staphylococcus aureus* strain Newman leads to deficient adherence and host cell invasion due to loss of the cell wall anchor function. Infect Immun 72(12):7155–7163. https://doi.org/10.1128/IAI.72.12.7155-7163.2004
84. Fowler T, Wann ER, Joh D, Johansson S, Foster TJ, Hook M (2000) Cellular invasion by *Staphylococcus aureus* involves a fibronectin bridge between the bacterial fibronectin-binding MSCRAMMs and host cell beta1 integrins 1. Eur J Cell Biol 79(10):672–679. https://doi.org/10.1078/0171-9335-00104
85. Isberg RR, Leong JM (1990) Multiple β_1 chain integrins are receptors for invasin, a protein that promotes bacterial penetration into mammalian cells. Cell 60(5):861–871. https://doi.org/10.1016/0092-8674(90)90099-z
86. Bingham RJ, Rudiño-Piñera E, Meenan NA, Schwarz-Linek U, Turkenburg JP, Höök M et al (2008) Crystal structures of fibronectin-binding sites from *Staphylococcus aureus* FnBPA in complex with fibronectin domains. Proc Natl Acad Sci U S A 105(34):12254–12258. https://doi.org/10.1073/pnas.0803556105
87. Oviedo-Boyso J, Cortés-Vieyra R, Huante-Mendoza A, Yu HB, Valdez-Alarcón JJ, Bravo-Patiño A et al (2011) The phosphoinositide-3-kinase-Akt signaling pathway is important for

Staphylococcus aureus internalization by endothelial cells. Infect Immun 79(11):4569–4577. https://doi.org/10.1128/IAI.05303-11

88. Fowler T, Johansson S, Wary KK, Höök M (2003) Src kinase has a central role in in vitro cellular internalization of *Staphylococcus aureus*. Cell Microbiol 5(6):417–426. https://doi.org/10.1046/j.1462-5822.2003.00290.x
89. Agerer F, Lux S, Michel A, Rohde M, Ohlsen K, Hauck CR (2005) Cellular invasion by *Staphylococcus aureus* reveals a functional link between focal adhesion kinase and cortactin in integrin-mediated internalisation. J Cell Sci 118(Pt 10):2189–2200. https://doi.org/10.1242/jcs.02328
90. Wang JH, Zhang K, Wang N, Qiu XM, Wang YB, He P (2013) Involvement of phosphatidylinositol 3-kinase/Akt signaling pathway in β1 integrin-mediated internalization of *Staphylococcus aureus* by alveolar epithelial cells. J Microbiol 51(5):644–650. https://doi.org/10.1007/s12275-013-3040-x
91. Goldmann O, Tuchscherr L, Rohde M, Medina E (2016) α-Hemolysin enhances *Staphylococcus aureus* internalization and survival within mast cells by modulating the expression of β1 integrin. Cell Microbiol 18(6):807–819. https://doi.org/10.1111/cmi.12550
92. Askarian F, Ajayi C, Hanssen AM, van Sorge NM, Pettersen I, Diep DB et al (2016) The interaction between *Staphylococcus aureus* SdrD and desmoglein 1 is important for adhesion to host cells. Sci Rep 6:22134. https://doi.org/10.1038/srep22134
93. Claes J, Liesenborghs L, Peetermans M, Veloso TR, Missiakas D, Schneewind O et al (2017) Clumping factor a, von Willebrand factor-binding protein and von Willebrand factor anchor *Staphylococcus aureus* to the vessel wall. J Thromb Haemost 15(5):1009–1019. https://doi.org/10.1111/jth.13653
94. Janek D, Zipperer A, Kulik A, Krismer B, Peschel A (2016) High frequency and diversity of antimicrobial activities produced by nasal *staphylococcus* strains against bacterial competitors. PLoS Pathog 12(8):e1005812. https://doi.org/10.1371/journal.ppat.1005812
95. Wos-Oxley ML, Chaves-Moreno D, Jáuregui R, Oxley AP, Kaspar U, Plumeier I et al (2016) Exploring the bacterial assemblages along the human nasal passage. Environ Microbiol 18(7):2259–2271. https://doi.org/10.1111/1462-2920.13378
96. Eichelberger KR, Cassat JE (2021) Metabolic adaptations during *Staphylococcus aureus* and *Candida albicans* co-infection. Front Immunol 12:797550. https://doi.org/10.3389/fimmu.2021.797550
97. Beaume M, Kohler T, Fontana T, Tognon M, Renzoni A, van Delden C (2015) Metabolic pathways of *Pseudomonas aeruginosa* involved in competition with respiratory bacterial pathogens. Front Microbiol 6:321. https://doi.org/10.3389/fmicb.2015.00321
98. Fehrmann C, Jurk K, Bertling A, Seidel G, Fegeler W, Kehrel BE et al (2013) Role for the fibrinogen-binding proteins coagulase and Efb in the *Staphylococcus aureus-Candida* interaction. Int J Med Microbiol 303(5):230–238. https://doi.org/10.1016/j.ijmm.2013.02.011
99. Kiedrowski MR, Gaston JR, Kocak BR, Coburn SL, Lee S, Pilewski JM et al (2018) *Staphylococcus aureus* biofilm growth on cystic fibrosis airway epithelial cells is enhanced during respiratory syncytial virus coinfection. mSphere 3(4):e00341–e00318. https://doi.org/10.1128/mSphere.00341-18
100. Alves PM, Al-Badi E, Withycombe C, Jones PM, Purdy KJ, Maddocks SE (2018) Interaction between *Staphylococcus aureus* and *Pseudomonas aeruginosa* is beneficial for colonisation and pathogenicity in a mixed biofilm. Pathog Dis 76(1). https://doi.org/10.1093/femspd/fty003
101. Niemann S, Ehrhardt C, Medina E, Warnking K, Tuchscherr L, Heitmann V et al (2012) Combined action of influenza virus and *Staphylococcus aureus* panton-valentine leukocidin provokes severe lung epithelium damage. J Infect Dis 206(7):1138–1148. https://doi.org/10.1093/infdis/jis468
102. Tognon M, Kohler T, Gdaniec BG, Hao Y, Lam JS, Beaume M et al (2017) Co-evolution with *Staphylococcus aureus* leads to lipopolysaccharide alterations in *Pseudomonas aeruginosa*. ISME J 11(10):2233–2243. https://doi.org/10.1038/ismej.2017.83

103. Harriott MM, Noverr MC (2009) *Candida albicans* and *Staphylococcus aureus* form polymicrobial biofilms: effects on antimicrobial resistance. Antimicrob Agents Chemother 53(9):3914–3922. https://doi.org/10.1128/AAC.00657-09
104. Junge S, Görlich D, den Reijer M, Wiedemann B, Tümmler B, Ellemunter H et al (2016) Factors associated with worse lung function in cystic fibrosis patients with persistent *Staphylococcus aureus*. PloS One 11(11):e0166220. https://doi.org/10.1371/journal.pone.0166220
105. Burgess JL, Wyant WA, Abdo Abujamra B, Kirsner RS, Jozic I (2021) Diabetic wound-healing science. Medicina (Kaunas) 57(10):1072. https://doi.org/10.3390/medicina57101072
106. von Eiff C, Becker K, Machka K, Stammer H, Peters G (2001) Nasal carriage as a source of *Staphylococcus aureus* bacteremia. N Engl J Med 344(1):11–16. https://doi.org/10.1056/NEJM200101043440102
107. Luzar MA, Coles GA, Faller B, Slingeneyer A, Dah GD, Briat C et al (1990) *Staphylococcus aureus* nasal carriage and infection in patients on continuous ambulatory peritoneal dialysis. N Engl J Med 322(8):505–509. https://doi.org/10.1056/NEJM199002223220804
108. Kluytmans JA, Mouton JW, Ijzerman EP, Vandenbroucke-Grauls CM, Maat AW, Wagenvoort JH et al (1995) Nasal carriage of *Staphylococcus aureus* as a major risk factor for wound infections after cardiac surgery. J Infect Dis 171(1):216–219. https://doi.org/10.1093/infdis/171.1.216
109. Wertheim HF, Vos MC, Ott A, van Belkum A, Voss A, Kluytmans JA et al (2004) Risk and outcome of nosocomial *Staphylococcus aureus* bacteraemia in nasal carriers versus non-carriers. Lancet 364(9435):703–705. https://doi.org/10.1016/S0140-6736(04)16897-9
110. Pujol M, Peña C, Pallares R, Ariza J, Ayats J, Dominguez MA et al (1996) Nosocomial *Staphylococcus aureus* bacteremia among nasal carriers of methicillin-resistant and methicillin-susceptible strains. Am J Med 100(5):509–516. https://doi.org/10.1016/s0002-9343(96)00014-9
111. Moss B, Squire JR et al (1948) Nose and skin carriage of *Staphylococcus aureus* in patients receiving penicillin. Lancet 1(6496):320–325. https://doi.org/10.1016/s0140-6736(48)92088-1
112. Gillespie EH, Devenish EA, Cowan ST (1939) Pathogenic staphylococci. Their incidence in the nose and on the skin. Lancet 234(6060):870–873. https://doi.org/10.1016/S0140-6736(00)62957-4
113. Kloos WE, Musselwhite MS (1975) Distribution and persistence of *Staphylococcus* and *Micrococcus* species and other aerobic bacteria on human skin. Appl Microbiol 30(3):381–385. https://doi.org/10.1128/am.30.3.381-395.1975
114. Williams RE (1963) Healthy carriage of *Staphylococcus aureus*: its prevalence and importance. Bacteriol Rev 27:56–71. https://doi.org/10.1128/br.27.1.56-71.1963
115. Miles AA (1944) Epidemiology of wound infection. Lancet 243(6302):809–814. https://doi.org/10.1016/S0140-6736(00)75061-6
116. Sollid JU, Furberg AS, Hanssen AM, Johannessen M (2014) *Staphylococcus aureus*: determinants of human carriage. Infect Genet Evol 21:531–541. https://doi.org/10.1016/j.meegid.2013.03.020
117. Becker K, Schaumburg F, Fegeler C, Friedrich AW, Köck R, Prevalence of Multiresistant Microorganisms (PMM) Study Group (2017) *Staphylococcus aureus* from the German general population is highly diverse. Int J Med Microbiol 307(1):21–27. https://doi.org/10.1016/j.ijmm.2016.11.007
118. Tuazon CU, Sheagren JN (1974) Increased rate of carriage of *Staphylococcus aureus* among narcotic addicts. J Infect Dis 129(6):725–727. https://doi.org/10.1093/infdis/129.6.725
119. Chow JW, Yu VL (1989) *Staphylococcus aureus* nasal carriage in hemodialysis patients. Its role in infection and approaches to prophylaxis. Arch Intern Med 149(6):1258–1262. https://doi.org/10.1001/archinte.1989.00390060012003

120. Cespedes C, Miller M, Quagliarello B, Vavagiakis P, Klein RS, Lowy FD (2002) Differences between *Staphylococcus aureus* isolates from medical and nonmedical hospital personnel. J Clin Microbiol 40(7):2594–2597. https://doi.org/10.1128/JCM.40.7.2594-2597.2002
121. Boyko EJ, Lipsky BA, Sandoval R, Keane EM, Monahan JS, Pecoraro RE et al (1989) NIDDM and prevalence of nasal *Staphylococcus aureus* colonization. San Luis Valley Diabetes Study. Diabetes Care 12(3):189–192. https://doi.org/10.2337/diacare.12.3.189
122. Tuazon CU, Perez A, Kishaba T, Sheagren JN (1975) *Staphylococcus aureus* among insulin-injecting diabetic patients. An increased carrier rate. JAMA 231(12):1272. https://doi.org/10.1001/jama.1975.03240240042025
123. Mertz D, Frei R, Jaussi B, Tietz A, Stebler C, Flückiger U et al (2007) Throat swabs are necessary to reliably detect carriers of *Staphylococcus aureus*. Clin Infect Dis 45(4):475–477. https://doi.org/10.1086/520016
124. Köck R, Werner P, Friedrich AW, Fegeler C, Becker K, Prevalence of Multiresistant Microorganisms (PMM) Study Group (2016) Persistence of nasal colonization with human pathogenic bacteria and associated antimicrobial resistance in the German general population. New Microbes New Infect 9:24–34. https://doi.org/10.1016/j.nmni.2015.11.004
125. van Belkum A, Verkaik NJ, de Vogel CP, Boelens HA, Verveer J, Nouwen JL et al (2009) Reclassification of *Staphylococcus aureus* nasal carriage types. J Infect Dis 199(12):1820–1826. https://doi.org/10.1086/599119
126. Nakatsuji T, Chen TH, Narala S, Chun KA, Two AM, Yun T et al (2017) Antimicrobials from human skin commensal bacteria protect against *Staphylococcus aureus* and are deficient in atopic dermatitis. Sci Transl Med 9(378):eaah4680. https://doi.org/10.1126/scitranslmed.aah4680
127. Iwase T, Uehara Y, Shinji H, Tajima A, Seo H, Takada K et al (2010) *Staphylococcus epidermidis* Esp inhibits *Staphylococcus aureus* biofilm formation and nasal colonization. Nature 465(7296):346–349. https://doi.org/10.1038/nature09074
128. Sugimoto S, Iwamoto T, Takada K, Okuda K, Tajima A, Iwase T et al (2013) *Staphylococcus epidermidis* Esp degrades specific proteins associated with *Staphylococcus aureus* biofilm formation and host-pathogen interaction. J Bacteriol 195(8):1645–1655. https://doi.org/10.1128/JB.01672-12
129. Zipperer A, Konnerth MC, Laux C, Berscheid A, Janek D, Weidenmaier C et al (2016) Human commensals producing a novel antibiotic impair pathogen colonization. Nature 535(7613):511–516. https://doi.org/10.1038/nature18634
130. Wos-Oxley ML, Plumeier I, von Eiff C, Taudien S, Platzer M, Vilchez-Vargas R et al (2010) A poke into the diversity and associations within human anterior nare microbial communities. ISME J 4(7):839–851. https://doi.org/10.1038/ismej.2010.15
131. Bogaert D, van Belkum A, Sluijter M, Luijendijk A, de Groot R, Rumke HC et al (2004) Colonisation by *Streptococcus pneumoniae* and *Staphylococcus aureus* in healthy children. Lancet 363(9424):1871–1872. https://doi.org/10.1016/S0140-6736(04)16357-5
132. Uehara Y, Nakama H, Agematsu K, Uchida M, Kawakami Y, Abdul Fattah AS et al (2000) Bacterial interference among nasal inhabitants: eradication of *Staphylococcus aureus* from nasal cavities by artificial implantation of *Corynebacterium* sp. J Hosp Infect 44(2):127–133. https://doi.org/10.1053/jhin.1999.0680
133. Lina G, Boutite F, Tristan A, Bes M, Etienne J, Vandenesch F (2003) Bacterial competition for human nasal cavity colonization: role of staphylococcal *agr* alleles. Appl Environ Microbiol 69(1):18–23. https://doi.org/10.1128/AEM.69.1.18-23.2003
134. Danbolt N (1932) Undersøkelser over stafylokokker. Det Norske Vitenskeps-Akademi, Oslo
135. Weinstein HJ (1959) The relation between the nasal-staphylococcal-carrier state and the incidence of postoperative complications. N Engl J Med 260(26):1303–1308. https://doi.org/10.1056/NEJM195906252602601
136. Bode LG, Kluytmans JA, Wertheim HF, Bogaers D, Vandenbroucke-Grauls CM, Roosendaal R et al (2010) Preventing surgical-site infections in nasal carriers of *Staphylococcus aureus*. N Engl J Med 362(1):9–17. https://doi.org/10.1056/NEJMoa0808939

137. Muto CA, Jernigan JA, Ostrowsky BE, Richet HM, Jarvis WR, Boyce JM et al (2003) SHEA guideline for preventing nosocomial transmission of multidrug-resistant strains of *Staphylococcus aureus* and *Enterococcus*. Infect Control Hosp Epidemiol 24(5):362–386. https://doi.org/10.1086/502213
138. Kommission für Krankenhaushygiene und Infektionsprävention beim Robert Koch-Institut (2014) Empfehlungen zur Prävention und Kontrolle von Methicillinresistenten *Staphylococcus aureus*-Stämmen (MRSA) in medizinischen und pflegerischen Einrichtungen. Empfehlung der Kommission für Krankenhaushygiene und Infektionsprävention (KRINKO) beim Robert Koch-Institut. BundesgesundheitsblGesundheitsforschGesundheitsschutz 57(6):696–732. https://doi.org/10.1007/s00103-014-1980-x
139. Calfee DP, Salgado CD, Milstone AM, Harris AD, Kuhar DT, Moody J et al (2014) Strategies to prevent methicillin-resistant *Staphylococcus aureus* transmission and infection in acute care hospitals: 2014 update. Infect Control Hosp Epidemiol 35(Suppl 2):S108–S132. https://doi.org/10.1017/s0899823x00193882
140. Jain R, Kralovic SM, Evans ME, Ambrose M, Simbartl LA, Obrosky DS et al (2011) Veterans affairs initiative to prevent methicillin-resistant *Staphylococcus aureus* infections. N Engl J Med 364(15):1419–1430. https://doi.org/10.1056/NEJMoa1007474
141. Köck R, Becker K, Cookson B, van Gemert-Pijnen JE, Harbarth S, Kluytmans J et al (2014) Systematic literature analysis and review of targeted preventive measures to limit healthcare-associated infections by meticillin-resistant *Staphylococcus aureus*. Euro Surveill 19(29):20860. https://doi.org/10.2807/1560-7917.es2014.19.29.20860
142. Ammerlaan HS, Kluytmans JA, Berkhout H, Buiting A, de Brauwer EI, van den Broek PJ et al (2011) Eradication of carriage with methicillin-resistant *Staphylococcus aureus*: effectiveness of a national guideline. J Antimicrob Chemother 66(10):2409–2417. https://doi.org/10.1093/jac/dkr243
143. Awad SS, Palacio CH, Subramanian A, Byers PA, Abraham P, Lewis DA et al (2009) Implementation of a methicillin-resistant *Staphylococcus aureus* (MRSA) prevention bundle results in decreased MRSA surgical site infections. Am J Surg 198(5):607–610. https://doi.org/10.1016/j.amjsurg.2009.07.010
144. Jurke A, Köck R, Becker K, Thole S, Hendrix R, Rossen J et al (2013) Reduction of the nosocomial meticillin-resistant *Staphylococcus aureus* incidence density by a region-wide search and follow-strategy in forty German hospitals of the EUREGIO, 2009 to 2011. Euro Surveill 18(36):pii=20579. https://doi.org/10.2807/1560-7917.es2013.18.36.20579
145. Webb RM, Denton C, Spruill E, Henson G, Bruce L, Woods GL et al (2016) Multiclonal methicillin-resistant *Staphylococcus aureus* (MRSA) outbreak and its control after use of the veterans affairs (VA) MRSA bundle in a VA long-term care facility, 2004-2014. Am J Infect Control 44(6):727–729. https://doi.org/10.1016/j.ajic.2015.12.041
146. Levy PY, Ollivier M, Drancourt M, Raoult D, Argenson JN (2013) Relation between nasal carriage of *Staphylococcus aureus* and surgical site infection in orthopedic surgery: the role of nasal contamination. A systematic literature review and meta-analysis. Orthop Traumatol Surg Res 99(6):645–651. https://doi.org/10.1016/j.otsr.2013.03.030
147. Bode LG, van Rijen MM, Wertheim HF, Vandenbroucke-Grauls CM, Troelstra A, Voss A et al (2016) Long-term mortality after rapid screening and decolonization of *Staphylococcus aureus* carriers: observational follow-up study of a randomized, placebo-controlled trial. Ann Surg 263(3):511–515. https://doi.org/10.1097/SLA.0000000000001060
148. Campoccia D, Montanaro L, Arciola CR (2006) The significance of infection related to orthopedic devices and issues of antibiotic resistance. Biomaterials 27(11):2331–2339. https://doi.org/10.1016/j.biomaterials.2005.11.044
149. von Eiff C, Arciola CR, Montanaro L, Becker K, Campoccia D (2006) Emerging *Staphylococcus* species as new pathogens in implant infections. Int J Artif Organs 29(4):360–367. https://doi.org/10.1177/039139880602900405

150. Arciola CR, An YH, Campoccia D, Donati ME, Montanaro L (2005) Etiology of implant orthopedic infections: a survey on 1027 clinical isolates. Int J Artif Organs 28(11):1091–1100. https://doi.org/10.1177/039139880502801106
151. von Eiff C, Jansen B, Kohnen W, Becker K (2005) Infections associated with medical devices: pathogenesis, management and prophylaxis. Drugs 65(2):179–214. https://doi.org/10.2165/00003495-200565020-00003
152. Santarpia L, Buonomo A, Pagano MC, Alfonsi L, Foggia M, Mottola M et al (2016) Central venous catheter related bloodstream infections in adult patients on home parenteral nutrition: prevalence, predictive factors, therapeutic outcome. Clin Nutr 35(6):1394–1398. https://doi.org/10.1016/j.clnu.2016.03.009
153. Tuchscherr L, Löffler B (2016) *Staphylococcus aureus* dynamically adapts global regulators and virulence factor expression in the course from acute to chronic infection. Curr Genet 62(1):15–17. https://doi.org/10.1007/s00294-015-0503-0
154. Hale JH (1947) Studies on *Staphylococcus* mutation; characteristics of the "G" (gonidial) variant and factors concerned in its production. Br J Exp Pathol 28(3):202–210
155. Goudie JB, Goudie RB (1955) Recurrent infection by a stable-dwarf-colony variant of *Staphylococcus aureus*. J Clin Pathol 8:284–287
156. Swingle EL (1935) Studies on small colony variants of *Staphylococcus aureus*. J Bacteriol 29(5):467–489. https://doi.org/10.1128/jb.29.5.467-489.1935
157. Proctor RA, van Langevelde P, Kristjansson M, Maslow JN, Arbeit RD (1995) Persistent and relapsing infections associated with small-colony variants of *Staphylococcus aureus*. Clin Infect Dis 20(1):95–102. https://doi.org/10.1093/clinids/20.1.95
158. Rolauffs B, Bernhardt TM, von Eiff C, Hart ML, Bettin D (2002) Osteopetrosis, femoral fracture, and chronic osteomyelitis caused by *Staphylococcus aureus* small colony variants (SCV) treated by girdlestone resection—6-year follow-up. Arch Orthop Trauma Surg 122(9–10):547–550. https://doi.org/10.1007/s00402-002-0435-2
159. von Eiff C, Bettin D, Proctor RA, Rolauffs B, Lindner N, Winkelmann W et al (1997) Recovery of small colony variants of *Staphylococcus aureus* following gentamicin bead placement for osteomyelitis. Clin Infect Dis 25(5):1250–1251. https://doi.org/10.1086/516962
160. Kalinka J, Hachmeister M, Geraci J, Sordelli D, Hansen U, Niemann S et al (2014) *Staphylococcus aureus* isolates from chronic osteomyelitis are characterized by high host cell invasion and intracellular adaptation, but still induce inflammation. Int J Med Microbiol 304(8):1038–1049. https://doi.org/10.1016/j.ijmm.2014.07.013
161. Spearman P, Lakey D, Jotte S, Chernowitz A, Claycomb S, Stratton C (1996) Sternoclavicular joint septic arthritis with small-colony variant *Staphylococcus aureus*. Diagn Microbiol Infect Dis 26(1):13–15. https://doi.org/10.1016/s0732-8893(96)00145-9
162. Piffaut C, Lustig S, Laurent F, Chidiac C, Ferry T, Lyon BJI Study Group (2013) Small colony variant-producing *S aureus* prosthesis joint infection highlighted by sonication and treated with prolonged high doses of daptomycin. BMJ Case Rep 2013:bcr2013008637. https://doi.org/10.1136/bcr-2013-008637
163. Kipp F, Ziebuhr W, Becker K, Krimmer V, Höß N, Peters G et al (2003) Detection of *Staphylococcus aureus* by 16S rRNA directed in situ hybridisation in a patient with a brain abscess caused by small colony variants. J Neurol Neurosurg Psychiatry 74(7):1000–1002. https://doi.org/10.1136/jnnp.74.7.1000
164. Seifert H, von Eiff C, Fätkenheuer G (1999) Fatal case due to methicillin-resistant *Staphylococcus aureus* small colony variants in an AIDS patient. Emerg Infect Dis 5(3):450–453. https://doi.org/10.3201/eid0503.990319
165. Slifkin M, Merkow LP, Kreuzberger SA, Engwall C, Pardo M (1971) Characterization of CO_2 dependent microcolony variants of *Staphylococcus aureus*. Am J Clin Pathol 56(5):584–592. https://doi.org/10.1093/ajcp/56.5.584
166. Agarwal H, Verrall R, Singh SP, Tang YW, Wilson G (2007) Small colony variant *Staphylococcus aureus* multiorgan infection. Pediatr Infect Dis J 26(3):269–271. https://doi.org/10.1097/01.inf.0000256749.29244.67

167. Cervantes-García E, García-Gonzalez R, Reyes-Torres A, Resendiz-Albor AA, Salazar-Schettino PM (2015) *Staphylococcus aureus* small colony variants in diabetic foot infections. Diabetic Foot Ankle 6:26431. https://doi.org/10.3402/dfa.v6.26431
168. Abele-Horn M, Schupfner B, Emmerling P, Waldner H, Göring H (2000) Persistent wound infection after herniotomy associated with small-colony variants of *Staphylococcus aureus*. Infection 28(1):53–54. https://doi.org/10.1007/s150100050014
169. Gómez-González C, Acosta J, Villa J, Barrado L, Sanz F, Orellana MÁ et al (2010) Clinical and molecular characteristics of infections with CO_2-dependent small-colony variants of *Staphylococcus aureus*. J Clin Microbiol 48(8):2878–2884. https://doi.org/10.1128/JCM.00520-10
170. Baddour LM, Christensen GD (1987) Prosthetic valve endocarditis due to small-colony staphylococcal variants. Rev Infect Dis 9(6):1168–1174. https://doi.org/10.1093/clinids/9.6.1168
171. Chen H, Wang Q, Yin Y, Li S, Niu DK, Wang H (2018) Genotypic variations between wild-type and small colony variant of *Staphylococcus aureus* in prosthetic valve infectious endocarditis: a comparative genomic and transcriptomic analysis. Int J Antimicrob Agents 51(4):655–658. https://doi.org/10.1016/j.ijantimicag.2017.12.006
172. Tan NC, Cooksley CM, Roscioli E, Drilling AJ, Douglas R, Wormald PJ et al (2014) Small-colony variants and phenotype switching of intracellular *Staphylococcus aureus* in chronic rhinosinusitis. Allergy 69(10):1364–1371. https://doi.org/10.1111/all.12457
173. Spanu T, Romano L, D'Inzeo T, Masucci L, Albanese A, Papacci F et al (2005) Recurrent ventriculoperitoneal shunt infection caused by small-colony variants of *Staphylococcus aureus*. Clin Infect Dis 41(5):e48–e52. https://doi.org/10.1086/432577
174. Seifert H, Wisplinghoff H, Schnabel P, von Eiff C (2003) Small colony variants of *Staphylococcus aureus* and pacemaker-related infection. Emerg Infect Dis 9(10):1316–1318. https://doi.org/10.3201/eid0910.0302000
175. Tande AJ, Osmon DR, Greenwood-Quaintance KE, Mabry TM, Hanssen AD, Patel R (2014) Clinical characteristics and outcomes of prosthetic joint infection caused by small colony variant staphylococci. MBio 5(5):e01910–e01914. https://doi.org/10.1128/mBio.01910-14
176. Wolter DJ, Emerson JC, McNamara S, Buccat AM, Qin X, Cochrane E et al (2013) *Staphylococcus aureus* small-colony variants are independently associated with worse lung disease in children with cystic fibrosis. Clin Infect Dis 57(3):384–391. https://doi.org/10.1093/cid/cit270
177. Kahl B, Herrmann M, Everding AS, Koch HG, Becker K, Harms E et al (1998) Persistent infection with small colony variant strains of *Staphylococcus aureus* in patients with cystic fibrosis. J Infect Dis 177(4):1023–1029. https://doi.org/10.1086/515238
178. Schneider M, Mühlemann K, Droz S, Couzinet S, Casaulta C, Zimmerli S (2008) Clinical characteristics associated with isolation of small-colony variants of *Staphylococcus aureus* and *Pseudomonas aeruginosa* from respiratory secretions of patients with cystic fibrosis. J Clin Microbiol 46(5):1832–1834. https://doi.org/10.1128/JCM.00361-08
179. Lernau H, Sompolinsky D (1962) Bovine mastitis due to dwarf-colony variants of *Staphylococcus aureus*. Cornell Vet 52:445–451
180. Atalla H, Gyles C, Jacob CL, Moisan H, Malouin F, Mallard B (2008) Characterization of a *Staphylococcus aureus* small colony variant (SCV) associated with persistent bovine mastitis. Foodborne Pathog Dis 5(6):785–799. https://doi.org/10.1089/fpd.2008.0110
181. Becker K, Laham NA, Fegeler W, Proctor RA, Peters G, von Eiff C (2006) Fourier-transform infrared spectroscopic analysis is a powerful tool for studying the dynamic changes in *Staphylococcus aureus* small-colony variants. J Clin Microbiol 44(9):3274–3278. https://doi.org/10.1128/JCM.00847-06
182. Edwards AM (2012) Phenotype switching is a natural consequence of *Staphylococcus aureus* replication. J Bacteriol 194(19):5404–5412. https://doi.org/10.1128/JB.00948-12
183. Proctor RA, Kriegeskorte A, Kahl BC, Becker K, Löffler B, Peters G (2014) *Staphylococcus aureus* small colony variants (SCVs): a road map for the metabolic pathways involved in persistent infections. Front Cell Infect Microbiol 4:99. https://doi.org/10.3389/fcimb.2014.00099

184. Tuchscherr L, Heitmann V, Hussain M, Viemann D, Roth J, von Eiff C et al (2010) *Staphylococcus aureus* small-colony variants are adapted phenotypes for intracellular persistence. J Infect Dis 202(7):1031–1040. https://doi.org/10.1086/656047
185. Musher DM, Baughn RE, Templeton GB, Minuth JN (1977) Emergence of variant forms of *Staphylococcus aureus* after exposure to gentamicin and infectivity of the variants in experimental animals. J Infect Dis 136(3):360–369. https://doi.org/10.1093/infdis/136.3.360
186. Balwit JM, van Langevelde P, Vann JM, Proctor RA (1994) Gentamicin-resistant menadione and hemin auxotrophic *Staphylococcus aureus* persist within cultured endothelial cells. J Infect Dis 170(4):1033–1037. https://doi.org/10.1093/infdis/170.4.1033
187. Bui LM, Hoffmann P, Turnidge JD, Zilm PS, Kidd SP (2015) Prolonged growth of a clinical *Staphylococcus aureus* strain selects for a stable small-colony-variant cell type. Infect Immun 83(2):470–481. https://doi.org/10.1128/IAI.02702-14
188. Bui LM, Turnidge JD, Kidd SP (2015) The induction of *Staphylococcus aureus* biofilm formation or small Colony variants is a strain-specific response to host-generated chemical stresses. Microbes Infect 17(1):77–82. https://doi.org/10.1016/j.micinf.2014.09.009
189. Lattar SM, Tuchscherr LP, Caccuri RL, Centron D, Becker K, Alonso CA et al (2009) Capsule expression and genotypic differences among *Staphylococcus aureus* isolates from patients with chronic or acute osteomyelitis. Infect Immun 77(5):1968–1975. https://doi.org/10.1128/IAI.01214-08
190. Lannergård J, von Eiff C, Sander G, Cordes T, Seggewiß J, Peters G et al (2008) Identification of the genetic basis for clinical menadione-auxotrophic small-colony variant isolates of *Staphylococcus aureus*. Antimicrob Agents Chemother 52(11):4017–4022. https://doi.org/10.1128/AAC.00668-08
191. Gao W, Chua K, Davies JK, Newton HJ, Seemann T, Harrison PF et al (2010) Two novel point mutations in clinical *Staphylococcus aureus* reduce linezolid susceptibility and switch on the stringent response to promote persistent infection. PLoS Pathog 6(6):e1000944. https://doi.org/10.1371/journal.ppat.1000944
192. Chatterjee I, Kriegeskorte A, Fischer A, Deiwick S, Theimann N, Proctor RA et al (2008) In vivo mutations of thymidylate synthase (encoded by *thyA*) are responsible for thymidine dependency in clinical small-colony variants of *Staphylococcus aureus*. J Bacteriol 190(3):834–842. https://doi.org/10.1128/JB.00912-07
193. von Eiff C, Heilmann C, Proctor RA, Woltz C, Peters G, Götz F (1997) A site-directed *Staphylococcus aureus hemB* mutant is a small-colony variant which persists intracellularly. J Bacteriol 179(15):4706–4712. https://doi.org/10.1128/jb.179.15.4706-4712.1997
194. Kriegeskorte A, König S, Sander G, Pirkl A, Mahabir E, Proctor RA et al (2011) Small colony variants of *Staphylococcus aureus* reveal distinct protein profiles. Proteomics 11(12):2476–2490. https://doi.org/10.1002/pmic.201000796
195. Bates DM, von Eiff C, McNamara PJ, Peters G, Yeaman MR, Bayer AS et al (2003) *Staphylococcus aureus menD* and *hemB* mutants are as infective as the parent strains, but the menadione biosynthetic mutant persists within the kidney. J Infect Dis 187(10):1654–1661. https://doi.org/10.1086/374642
196. Kriegeskorte A, Block D, Drescher M, Windmüller N, Mellmann A, Baum C et al (2014) Inactivation of *thyA* in *Staphylococcus aureus* attenuates virulence and has a strong impact on metabolism and virulence gene expression. MBio 5(4):e01447–e01414. https://doi.org/10.1128/mBio.01447-14
197. Tuchscherr L, Medina E, Hussain M, Volker W, Heitmann V, Niemann S et al (2011) *Staphylococcus aureus* phenotype switching: an effective bacterial strategy to escape host immune response and establish a chronic infection. EMBO Mol Med 3(3):129–141. https://doi.org/10.1002/emmm.201000115
198. Proctor R (2019) Respiration and small colony variants of *Staphylococcus aureus*. Microbiol Spectr 7(3). https://doi.org/10.1128/microbiolspec.GPP3-0069-2019
199. Tuchscherr L, Kreis CA, Hoerr V, Flint L, Hachmeister M, Geraci J et al (2016) *Staphylococcus aureus* develops increased resistance to antibiotics by forming dynamic small colony vari-

ants during chronic osteomyelitis. J Antimicrob Chemother 71(2):438–448. https://doi.org/10.1093/jac/dkv371
200. Abu-Qatouseh LF, Chinni SV, Seggewiß J, Proctor RA, Brosius J, Rozhdestvensky TS et al (2010) Identification of differentially expressed small non-protein-coding RNAs in *Staphylococcus aureus* displaying both the normal and the small-colony variant phenotype. J Mol Med (Berl) 88(6):565–575. https://doi.org/10.1007/s00109-010-0597-2
201. Cui L, Neoh HM, Iwamoto A, Hiramatsu K (2012) Coordinated phenotype switching with large-scale chromosome flip-flop inversion observed in bacteria. Proc Natl Acad Sci U S A 109(25):E1647–E1656. https://doi.org/10.1073/pnas.1204307109
202. Guerillot R, Kostoulias X, Donovan L, Li L, Carter GP, Hachani A et al (2019) Unstable chromosome rearrangements in *Staphylococcus aureus* cause phenotype switching associated with persistent infections. Proc Natl Acad Sci U S A 116(40):20135–20140. https://doi.org/10.1073/pnas.1904861116
203. Gao W, Monk IR, Tobias NJ, Gladman SL, Seemann T, Stinear TP et al (2015) Large tandem chromosome expansions facilitate niche adaptation during persistent infection with drug-resistant *Staphylococcus aureus*. Microb Genom 1(2):e000026. https://doi.org/10.1099/mgen.0.000026
204. Kahl BC, Belling G, Reichelt R, Herrmann M, Proctor RA, Peters G (2003) Thymidine-dependent small-colony variants of *Staphylococcus aureus* exhibit gross morphological and ultrastructural changes consistent with impaired cell separation. J Clin Microbiol 41(1):410–413. https://doi.org/10.1128/JCM.41.1.410-413.2003
205. Dean MA, Olsen RJ, Long SW, Rosato AE, Musser JM (2014) Identification of point mutations in clinical *Staphylococcus aureus* strains that produce small-colony variants auxotrophic for menadione. Infect Immun 82(4):1600–1605. https://doi.org/10.1128/IAI.01487-13
206. Besier S, Ludwig A, Ohlsen K, Brade V, Wichelhaus TA (2007) Molecular analysis of the thymidine-auxotrophic small colony variant phenotype of *Staphylococcus aureus*. Int J Med Microbiol 297(4):217–225. https://doi.org/10.1016/j.ijmm.2007.02.003
207. Kriegeskorte A, Lore NI, Bragonzi A, Riva C, Kelkenberg M, Becker K et al (2015) Thymidine-dependent *Staphylococcus aureus* small-colony variants are induced by trimethoprim-sulfamethoxazole (SXT) and have increased fitness during SXT challenge. Antimicrob Agents Chemother 59(12):7265–7272. https://doi.org/10.1128/AAC.00742-15
208. Thomas ME (1955) Studies on a CO_2-dependent staphylococcus. J Clin Pathol 8(4):288–291. https://doi.org/10.1136/jcp.8.4.288
209. Schleimer N, Kaspar U, Drescher M, Seggewiß J, von Eiff C, Proctor RA et al (2018) The energy-coupling factor transporter module EcfAA'T, a novel candidate for the genetic basis of fatty acid-auxotrophic small-colony variants of *Staphylococcus aureus*. Front Microbiol 9:1863. https://doi.org/10.3389/fmicb.2018.01863
210. Baumert N, von Eiff C, Schaaff F, Peters G, Proctor RA, Sahl HG (2002) Physiology and antibiotic susceptibility of *Staphylococcus aureus* small colony variants. Microb Drug Resist 8(4):253–260. https://doi.org/10.1089/10766290260469507
211. Vaudaux P, Francois P, Bisognano C, Kelley WL, Lew DP, Schrenzel J et al (2002) Increased expression of clumping factor and fibronectin-binding proteins by *hemB* mutants of *Staphylococcus aureus* expressing small colony variant phenotypes. Infect Immun 70(10):5428–5437. https://doi.org/10.1128/IAI.70.10.5428-5437.2002
212. Jonsson IM, von Eiff C, Proctor RA, Peters G, Ryden C, Tarkowski A (2003) Virulence of a *hemB* mutant displaying the phenotype of a *Staphylococcus aureus* small colony variant in a murine model of septic arthritis. Microb Pathog 34(2):73–79. https://doi.org/10.1016/s0882-4010(02)00208-5
213. Samuelsen O, Haukland HH, Kahl BC, von Eiff C, Proctor RA, Ulvatne H et al (2005) *Staphylococcus aureus* small colony variants are resistant to the antimicrobial peptide lactoferricin B. J Antimicrob Chemother 56(6):1126–1129. https://doi.org/10.1093/jac/dki385

214. Sifri CD, Baresch-Bernal A, Calderwood SB, von Eiff C (2006) Virulence of *Staphylococcus aureus* small colony variants in the *Caenorhabditis elegans* infection model. Infect Immun 74(2):1091–1096. https://doi.org/10.1128/IAI.74.2.1091-1096.2006
215. Werth N, Beerlage C, Rosenberger C, Yazdi AS, Edelmann M, Amr A et al (2010) Activation of hypoxia inducible factor 1 is a general phenomenon in infections with human pathogens. PloS One 5(7):e11576. https://doi.org/10.1371/journal.pone.0011576
216. Gläser R, Becker K, von Eiff C, Meyer-Hoffert U, Harder J (2014) Decreased susceptibility of *Staphylococcus aureus* small-colony variants toward human antimicrobial peptides. J Invest Dermatol 134(9):2347–2350. https://doi.org/10.1038/jid.2014.176
217. Idelevich EA, Kriegeskorte A, Schleimer N, Peters G, von Eiff C, Becker K (2018) *In vitro* susceptibility of clinical *Staphylococcus aureus* small-colony variants to beta-lactam and non-beta-lactam antibiotics. Antimicrob Agents Chemother 62(4):e02532–e02517. https://doi.org/10.1128/AAC.02532-17
218. Garcia LG, Lemaire S, Kahl BC, Becker K, Proctor RA, Denis O et al (2012) Pharmacodynamic evaluation of the activity of antibiotics against hemin- and menadione-dependent small-colony variants of *Staphylococcus aureus* in models of extracellular (broth) and intracellular (THP-1 monocytes) infections. Antimicrob Agents Chemother 56(7):3700–3711. https://doi.org/10.1128/AAC.00285-12
219. Garcia LG, Lemaire S, Kahl BC, Becker K, Proctor RA, Denis O et al (2013) Antibiotic activity against small-colony variants of *Staphylococcus aureus*: review of in vitro, animal and clinical data. J Antimicrob Chemother 68(7):1455–1464. https://doi.org/10.1093/jac/dkt072
220. McNamara PJ, Proctor RA (2000) *Staphylococcus aureus* small colony variants, electron transport and persistent infections. Int J Antimicrob Agents 14(2):117–122. https://doi.org/10.1016/s0924-8579(99)00170-3
221. Seggewiß J, Becker K, Kotte O, Eisenacher M, Yazdi MR, Fischer A et al (2006) Reporter metabolite analysis of transcriptional profiles of a *Staphylococcus aureus* strain with normal phenotype and its isogenic *hemB* mutant displaying the small-colony-variant phenotype. J Bacteriol 188(22):7765–7777. https://doi.org/10.1128/JB.00774-06
222. von Eiff C, McNamara P, Becker K, Bates D, Lei XH, Ziman M et al (2006) Phenotype microarray profiling of *Staphylococcus aureus menD* and *hemB* mutants with the small-colony-variant phenotype. J Bacteriol 188(2):687–693. https://doi.org/10.1128/JB.188.2.687-693.2006
223. Kriegeskorte A, Grubmuller S, Huber C, Kahl BC, von Eiff C, Proctor RA et al (2014) *Staphylococcus aureus* small colony variants show common metabolic features in central metabolism irrespective of the underlying auxotrophism. Front Cell Infect Microbiol 4:141. https://doi.org/10.3389/fcimb.2014.00141
224. Kohler C, von Eiff C, Peters G, Proctor RA, Hecker M, Engelmann S (2003) Physiological characterization of a heme-deficient mutant of *Staphylococcus aureus* by a proteomic approach. J Bacteriol 185(23):6928–6937. https://doi.org/10.1128/JB.185.23.6928-6937.2003
225. Vann JM, Proctor RA (1988) Cytotoxic effects of ingested *Staphylococcus aureus* on bovine endothelial cells: role of *S. aureus* alpha-hemolysin. Microb Pathog 4(6):443–453. https://doi.org/10.1016/0882-4010(88)90029-0
226. Goerke C, Wolz C (2004) Regulatory and genomic plasticity of *Staphylococcus aureus* during persistent colonization and infection. Int J Med Microbiol 294(2–3):195–202. https://doi.org/10.1016/j.ijmm.2004.06.013
227. Kahl BC, Belling G, Becker P, Chatterjee I, Wardecki K, Hilgert K et al (2005) Thymidine-dependent *Staphylococcus aureus* small-colony variants are associated with extensive alterations in regulator and virulence gene expression profiles. Infect Immun 73(7):4119–4126. https://doi.org/10.1128/IAI.73.7.4119-4126.2005
228. Novick RP, Ross HF, Projan SJ, Kornblum J, Kreiswirth B, Moghazeh S (1993) Synthesis of staphylococcal virulence factors is controlled by a regulatory RNA molecule. EMBO J 12(10):3967–3975. https://doi.org/10.1002/j.1460-2075.1993.tb06074.x

229. Vesga O, Groeschel MC, Otten MF, Brar DW, Vann JM, Proctor RA (1996) *Staphylococcus aureus* small colony variants are induced by the endothelial cell intracellular milieu. J Infect Dis 173(3):739–742. https://doi.org/10.1093/infdis/173.3.739
230. Gilligan PH, Gage PA, Welch DF, Muszynski MJ, Wait KR (1987) Prevalence of thymidine-dependent *Staphylococcus aureus* in patients with cystic fibrosis. J Clin Microbiol 25(7):1258–1261. https://doi.org/10.1128/jcm.25.7.1258-1261.1987
231. Besier S, Smaczny C, von Mallinckrodt C, Krahl A, Ackermann H, Brade V et al (2007) Prevalence and clinical significance of *Staphylococcus aureus* small-colony variants in cystic fibrosis lung disease. J Clin Microbiol 45(1):168–172. https://doi.org/10.1128/JCM.01510-06
232. Cucarella C, Tormo MA, Knecht E, Amorena B, Lasa I, Foster TJ et al (2002) Expression of the biofilm-associated protein interferes with host protein receptors of *Staphylococcus aureus* and alters the infective process. Infect Immun 70(6):3180–3186. https://doi.org/10.1128/IAI.70.6.3180-3186.2002
233. Valle J, Latasa C, Gil C, Toledo-Arana A, Solano C, Penadés JR et al (2012) Bap, a biofilm matrix protein of *Staphylococcus aureus* prevents cellular internalization through binding to GP96 host receptor. PLoS Pathog 8(8):e1002843. https://doi.org/10.1371/journal.ppat.1002843
234. Taglialegna A, Navarro S, Ventura S, Garnett JA, Matthews S, Penades JR et al (2016) Staphylococcal bap proteins build amyloid scaffold biofilm matrices in response to environmental signals. PLoS Pathog 12(6):e1005711. https://doi.org/10.1371/journal.ppat.1005711
235. Vazquez V, Liang X, Horndahl JK, Ganesh VK, Smeds E, Foster TJ et al (2011) Fibrinogen is a ligand for the *Staphylococcus aureus* microbial surface components recognizing adhesive matrix molecules (MSCRAMM) bone sialoprotein-binding protein (Bbp). J Biol Chem 286(34):29797–29805. https://doi.org/10.1074/jbc.M110.214981
236. Tung H, Guss B, Hellman U, Persson L, Rubin K, Rydén C (2000) A bone sialoprotein-binding protein from *Staphylococcus aureus*: a member of the staphylococcal Sdr family. Biochem J 345(Pt 3):611–619
237. Zhang X, Wu M, Zhuo W, Gu J, Zhang S, Ge J et al (2015) Crystal structures of Bbp from *Staphylococcus aureus* reveal the ligand binding mechanism with fibrinogen α. Protein Cell 6(10):757–766. https://doi.org/10.1007/s13238-015-0205-x
238. Yacoub A, Lindahl P, Rubin K, Wendel M, Heinegård D, Rydén C (1994) Purification of a bone sialoprotein-binding protein from *Staphylococcus aureus*. Eur J Biochem 222(3):919–925. https://doi.org/10.1111/j.1432-1033.1994.tb18940.x
239. Ganesh VK, Rivera JJ, Smeds E, Ko YP, Bowden MG, Wann ER et al (2008) A structural model of the *Staphylococcus aureus* ClfA-fibrinogen interaction opens new avenues for the design of anti-staphylococcal therapeutics. PLoS Pathog 4(11):e1000226. https://doi.org/10.1371/journal.ppat.1000226
240. McDevitt D, Francois P, Vaudaux P, Foster TJ (1994) Molecular characterization of the clumping factor (fibrinogen receptor) of *Staphylococcus aureus*. Mol Microbiol 11(2):237–248. https://doi.org/10.1111/j.1365-2958.1994.tb00304.x
241. Hair PS, Ward MD, Semmes OJ, Foster TJ, Cunnion KM (2008) *Staphylococcus aureus* clumping factor a binds to complement regulator factor I and increases factor I cleavage of C3b. J Infect Dis 198(1):125–133. https://doi.org/10.1086/588825
242. Ashraf S, Cheng J, Zhao X (2017) Clumping factor a of *Staphylococcus aureus* interacts with AnnexinA2 on mammary epithelial cells. Sci Rep 7:40608. https://doi.org/10.1038/srep40608
243. Ganesh VK, Barbu EM, Deivanayagam CC, Le B, Anderson AS, Matsuka YV et al (2011) Structural and biochemical characterization of *Staphylococcus aureus* clumping factor B/ligand interactions. J Biol Chem 286(29):25963–25972. https://doi.org/10.1074/jbc.M110.217414
244. Walsh EJ, Miajlovic H, Gorkun OV, Foster TJ (2008) Identification of the *Staphylococcus aureus* MSCRAMM clumping factor B (ClfB) binding site in the alphaC-domain of human fibrinogen. Microbiology 154(Pt 2):550–558. https://doi.org/10.1099/mic.0.2007/010868-0

245. Xiang H, Feng Y, Wang J, Liu B, Chen Y, Liu L et al (2012) Crystal structures reveal the multi-ligand binding mechanism of *Staphylococcus aureus* ClfB. PLoS Pathog 8(6):e1002751. https://doi.org/10.1371/journal.ppat.1002751
246. Mulcahy ME, Geoghegan JA, Monk IR, O'Keeffe KM, Walsh EJ, Foster TJ et al (2012) Nasal colonisation by *Staphylococcus aureus* depends upon clumping factor B binding to the squamous epithelial cell envelope protein loricrin. PLoS Pathog 8(12):e1003092. https://doi.org/10.1371/journal.ppat.1003092
247. Haim M, Trost A, Maier CJ, Achatz G, Feichtner S, Hintner H et al (2010) Cytokeratin 8 interacts with clumping factor B: a new possible virulence factor target. Microbiology 156(Pt 12):3710–3721. https://doi.org/10.1099/mic.0.034413-0
248. Roche FM, Massey R, Peacock SJ, Day NP, Visai L, Speziale P et al (2003) Characterization of novel LPXTG-containing proteins of *Staphylococcus aureus* identified from genome sequences. Microbiology 149(Pt 3):643–654. https://doi.org/10.1099/mic.0.25996-0
249. Thammavongsa V, Schneewind O, Missiakas DM (2011) Enzymatic properties of *Staphylococcus aureus* adenosine synthase (AdsA). BMC Biochem 12:56. https://doi.org/10.1186/1471-2091-12-56
250. Thammavongsa V, Kern JW, Missiakas DM, Schneewind O (2009) *Staphylococcus aureus* synthesizes adenosine to escape host immune responses. J Exp Med 206(11):2417–2427. https://doi.org/10.1084/jem.20090097
251. Espersen F, Clemmensen I (1982) Isolation of a fibronectin-binding protein from *Staphylococcus aureus*. Infect Immun 37(2):526–531. https://doi.org/10.1128/iai.37.2.526-531.1982
252. Flock JI, Fröman G, Jönsson K, Guss B, Signäs C, Nilsson B et al (1987) Cloning and expression of the gene for a fibronectin-binding protein from *Staphylococcus aureus*. EMBO J 6(8):2351–2357. https://doi.org/10.1002/j.1460-2075.1987.tb02511.x
253. Fitzgerald JR, Loughman A, Keane F, Brennan M, Knobel M, Higgins J et al (2006) Fibronectin-binding proteins of *Staphylococcus aureus* mediate activation of human platelets via fibrinogen and fibronectin bridges to integrin GPIIb/IIIa and IgG binding to the FcγRIIa receptor. Mol Microbiol 59(1):212–230. https://doi.org/10.1111/j.1365-2958.2005.04922.x
254. Keane FM, Loughman A, Valtulina V, Brennan M, Speziale P, Foster TJ (2007) Fibrinogen and elastin bind to the same region within the a domain of fibronectin binding protein a, an MSCRAMM of *Staphylococcus aureus*. Mol Microbiol 63(3):711–723. https://doi.org/10.1111/j.1365-2958.2006.05552.x
255. Schwarz-Linek U, Werner JM, Pickford AR, Gurusiddappa S, Kim JH, Pilka ES et al (2003) Pathogenic bacteria attach to human fibronectin through a tandem β-zipper. Nature 423(6936):177–181. https://doi.org/10.1038/nature01589
256. Peacock SJ, Foster TJ, Cameron BJ, Berendt AR (1999) Bacterial fibronectin-binding proteins and endothelial cell surface fibronectin mediate adherence of *Staphylococcus aureus* to resting human endothelial cells. Microbiology 145(Pt 12):3477–3486. https://doi.org/10.1099/00221287-145-12-3477
257. Jönsson K, Signäs C, Müller HP, Lindberg M (1991) Two different genes encode fibronectin binding proteins in *Staphylococcus aureus*. The complete nucleotide sequence and characterization of the second gene. Eur J Biochem 202(3):1041–1048. https://doi.org/10.1111/j.1432-1033.1991.tb16468.x
258. Burke FM, Di Poto A, Speziale P, Foster TJ (2011) The a domain of fibronectin-binding protein B of *Staphylococcus aureus* contains a novel fibronectin binding site. FEBS J 278(13):2359–2371. https://doi.org/10.1111/j.1742-4658.2011.08159.x
259. Clarke SR, Brummell KJ, Horsburgh MJ, McDowell PW, Mohamad SA, Stapleton MR et al (2006) Identification of in vivo-expressed antigens of *Staphylococcus aureus* and their use in vaccinations for protection against nasal carriage. J Infect Dis 193(8):1098–1108. https://doi.org/10.1086/501471

260. Clarke SR, Andre G, Walsh EJ, Dufrene YF, Foster TJ, Foster SJ (2009) Iron-regulated surface determinant protein a mediates adhesion of *Staphylococcus aureus* to human corneocyte envelope proteins. Infect Immun 77(6):2408–2416. https://doi.org/10.1128/IAI.01304-08
261. Clarke SR, Wiltshire MD, Foster SJ (2004) IsdA of *Staphylococcus aureus* is a broad spectrum, iron-regulated adhesin. Mol Microbiol 51(5):1509–1519. https://doi.org/10.1111/j.1365-2958.2003.03938.x
262. Clarke SR, Foster SJ (2008) IsdA protects *Staphylococcus aureus* against the bactericidal protease activity of apolactoferrin. Infect Immun 76(4):1518–1526. https://doi.org/10.1128/IAI.01530-07
263. Wiltshire MD, Foster SJ (2001) Identification and analysis of *Staphylococcus aureus* components expressed by a model system of growth in serum. Infect Immun 69(8):5198–5202. https://doi.org/10.1128/IAI.69.8.5198-5202.2001
264. Mazmanian SK, Skaar EP, Gaspar AH, Humayun M, Gornicki P, Jelenska J et al (2003) Passage of heme-iron across the envelope of *Staphylococcus aureus*. Science 299(5608):906–909. https://doi.org/10.1126/science.1081147
265. Taylor JM, Heinrichs DE (2002) Transferrin binding in *Staphylococcus aureus*: involvement of a cell wall-anchored protein. Mol Microbiol 43(6):1603–1614. https://doi.org/10.1046/j.1365-2958.2002.02850.x
266. Cheng AG, Kim HK, Burts ML, Krausz T, Schneewind O, Missiakas DM (2009) Genetic requirements for *Staphylococcus aureus* abscess formation and persistence in host tissues. FASEB J 23(10):3393–3404. https://doi.org/10.1096/fj.09-135467
267. Pluym M, Muryoi N, Heinrichs DE, Stillman MJ (2008) Heme binding in the NEAT domains of IsdA and IsdC of *Staphylococcus aureus*. J Inorg Biochem 102(3):480–488. https://doi.org/10.1016/j.jinorgbio.2007.11.011
268. Grigg JC, Ukpabi G, Gaudin CF, Murphy ME (2010) Structural biology of heme binding in the *Staphylococcus aureus* Isd system. J Inorg Biochem 104(3):341–348. https://doi.org/10.1016/j.jinorgbio.2009.09.012
269. Pilpa RM, Robson SA, Villareal VA, Wong ML, Phillips M, Clubb RT (2009) Functionally distinct NEAT (NEAr transporter) domains within the *Staphylococcus aureus* IsdH/HarA protein extract heme from methemoglobin. J Biol Chem 284(2):1166–1176. https://doi.org/10.1074/jbc.M806007200
270. Zapotoczna M, Jevnikar Z, Miajlovic H, Kos J, Foster TJ (2013) Iron-regulated surface determinant B (IsdB) promotes *Staphylococcus aureus* adherence to and internalization by non-phagocytic human cells. Cell Microbiol 15(6):1026–1041. https://doi.org/10.1111/cmi.12097
271. Miajlovic H, Zapotoczna M, Geoghegan JA, Kerrigan SW, Speziale P, Foster TJ (2010) Direct interaction of iron-regulated surface determinant IsdB of *Staphylococcus aureus* with the GPIIb/IIIa receptor on platelets. Microbiology 156(Pt 3):920–928. https://doi.org/10.1099/mic.0.036673-0
272. Alfeo MJ, Pagotto A, Barbieri G, Foster TJ, Vanhoorelbeke K, De Filippis V et al (2021) *Staphylococcus aureus* iron-regulated surface determinant B (IsdB) protein interacts with von Willebrand factor and promotes adherence to endothelial cells. Sci Rep 11(1):22799. https://doi.org/10.1038/s41598-021-02065-w
273. Visai L, Yanagisawa N, Josefsson E, Tarkowski A, Pezzali I, Rooijakkers SH et al (2009) Immune evasion by *Staphylococcus aureus* conferred by iron-regulated surface determinant protein IsdH. Microbiology 155(Pt 3):667–679. https://doi.org/10.1099/mic.0.025684-0
274. Sæderup KL, Stødkilde K, Graversen JH, Dickson CF, Etzerodt A, Hansen SW et al (2016) The *Staphylococcus aureus* protein IsdH inhibits host hemoglobin scavenging to promote heme acquisition by the pathogen. J Biol Chem 291(46):23989–23998. https://doi.org/10.1074/jbc.M116.755934
275. Pietrocola G, Pellegrini A, Alfeo MJ, Marchese L, Foster TJ, Speziale P (2020) The iron-regulated surface determinant B (IsdB) protein from *Staphylococcus aureus* acts as a receptor for the host protein vitronectin. J Biol Chem 295(29):10008–10022. https://doi.org/10.1074/jbc.RA120.013510

276. Dryla A, Gelbmann D, von Gabain A, Nagy E (2003) Identification of a novel iron regulated staphylococcal surface protein with haptoglobin-haemoglobin binding activity. Mol Microbiol 49(1):37–53. https://doi.org/10.1046/j.1365-2958.2003.03542.x
277. Nguyen T, Ghebrehiwet B, Peerschke EI (2000) *Staphylococcus aureus* protein a recognizes platelet gC1qR/p33: a novel mechanism for staphylococcal interactions with platelets. Infect Immun 68(4):2061–2068. https://doi.org/10.1128/IAI.68.4.2061-2068.2000
278. Deisenhofer J (1981) Crystallographic refinement and atomic models of a human fc fragment and its complex with fragment B of protein a from *Staphylococcus aureus* at 2.9- and 2.8-a resolution. Biochemistry 20(9):2361–2370. https://doi.org/10.1021/bi00512a001
279. Jansson B, Uhlen M, Nygren PA (1998) All individual domains of staphylococcal protein a show fab binding. FEMS Immunol Med Microbiol 20(1):69–78. https://doi.org/10.1111/j.1574-695X.1998.tb01112.x
280. Inganäs M (1981) Comparison of mechanisms of interaction between protein a from *Staphylococcus aureus* and human monoclonal IgG, IgA and IgM in relation to the classical FC gamma and the alternative F(ab′)2 epsilon protein a interactions. Scand J Immunol 13(4):343–352. https://doi.org/10.1111/j.1365-3083.1981.tb00143.x
281. Falugi F, Kim HK, Missiakas DM, Schneewind O (2013) Role of protein a in the evasion of host adaptive immune responses by *Staphylococcus aureus*. MBio 4(5):e00575–e00513. https://doi.org/10.1128/mBio.00575-13
282. Goodyear CS, Silverman GJ (2003) Death by a B cell superantigen: in vivo VH-targeted apoptotic supraclonal B cell deletion by a staphylococcal toxin. J Exp Med 197(9):1125–1139. https://doi.org/10.1084/jem.20020552
283. O'Brien L, Kerrigan SW, Kaw G, Hogan M, Penades J, Litt D et al (2002) Multiple mechanisms for the activation of human platelet aggregation by *Staphylococcus aureus*: roles for the clumping factors ClfA and ClfB, the serine-aspartate repeat protein SdrE and protein A. Mol Microbiol 44(4):1033–1144. https://doi.org/10.1046/j.1365-2958.2002.02935.x
284. Uhlén M, Guss B, Nilsson B, Gatenbeck S, Philipson L, Lindberg M (1984) Complete sequence of the staphylococcal gene encoding protein A. a gene evolved through multiple duplications. J Biol Chem 259(3):1695–1702. https://doi.org/10.1016/S0021-9258(17)43463-6
285. Gómez MI, Lee A, Reddy B, Muir A, Soong G, Pitt A et al (2004) *Staphylococcus aureus* protein a induces airway epithelial inflammatory responses by activating TNFR1. Nat Med 10(8):842–848. https://doi.org/10.1038/nm1079
286. Hartleib J, Köhler N, Dickinson RB, Chhatwal GS, Sixma JJ, Hartford OM et al (2000) Protein a is the von Willebrand factor binding protein on *Staphylococcus aureus*. Blood 96(6):2149–2156. https://doi.org/10.1182/blood.V96.6.2149
287. Peterson PK, Verhoef J, Sabath LD, Quie PG (1977) Effect of protein a on staphylococcal opsonization. Infect Immun 15(3):760–764. https://doi.org/10.1128/iai.15.3.760-764.1977
288. Patti JM, Jonsson H, Guss B, Switalski LM, Wiberg K, Lindberg M et al (1992) Molecular characterization and expression of a gene encoding a *Staphylococcus aureus* collagen adhesin. J Biol Chem 267(7):4766–4772. https://doi.org/10.1016/S0021-9258(18)42898-0
289. Zong Y, Xu Y, Liang X, Keene DR, Hook A, Gurusiddappa S et al (2005) A ‘Collagen Hug’ model for *Staphylococcus aureus* CNA binding to collagen. EMBO J 24(24):4224–4236. https://doi.org/10.1038/sj.emboj.7600888
290. Kang M, Ko YP, Liang X, Ross CL, Liu Q, Murray BE et al (2013) Collagen-binding microbial surface components recognizing adhesive matrix molecule (MSCRAMM) of gram-positive bacteria inhibit complement activation via the classical pathway. J Biol Chem 288(28):20520–20531. https://doi.org/10.1074/jbc.M113.454462
291. Valotteau C, Prystopiuk V, Pietrocola G, Rindi S, Peterle D, De Filippis V et al (2017) Single-cell and single-molecule analysis unravels the multifunctionality of the *Staphylococcus aureus* collagen-binding protein Cna. ACS Nano 11(2):2160–2170. https://doi.org/10.1021/acsnano.6b08404
292. Schroeder K, Jularic M, Horsburgh SM, Hirschhausen N, Neumann C, Bertling A et al (2009) Molecular characterization of a novel *Staphylococcus aureus* surface protein (SasC) involved

in cell aggregation and biofilm accumulation. PloS One 4(10):e7567. https://doi.org/10.1371/journal.pone.0007567
293. Roche FM, Meehan M, Foster TJ (2003) The *Staphylococcus aureus* surface protein SasG and its homologues promote bacterial adherence to human desquamated nasal epithelial cells. Microbiology 149(Pt 10):2759–2767. https://doi.org/10.1099/mic.0.26412-0
294. Geoghegan JA, Corrigan RM, Gruszka DT, Speziale P, O'Gara JP, Potts JR et al (2010) Role of surface protein SasG in biofilm formation by *Staphylococcus aureus*. J Bacteriol 192(21):5663–5673. https://doi.org/10.1128/JB.00628-10
295. Huesca M, Peralta R, Sauder DN, Simor AE, McGavin MJ (2002) Adhesion and virulence properties of epidemic Canadian methicillin-resistant *Staphylococcus aureus* strain 1: identification of novel adhesion functions associated with plasmin-sensitive surface protein. J Infect Dis 185(9):1285–1296. https://doi.org/10.1086/340123
296. Hussain M, Schäfer D, Juuti KM, Peters G, Haslinger-Löffler B, Kuusela PI et al (2009) Expression of Pls (plasmin sensitive) in *Staphylococcus aureus* negative for *pls* reduces adherence and cellular invasion and acts by steric hindrance. J Infect Dis 200(1):107–117. https://doi.org/10.1086/599359
297. Li M, Du X, Villaruz AE, Diep BA, Wang D, Song Y et al (2012) MRSA epidemic linked to a quickly spreading colonization and virulence determinant. Nat Med 18(5):816–819. https://doi.org/10.1038/nm.2692
298. Barbu EM, Mackenzie C, Foster TJ, Hook M (2014) SdrC induces staphylococcal biofilm formation through a homophilic interaction. Mol Microbiol 94(1):172–185. https://doi.org/10.1111/mmi.12750
299. Barbu EM, Ganesh VK, Gurusiddappa S, Mackenzie RC, Foster TJ, Sudhof TC et al (2010) β-Neurexin is a ligand for the *Staphylococcus aureus* MSCRAMM SdrC. PLoS Pathog 6(1):e1000726. https://doi.org/10.1371/journal.ppat.1000726
300. Josefsson E, McCrea KW, Ní Eidhin D, O'Connell D, Cox J, Höök M et al (1998) Three new members of the serine-aspartate repeat protein multigene family of *Staphylococcus aureus*. Microbiology 144(Pt 12):3387–3395. https://doi.org/10.1099/00221287-144-12-3387
301. Feuillie C, Formosa-Dague C, Hays LM, Vervaeck O, Derclaye S, Brennan MP et al (2017) Molecular interactions and inhibition of the staphylococcal biofilm-forming protein SdrC. Proc Natl Acad Sci U S A 114(14):3738–3743. https://doi.org/10.1073/pnas.1616805114
302. Askarian F, Uchiyama S, Valderrama JA, Ajayi C, Sollid JUE, van Sorge NM et al (2017) Serine-aspartate repeat protein D increases *Staphylococcus aureus* virulence and survival in blood. Infect Immun 85(1):e00559–e00516. https://doi.org/10.1128/IAI.00559-16
303. Josefsson E, O'Connell D, Foster TJ, Durussel I, Cox JA (1998) The binding of calcium to the B-repeat segment of SdrD, a cell surface protein of *Staphylococcus aureus*. J Biol Chem 273(47):31145–31152. https://doi.org/10.1074/jbc.273.47.31145
304. Zhang Y, Wu M, Hang T, Wang C, Yang Y, Pan W et al (2017) *Staphylococcus aureus* SdrE captures complement factor H's C-terminus via a novel 'close, dock, lock and latch' mechanism for complement evasion. Biochem J 474(10):1619–1631. https://doi.org/10.1042/BCJ20170085
305. Sharp JA, Echague CG, Hair PS, Ward MD, Nyalwidhe JO, Geoghegan JA et al (2012) *Staphylococcus aureus* surface protein SdrE binds complement regulator factor H as an immune evasion tactic. PloS One 7(5):e38407. https://doi.org/10.1371/journal.pone.0038407
306. Siboo IR, Chambers HF, Sullam PM (2005) Role of SraP, a serine-rich surface protein of *Staphylococcus aureus*, in binding to human platelets. Infect Immun 73(4):2273–2280. https://doi.org/10.1128/IAI.73.4.2273-2280.2005
307. Sanchez CJ, Shivshankar P, Stol K, Trakhtenbroit S, Sullam PM, Sauer K et al (2010) The pneumococcal serine-rich repeat protein is an intra-species bacterial adhesin that promotes bacterial aggregation in vivo and in biofilms. PLoS Pathog 6(8):e1001044. https://doi.org/10.1371/journal.ppat.1001044

308. Yang YH, Jiang YL, Zhang J, Wang L, Bai XH, Zhang SJ et al (2014) Structural insights into SraP-mediated *Staphylococcus aureus* adhesion to host cells. PLoS Pathog 10(6):e1004169. https://doi.org/10.1371/journal.ppat.1004169
309. Kajimura J, Fujiwara T, Yamada S, Suzawa Y, Nishida T, Oyamada Y et al (2005) Identification and molecular characterization of an N-acetylmuramyl-L-alanine amidase Sle1 involved in cell separation of *Staphylococcus aureus*. Mol Microbiol 58(4):1087–1101. https://doi.org/10.1111/j.1365-2958.2005.04881.x
310. Takahashi J, Komatsuzawa H, Yamada S, Nishida T, Labischinski H, Fujiwara T et al (2002) Molecular characterization of an atl null mutant of *Staphylococcus aureus*. Microbiol Immunol 46(9):601–612. https://doi.org/10.1111/j.1348-0421.2002.tb02741.x
311. Pasztor L, Ziebandt AK, Nega M, Schlag M, Haase S, Franz-Wachtel M et al (2010) Staphylococcal major autolysin (Atl) is involved in excretion of cytoplasmic proteins. J Biol Chem 285(47):36794–36803. https://doi.org/10.1074/jbc.M110.167312
312. Bose JL, Lehman MK, Fey PD, Bayles KW (2012) Contribution of the *Staphylococcus aureus* Atl AM and GL murein hydrolase activities in cell division, autolysis, and biofilm formation. PloS One 7(7):e42244. https://doi.org/10.1371/journal.pone.0042244
313. Nega M, Tribelli PM, Hipp K, Stahl M, Götz F (2020) New insights in the coordinated amidase and glucosaminidase activity of the major autolysin (Atl) in *Staphylococcus aureus*. Commun Biol 3(1):695. https://doi.org/10.1038/s42003-020-01405-2
314. Hammel M, Sfyroera G, Pyrpassopoulos S, Ricklin D, Ramyar KX, Pop M et al (2007) Characterization of Ehp, a secreted complement inhibitory protein from *Staphylococcus aureus*. J Biol Chem 282(41):30051–30061. https://doi.org/10.1074/jbc.M704247200
315. Jongerius I, von Kockritz-Blickwede M, Horsburgh MJ, Ruyken M, Nizet V, Rooijakkers SH (2012) *Staphylococcus aureus* virulence is enhanced by secreted factors that block innate immune defenses. J Innate Immun 4(3):301–311. https://doi.org/10.1159/000334604
316. Sobke AC, Selimovic D, Orlova V, Hassan M, Chavakis T, Athanasopoulos AN et al (2006) The extracellular adherence protein from *Staphylococcus aureus* abrogates angiogenic responses of endothelial cells by blocking Ras activation. FASEB J 20(14):2621–2623. https://doi.org/10.1096/fj.06-5764fje
317. Bodén MK, Flock JI (1992) Evidence for three different fibrinogen-binding proteins with unique properties from *Staphylococcus aureus* strain Newman. Microb Pathog 12(4):289–298. https://doi.org/10.1016/0882-4010(92)90047-r
318. Palma M, Haggar A, Flock JI (1999) Adherence of *Staphylococcus aureus* is enhanced by an endogenous secreted protein with broad binding activity. J Bacteriol 181(9):2840–2845. https://doi.org/10.1128/JB.181.9.2840-2845.1999
319. Kreikemeyer B, McDevitt D, Podbielski A (2002) The role of the map protein in *Staphylococcus aureus* matrix protein and eukaryotic cell adherence. Int J Med Microbiol 292(3–4):283–295. https://doi.org/10.1078/1438-4221-00212
320. Jönsson K, McDevitt D, McGavin MH, Patti JM, Höök M (1995) *Staphylococcus aureus* expresses a major histocompatibility complex class II analog. J Biol Chem 270(37):21457–21460. https://doi.org/10.1074/jbc.270.37.21457
321. Hussain M, Haggar A, Heilmann C, Peters G, Flock JI, Herrmann M (2002) Insertional inactivation of Eap in *Staphylococcus aureus* strain Newman confers reduced staphylococcal binding to fibroblasts. Infect Immun 70(6):2933–2940. https://doi.org/10.1128/IAI.70.6.2933-2940.2002
322. Hansen U, Hussain M, Villone D, Herrmann M, Robenek H, Peters G et al (2006) The anchorless adhesin Eap (extracellular adherence protein) from *Staphylococcus aureus* selectively recognizes extracellular matrix aggregates but binds promiscuously to monomeric matrix macromolecules. Matrix Biol 25(4):252–260. https://doi.org/10.1016/j.matbio.2006.01.005
323. Chavakis T, Hussain M, Kanse SM, Peters G, Bretzel RG, Flock JI et al (2002) *Staphylococcus aureus* extracellular adherence protein serves as anti-inflammatory factor by inhibiting the recruitment of host leukocytes. Nat Med 8(7):687–693. https://doi.org/10.1038/nm728

324. Posner MG, Upadhyay A, Abubaker AA, Fortunato TM, Vara D, Canobbio I et al (2016) Extracellular fibrinogen-binding protein (Efb) from *Staphylococcus aureus* inhibits the formation of platelet-leukocyte complexes. J Biol Chem 291(6):2764–2776. https://doi.org/10.1074/jbc.M115.678359
325. Bodén MK, Flock JI (1994) Cloning and characterization of a gene for a 19 kDa fibrinogen-binding protein from *Staphylococcus aureus*. Mol Microbiol 12(4):599–606. https://doi.org/10.1111/j.1365-2958.1994.tb01046.x
326. Heilmann C, Herrmann M, Kehrel BE, Peters G (2002) Platelet-binding domains in 2 fibrinogen-binding proteins of *Staphylococcus aureus* identified by phage display. J Infect Dis 186(1):32–39. https://doi.org/10.1086/341081
327. Ko YP, Liang X, Smith CW, Degen JL, Höök M (2011) Binding of Efb from *Staphylococcus aureus* to fibrinogen blocks neutrophil adherence. J Biol Chem 286(11):9865–9874. https://doi.org/10.1074/jbc.M110.199687
328. Hussain M, Becker K, von Eiff C, Schrenzel J, Peters G, Herrmann M (2001) Identification and characterization of a novel 38.5-kilodalton cell surface protein of *Staphylococcus aureus* with extended-spectrum binding activity for extracellular matrix and plasma proteins. J Bacteriol 183(23):6778–6786. https://doi.org/10.1128/JB.183.23.6778-6786.2001
329. Geraci J, Neubauer S, Pöllath C, Hansen U, Rizzo F, Krafft C et al (2017) The *Staphylococcus aureus* extracellular matrix protein (Emp) has a fibrous structure and binds to different extracellular matrices. Sci Rep 7(1):13665. https://doi.org/10.1038/s41598-017-14168-4
330. Johnson M, Cockayne A, Morrissey JA (2008) Iron-regulated biofilm formation in *Staphylococcus aureus* Newman requires Ica and the secreted protein Emp. Infect Immun 76(4):1756–1765. https://doi.org/10.1128/IAI.01635-07
331. Park PW, Broekelmann TJ, Mecham BR, Mecham RP (1999) Characterization of the elastin binding domain in the cell-surface 25-kDa elastin-binding protein of *Staphylococcus aureus* (EbpS). J Biol Chem 274(5):2845–2850. https://doi.org/10.1074/jbc.274.5.2845
332. Park PW, Rosenbloom J, Abrams WR, Rosenbloom J, Mecham RP (1996) Molecular cloning and expression of the gene for elastin-binding protein (ebpS) in *Staphylococcus aureus*. J Biol Chem 271(26):15803–15809. https://doi.org/10.1074/jbc.271.26.15803

Chapter 2
Cell Wall-Anchored Surface Proteins of *Staphylococcus aureus*

Timothy J. Foster

Abstract *Staphylococcus aureus* expresses a plethora of proteins that are covalently anchored to cell wall peptidoglycan. They are classified on the basis of structural and functional properties. Several proteins are multifunctional. They promote adhesion to host cells and tissue, invasion of non-phagocytic host cells, evasion of innate and adaptive immune responses, biofilm formation and iron acquisition. Their roles in host colonization and pathogenesis of infectious diseases have been demonstrated in animal models.

Keywords Adhesion · Invasion · Immune evasion · Biofilm · Animal models · Fibrinogen · Fibronectin · Collagen · Complement · Neutrophils · Nasal colonization · Virulence

Abbreviations

Aap	Accumulation associated protein
AD	Atopic dermatitis
AdsA	Adenosine synthase
AFM	Atomic force microscopy
CA-MRSA	Community associated MRSA
CAUTI	Catheter associated urinary tract infection
Clf	Clumping factor
Cna	Collagen binding adhesin
CWA	Cell wall-anchored
dAMP	2′-Deoxyadenosine 3′ monophosphate
DEv-IgG	DE variant-immunoglobulin G fold
DLL	Dock lock latch
ECM	Extracellular matrix
Fg	Fibrinogen
FIVAR	Found in various architectures

T. J. Foster (✉)
Microbiology Department, Trinity College Dublin, Dublin, Ireland
e-mail: tfoster@tcd.ie

A. Nakane, K. Asano (eds.), *Staphylococcus aureus*,
https://doi.org/10.1007/978-981-99-9428-1_2

Fn	Fibronectin
FnBP	Fibronectin binding protein
FnBR	Fibronectin binding repeats
Gtf	Glycosyltransferase
Isd	Iron-regulated surface determinant
LPXTG	Leucine, proline, any residue, threonine, glycine
MRSA	Methicillin resistant *Staphylococcus aureus*
MSCRAMM	Microbial surface component recognizing adhesive matrix molecules
NEAT	Near transporter
NET	Neutrophil extracellular trap
Neu5AC	N-acetyl neuraminic acid
Plg	Plasminogen
RGD	Arginine glycine aspartate
Sbi	Second immunoglobulin binding protein
SCCmec	Staphylococcal chromosome cassette, methicillin resistance
SCV	Small colony variant
Sdr	Serine rich repeat protein
Spa	Staphylococcal protein A
SraP	Serine rich adhesin for platelets
SSR1	Short serine-rich repeat region
WTA	Wall teichoic acid

2.1 Introduction

Staphylococcus aureus is a common commensal of humans. Its primary site of colonization is the nares where it attaches to the ciliated epithelia of the posterior part of the cavity and to moist squamous epithelia in the anterior region [1–3]. Approximately 20–30% of the population are persistent carriers and carriage is a risk factor for infection [2]. *S. aureus* can cause a wide array of infections that can be superficial (eg skin abscesses) of invasive (eg endocarditis, osteomyelitis, septic arthritis) [4]. The bacterium is a major cause of bacteraemia in hospitalized patients. Most hospital acquired infections are caused by strains that are resistant to multiple antibiotics including most β-lactams while hypervirulent MRSA causing serious skin and soft tissue infections have spread in the community [5–7].

S. aureus can express a wide range of virulence factors including surface proteins that are covalently anchored to the cell wall peptidoglycan (cell wall-anchored CWA surface protein). The precise repertoire of CWA proteins on the surface varies from strain-to-strain. Up to 24 different CWA proteins have been described whereas the less virulent coagulase negative staphylococci such as *S. epidermidis* and *S. lugdunensis* express a smaller number [8, 9]. CWA protein expression depends on the growth conditions with some only being elaborated when iron is limited whereas others are found predominantly on cells from the exponential or stationary phases of growth [10–12].

CWA proteins are anchored to the cell surface by a process called sorting [13]. Each has an N-terminal secretory signal sequence which directs the newly translated protein to the secretion apparatus in the membrane where it is cleaved. At the C-terminus a sorting signal including an LPXTG motif is recognized by sortase. The nascent CWA protein is transferred to the peptidoglycan precursor lipid II and is then introduced into peptidoglycan. Thus, a CWA protein is modified posttranslationally both at the N-terminus and the C-terminus.

The focus of this article is to show how CWA proteins of staphylococci have been classified into groups based upon their structure and functions. Some proteins are multifunctional with the same domain binding to several different ligands. Functional redundancy occurs with the same ligand being recognized by two or more CWA proteins. They can promote adhesion to the extracellular matrix and to host cells, invasion of host cells that are not normally phagocytic, evasion of innate and adaptive immune responses, biofilm formation and acquisition of iron from haemoglobin.

2.2 Cell Wall Anchored Proteins: Structure and Function

Several distinct structural and/or functional classes of CWA protein occur in *S. aureus* (Fig. 2.1, Table 2.1). N-terminal domains are projected away from the cell surface and are the part of the protein that is usually engaged in interactions with the host. An important exception is the fibronectin binding repeats of fibronectin binding proteins (FnBPs) which are located towards the C-terminus.

2.2.1 The MSCRAMM Family

The acronym microbial surface components recognizing adhesive matrix molecules (MSCRAMMs) was originally applied to CWA proteins that bound to host extracellular matrix (ECM) proteins such a fibrinogen and fibronectin [14]. It is often used to describe all CWA proteins irrespectively of whether or not they bind the ECM. It is recommended that MSCRAMM should be reserved for proteins with a common structure motif comprising two adjacent IgG-like folds and which employ similar mechanisms of ligand binding exemplified by dock lock latch (DLL) and the collagen hug [15].

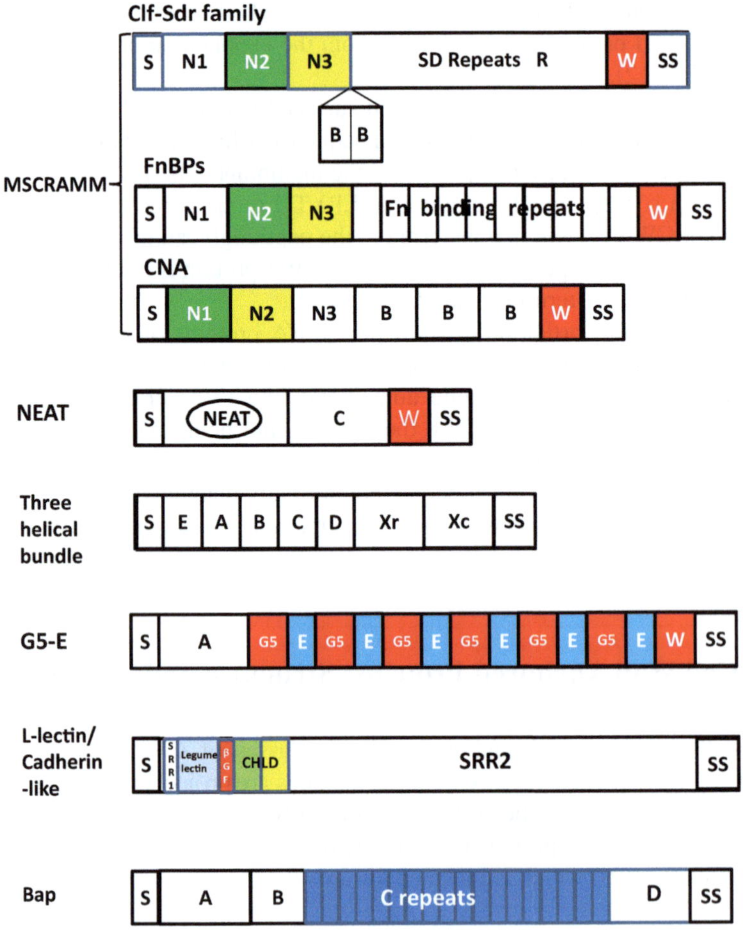

Fig. 2.1 Cell wall-anchored surface proteins classified based on structural motifs. The primary translation products of all cell wall-anchored (CWA) proteins contain a signal sequence (S) at the amino terminus and a wall-spanning region (W, Xc) and sorting signal (SS) at the carboxyl terminus. The CWA proteins that are depicted are those for which structural analysis has facilitated classification into distinct groups. (**a**) Microbial surface component recognizing adhesive matrix molecules (MSCRAMMs). The clumping factor (Clf) – serine aspartate repeat (Sdr) group comprises proteins that are closely related to ClfA. ClfA and ClfB have a similar domain organization, whereas SdrC, SdrD and SdrE contain additional B_{SDR} repeats that are located between the A domain and the serine-aspartate SD repeat R region. The N-terminal A region contains three separately folded domains, called N1, N2 and N3. Structurally, N2 and N3 form IgG-like folds that bind ligands by the 'dock, lock and latch' mechanism. Fibronectin-binding protein A (FnBPA) and FnBPB have A domains that are structurally and functionally similar to the A domain of the Clf–Sdr proteins. Located in place of the serine–aspartate repeat region are tandemly repeated fibronectin-binding domains (11 in FnBPA, 10 in FnBPB). The A region of the collagen adhesin

(continued)

(Cna) protein is organized differently to other MSCRAMMs, with N1 and N2 comprising IgG-like folds that bind to ligands using the 'collagen hug' mechanism. The A region is linked to the wall-spanning and anchorage domains by variable numbers of B_{CNA} repeats. (**b**) Near iron transporter (NEAT) motif protein family. The iron-regulated surface determinant (Isd) proteins have one (for IsdA), two (for IsdB) or three (for IsdH) NEAT motifs that bind to haemoglobin or haem. The figure depicts IsdA, which has a C-terminal hydrophilic stretch that reduces cell surface hydrophobicity and contributes to resistance to bactericidal lipids and antimicrobial peptides. (**c**) Three-helical bundle motif protein A. The five N-terminal tandemly linked triple-helical bundle domains (EABCD) that bind to IgG and other ligands are followed by the repeat-containing Xr region and the non-repetitive Xc region. (**d**) G5–E repeat family. The alternating repeats of the G5 and E domains of *S. aureus* surface protein G (SasG) from *S. aureus* [and the accumulation-associated protein (Aap) from *S. epidermidis*] link the N-terminal A region to the wall-spanning and anchorage domains. If the A domain is removed, the G5–E region can promote cell aggregation. (**e**) Legume-lectin, cadherin-like domain protein. The BR region of the serine rich adhesin of platelets (SraP) protein is flanked by serine-rich repeat domains. The BR region comprises three distinct structural domains, the legume lectin-like, the β-grasp fold (β-GF) and the cadherin-like (CHLD) domains

Table 2.1 Properties of cell wall-anchored surface proteins

Protein structural and functional group	Ligand and binding mechanism	Function	References
MSCRAMM family			
Clumping factor A (ClfA)	Fibrinogen γ chain C-terminus (DLL)	Adhesion to immobilized fibrinogen, immune evasion by binding soluble fibrinogen	[16, 24]
	Complement factor I. Mechanism unknown	Immune evasion, degradation of C3b	[152, 153]
	von Willebrand factor binding protein (vWBFBP). Mechanism unknown	Binds to endotheial cell αVβ3 integrin via vWBFBP and von Willebrand factor. Pathogenesis of endovascular infection	[174]
Clumping factor B (ClfB)	Fibrinogen α-chain repeat 5, keratin 10 and loricrin (DLL)	Adhesion to desquamated cells from nares and from skin of atopic dermatitis (AD) patients. Nasal colonization and colonization of AD skin	[18, 20, 27, 140, 175]
Serine aspartate repeat protein C (SdrC)	β-Neurexin (DLL)	Unknown	[36, 37]
	Unknown	Adhesion to desquamated epithelial cells. Nasal colonization?	[121]
	SdrC	Homophilic SdrC-SdrC interaction. Biofilm formation	[34, 36]

(continued)

Table 2.1 (continued)

Protein structural and functional group	Ligand and binding mechanism	Function	References
Serine aspartate repeat protein D (SdrD)	Desmoglein-1 Mechanism unknown	Adhesion to desquamated epithelial cells. Nasal colonization?	[121, 142]
Serine aspartate repeat protein E (SdrE)	Complement factor H. (Closed DLL)	Immune evasion, degradation of C3b	[26, 154]
Bone sialoprotein binding protein (isoform of SdrE)	Fibrinogen α-chain (DLL)	Adhesion to immobilized fibrinogen	[176]
Fibronectin binding protein A and B FnBPA and FnBPB A domains	Bind Fg γ-chain C-terminus and elastin (DLL)	Adhesion and invasion	[19, 29, 177]
	FnBPB A domain also binds fibronectin but not by DLL		[177]
	FnBPA, FnBPB	Homophilic interactions. Biofilm formation	[33, 38]
	FnBPA, FnBPB	Plasminogen	[39, 40]
	FnBPB (DLL)	Histones	[41]
Fibronectin binding repeats	Fibronectin, tandem β-zipper)	Adhesion to ECM, invasion of mammalian cells	[46–48, 60, 178]
Collagen binding protein (Cna)	Collagen triple helix (CH)	Adhesion to collagen rich tissue	[30]
	Complement protein C1q (CH)	Prevents classical pathway of complement activation	[118]
	Laminin	Adhesion to basement membrane	[32]
NEAT motif family			
Iron-regulated surface protein A (IsdA)	Haem, fibrinogen, fibronectin, cytokeratin 10, loricrin (N-terminal NEAT motif region)	Haem uptake and iron acquisition	[67]
		Adhesion to desquamated epithelial cells	[179] [68, 141]
	Unknown ligand(s). C-terminal domain NEAT motif region	Resistance to lactoferrin	[180]
		Resistance to bactericidal lipids and antimicrobial peptides Survival in neutrophils	[69]
IsdB	Haemoglobin, haem (N-terminal NEAT motif region)	Haem uptake and iron acquisition	[67, 181]
	β3 integrins (NEAT motif regions)	Invasion of non-phagocytic cells	[71]

Table 2.1 (continued)

Protein structural and functional group	Ligand and binding mechanism	Function	References
IsdH	Haem, haemoglobin (N-terminal and/or C-terminal NEAT motif region)	Haem uptake and iron acquisition	[181]
	Unknown ligand(s). N-terminal NEAT motif region)	Accelerated degradation of C3b	[72]
Three helical bundle			
Protein A	IgG Fc	Inhibits opsonophagocytosis	[148, 149]
	IgM Fab V_H3 subclass	B cell superantigen	[75]
	TNFR-1	Inflammation	[74]
	von Willebrand factor	Endovascular infection. Endocarditis	[76]
	Unknown ligand (region Xr)	Inflammation	[78]
G5-E domain family			
S. aureus surface protein G (SasG) and plasmin sensitive surface protein (Pls, SasG homolog in MRSA)	A domain no known ligand	A domain. Adhesion to desquamated epithelial cells. Primary attachment to unconditioned biomaterial	[86, 121] [89]
	G5E repeats of SasG. Zn^{2+}-dependent homophilic interaction	Biofilm formation	[92, 93, 95, 96]
Legume lectin-like cadherin-like			
Serine rich repeat protein (SraP). N-terminal legume-lectin domain	N-acetyl neuraminic acid (Neu5AC) Salivary agglutinin gp340	Adhesion to and invasion of mammalian cells	[98, 99]
SraP cadherin-like domain (CHDL)	SraP CHDL	Homophilic CHDL-CHDL interaction. Biofilm formation Projection of L-lectin domain	[98]
5′-Nucleotidase			
Adenosine synthase A (AdsA)	Converts ATP to adenosine	Survival in neutrophils. Inhibits oxidative burst	[100, 101]
	2′-deoxyadenosine 3′ monophosphate. Degradation product of DNA converted to 2′-deoxyadenosine	Macrophage and monocyte apoptosis	[103]
EF-hand motif			
Biofilm associated protein (BAP)	Biofilm Ca^{2+} dependent	Promotes persistence in ovine mastitis	[182]

(continued)

Table 2.1 (continued)

Protein structural and functional group	Ligand and binding mechanism	Function	References
Bovine/ovine mastitis isolates only	Binds GP96	Promotes adhesion to epithelial cells, reduces invasion	[183]
Other proteins. No known functional motifs			
S. aureus surface protein X (SasX)	Ligands unknown	Biofilm, cell aggregation, squamous cell adhesion, neutrophil evasion	[137]
S. aureus surface protein C (SasC) N-terminal FIVAR containing domain	Ligands unknown	Promotes primary attachment and accumulation phases of biofilm formation	[184]
S. aureus surface proteins SasF	Unknown	Resistance to bactericidal effects of long chain unsaturated fatty acids	[185]
SasB, SasD, SasJ, SasK, SasL		Putative LPXTG proteins identified from genome sequences. No known structure or function	[186]

DLL dock lock latch, *CH* collagen hug, *ECM* extracellular matrix, *NEAT* near iron transporter, *Pls* plasmin-sensitive, *MRSA* methicillin resistant *Staphylococcus aureus*, *C1q, C3a* complement components, *FIVAR* found in various architectures

2.2.2 The Dock Lock Latch Mechanism of Ligand Binding

The Clf-Sdr group of MSCRAMMs have an N-terminal A domain that comprises three subdomains called N1, N2 and N3, with N2 and N3 encompassing the minimum required for ligand binding by dock lock latch (DLL). Subdomains N2–N3 each comprise two β-sheets arranged in a variation of the IgG fold called DEv-IgG [16]. The DLL mechanism was developed from analysis of the X ray crystal structures of the SdrG protein of *S. epidermidis* in the unbound apo form and in complex with the β-chain peptide of fibrinogen (Fg) [17]. The same mechanism occurs (a) with clumping factor A (ClfA) and the A domains of FnBPA and FnBPB binding to the γ-chain peptide of Fg, and (b) ClfB binding to cytokeratin 10, loricrin and the α-chain of Fg [18–20].

In DLL an unfolded peptide ligand binds to a hydrophobic trench located between the N2 and N3 subdomains (Fig. 2.2). A conformational change results in an unfolded extension of N3 moving to cover the peptide to form the lock. The latch occurs when the peptide forms an extra β-strand in one of the β-sheets of N2 (β-strand complementation) [15, 17]. The ligand peptide binds β-strand G' in subdomain N3 through backbone interactions while side chains of amino acids in the peptide and residues in the binding trench determine specificity.

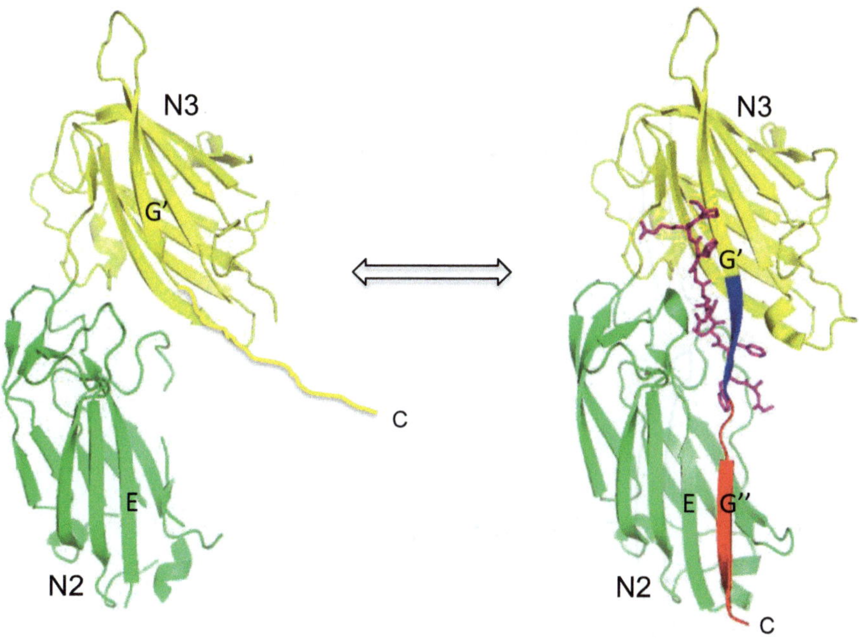

Fig. 2.2 MSCRAMM binding to ligand by dock lock latch. Ribbon diagram showing the structure of the N2 (green and N3 (yellow) subdomains of SdrG in the apo form and in complex with a peptide from the β-chain (purple) of fibrinogen following ligand binding by the dock lock latch mechanism. The C-terminal extension of N3 in the apo form undergoes a conformational change following ligand binding resulting in an addition β-strand in a β-sheet in subdomain N2 forming the latch (red) and lock (blue). The letters refer to β-strands

A significant development in understanding arose from recent studies using atomic force microscopy which showed that force required to separate SdrG from Fg is equivalent to that of a covalent bond [21]. Similar high rupture forces are required to separate cells expressing ClfB binding to loricrin [22] and FnBPA binding to Fg [23]. Another emerging theme is that shear forces induce the conformational changes involved in DLL.

With MSCRAMMss attached to the surface of a staphylococcal cell by a flexible stalk the conformational changes involved in DLL result in re-orientation of the protein with respect to the cell surface (Fig. 2.3). This could expose binding sites for other ligands such as plasminogen (see below).

2.2.3 Variations of Dock Lock Latch

There are several variants of the basic DLL mechanism. The ligand peptide can bind the trench in either a parallel (SdrG) or anti-parallel (ClfA, ClfB) orientation [16, 17, 20].

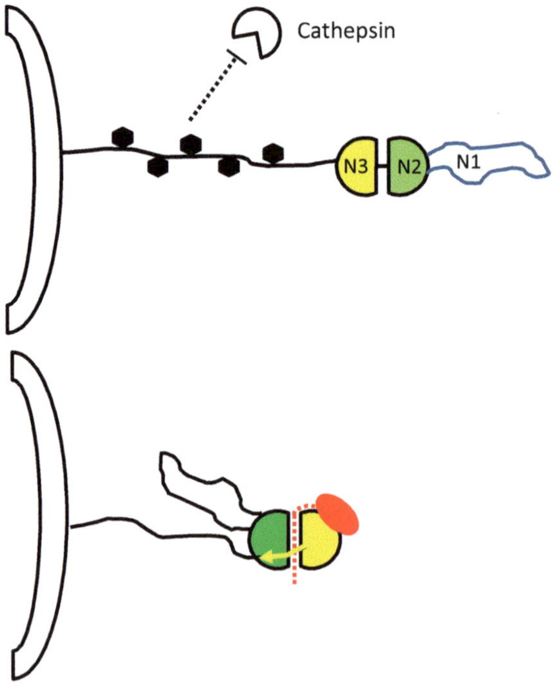

Fig. 2.3 Schematic diagrams of MSCRAMMs before and after ligand binding. The top figure depicts an MSCRAMM in the apo form with the N2 (green) and N3 (yellow) subdomains shown as semi-circles and the unstructured N1 subdomain. Serine residues in the flexible stalk are glycosylated which prevents degradation by cathepsin. The middle diagram shows the conformational change in an MSCRAMM with respect to the bacterial cell that occurs following ligand binding by dock lock latch. The C-terminal γ-chain peptide of Fg is depicted by the red dashed line and the gamma globule domain is in contact with the second ligand binding site in ClfA subdomain N3. It is not known if other MSCRAMMs have two binding sites on their ligands

The peptides bound by SdrG and ClfB have extensive C-terminal and N-terminal flanking residues and are only able to bind to the apo form of the MSCRAMM. In contrast residues in Fg bound by ClfA occur at the extreme C-terminus of the γ-chain which. These can penetrate the MSCRAMM when it is in the locked position [24].

ClfA has second binding site on Fg in addition to the γ-chain peptide [25]. The monoclonal antibody tefibazumab blocks binding of ClfA to Fg by binding to the "top" of subdomain N3, a site that is located well away from the DLL ligand binding trench. Tefibazumab is likely to block ClfA N3 binding to the Fg γ-globule, an interaction that this is required for high affinity binding (Fig. 2.3). The monoclonal antibody has a high affinity for ClfA but blocks ClfA binding to Fg poorly. High affinity binding of ClfA to Fg requires the two sites to act cooperatively in a complex multi-contact binding mechanism. Proof of such complicated two binding site interaction requires X ray crystal structures of the proteins in complex.

SdrE binds to C-terminal unit 20 of the complement control protein factor H by a variation of DLL called "closed dock lock latch" [26]. The ligand binding trench in SdrE is occluded by a loop that protrudes from subdomain N2 and forms four H bonds with β-strand G′ in subdomain N3. This loop must be displaced in order to open up the binding trench for the factor H C-terminal peptide to bind by DLL (Fig. 2.4).

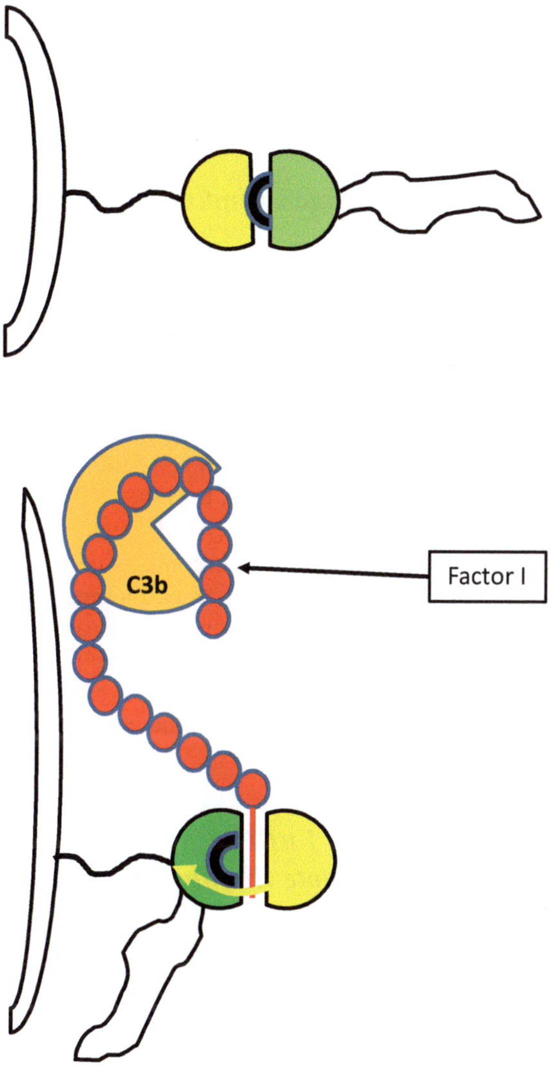

Fig. 2.4 SdrE binding to complement factor H. The top figure shows SdrE in the apo form with the unstructured N1 subdomain and the N2 and N3 subdomains (yellow and green semicircles, respectively). The loop that occludes the ligand binding trench is shown in blue. The bottom figure shows the conformational changes that occur when complement factor H binds by closed dock lock latch. Factor H (red) can then engage nearby C3b molecules (blue), and facilitate binding and activation of the protease factor I which cleaves C3b

2.2.4 MSCRAMMs Bind Different Ligands by DLL

The MSCRAMM ClfB can bind to cytokeratin 10, loricrin and the α-chain of Fg by DLL [18, 20, 27]. Several distinct ligands can be accommodated by the same binding trench. A consensus binding sequence was identified [20].

The A domains of FnBPA and FnBPB bind to elastin in addition to the γ-chain peptide of Fg [28, 29]. The conclusion that elastin binding occurred by DLL resulted from study of FnBP variants with substitutions in residues predicted to be involved in DLL, not from X ray crystal structures. It is possible that other MSCRAMMs bind several ligands by DLL.

2.2.5 The Collagen Binding Protein Cna

Binding to collagen by Cna involves variant DLL mechanism called the collagen hug [30]. The N1 and N2 subdomains of the A region of Cna are composed of IgG-like folds and perform similar roles to the N2 and N3 subdomains of Clf-like MSCRAMMs (Fig. 2.5). A long linker that separates N1 from N2 accomodates the thick collagen triple helix. The ligand binds to a shallow trench in N2. This stimulates a conformational change which allows the linker to wrap around the collagen. Latching occurs by β-strand complementation between the C-terminal extension of N2 and a β sheet in subdomain N1 [30].

Atomic force microscopy (AFM) studies showed that both weak and strong interactions occur when Cna encounters collagen. The initial interaction involves weak forces and molecules can be easily separated. Once the collagen hug has been completed the rupture force required to separate the molecules is of the same order as that required to separate SdrG from Fg [31, 32]. In contrast binding of Cna to complement protein C1q and laminin involves only weak forces of separation. The full collagen hug does not seem to be involved [32].

2.2.6 SdrC and FnBPs Engage in Homophilic Interactions and Promote Biofilm Formation

Biofilm formation occurs in several stages. First the bacteria attach to a surface which could be unconditioned biomaterial, biomaterial that has been conditioned with host proteins or to the surface of tissue or host cells. MSCRAMMs can promote primary attachment and several can promote accumulation of cells in the multi-layered biofilm [33–35]. The A domains of SdrC and FnBPs located on adjacent cells undergo homophilic interactions (Fig. 2.6). SdrC subdomain N2 has two interaction sites [34]. Molecular modelling indicated that a β-neurexin peptide which binds SdrC occludes one of the interaction sites and prevents biofilm

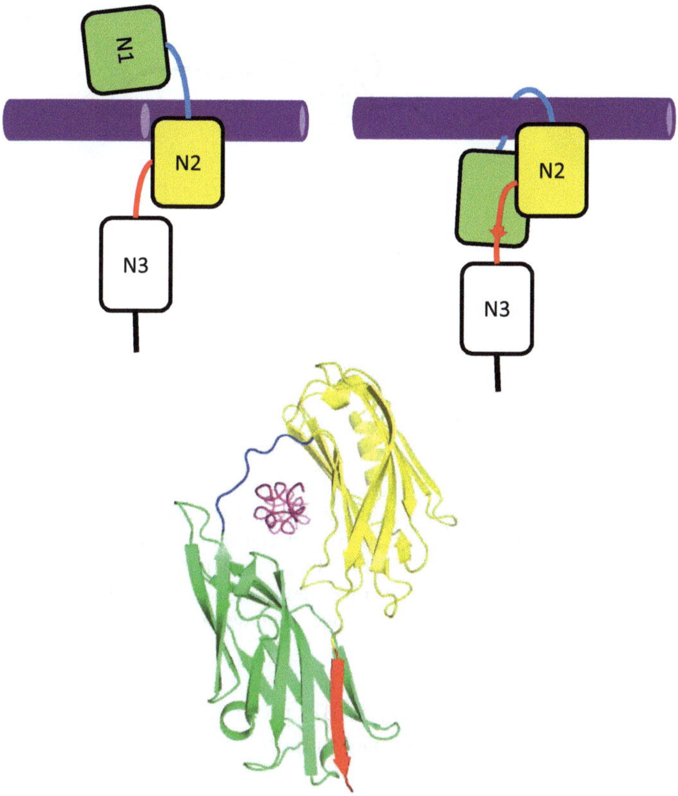

Fig. 2.5 Collagen binding protein and the collagen hug. The upper part shows a schematic diagram of the Cna protein in the apo form on the left and following binding by the collagen hug to the collagen triple helix on the right. Below is a ribbon diagram of the Cna protein in complex with the collagen triple helix. The N1 sub domain (green) and N2 subdomain (yellow) are separated by a long unfolded region (blue) that forms the lock around collagen (purple). β-strand complementation by the red strand completes the hug

formation [36, 37]. Studies with bacterial cells expressing variant FnBPs showed that the A domains are required for biofilm accumulation and that in the case of FnBPA DLL is not involved [33].

AFM has been employed to study forces involved in homophilic interactions between MSCRAMMs. It was concluded that cell-cell binding results from multiple weak interactions between proteins on individual cells [36, 38]. The rupture force required to separate a single homophilic protein-protein interaction was very low compared to that required to separate an MSCRAMM from a ligand bound by DLL. Weak interactions might facilitate detachment of bacterial cells from a mature biofilm and subsequent dissemination.

In the case of SdrC the A domain can also promote attachment to hydropholic surfaces and could thus be involved both in attachment to an indwelling medical

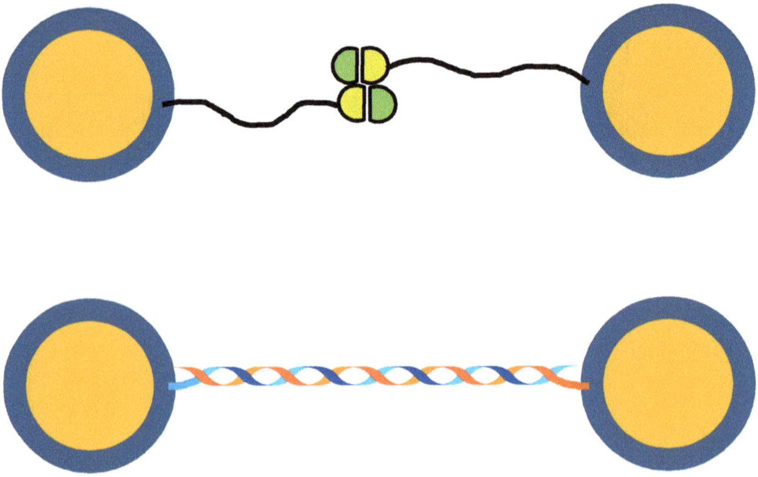

Fig. 2.6 Homophilic interactions and biofilm formation. The upper part shows a schematic diagram of the model for homophilic interactions between the A domains of the MSCRAMMs FnBPA, FnBPB and SdrC which promote cell-cell accumulation of staphylococcal cells (yellow spheres) during biofilm formation.. The lower part shows the extended fibrillar region of SdrG and Aap (orange and blue strands) which form extended zinc-dependent zipper interactions predicted to form a twisted rope-like structure

device and biofilm accumulation [36]. FnBPs promote adhesion to host proteins that coat biomaterial and can also participate in both stages.

2.2.7 FnBPs Bind Plasminogen and Histones

Several CWA proteins contribute to the ability of *S. aureus* cells to capture plasminogen (Plg) from serum [39]. Mutants lacking FnBPs are only partially defective in Plg binding whereas a sortase A defective mutant lacking all CWA proteins is almost completely defective. The full repertoire of CWA proteins that bind Plg remains to be discovered. To date studies have focussed on FnBPs [39, 40]. There are seven isoforms of the A domain of FnBP which vary in amino acid sequence by more than 10%. The A domain of all seven bind Plg with similar affinity [39]. Binding most likely occurs between lysine residues on the MSCRAMM and negatively charged residues on Plg. Alanine substitutions of two conserved patches of lysines in subdomain N2 of FnBPB resulted in reduced Plg binding. The ability of whole cells expressing FnBPs to bind to Plg was greatly enhanced by Fg [40]. This implies that the conformational changes that occur when Fg binds FnBPA by DLL reorients the cell-bound MSCRAMM and exposes a Plg binding site that might also involve contact with bound Fg (Fig. 2.7).

Activated neutrophils release nuclear DNA which forms large web-like structures called a neutrophil extracellular traps (NETs). The decondensed chromatin

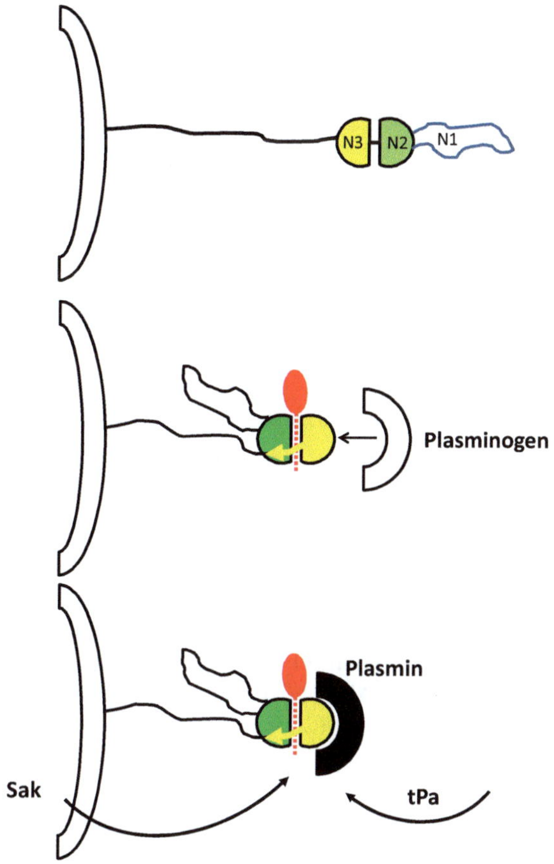

Fig. 2.7 Plasminogen binding by FnBPB. The upper figure shows FnBP in the apo form with its green and yellow N2 and N3 subdomains. The middle figure shows Fg (red) binding by DLL which facilitates plasminogen binding. This is converted to plasmin by host tissue plasminogen activator (tPa) or staphylokinase (Sak) expressed by *S. aureus* (lower figure)

contains granular proteins as well as histones. The latter have inhibitory activity by causing membrane damage like classical antimicrobial peptides. *S. aureus* is resistant to histones in vitro and to bactericidal activity of NETs [41]. FnBPB bind to histones most likely by DLL. The affinity of FnBPB for H3 is 20-fold higher than for Fg which indicates that it will bind histones preferentially when both proteins are present. FnBPB could bind both histone H3 and Plg simultaneously. After activation to plasmin the captured histone is cleaved and inactivated.

2.2.8 Functions Other MSCRAMM Regions

The SdrC, SdrD and SdrE proteins have two or more repeated domains of 110–113 residues (called B_{SDR}) that are located between the A region and the SD region (Fig. 2.1) [42]. The B_{SDR} repeats are arranged as a rigid rod that depends on Ca^{2+} for structural integrity and may project the A domain away from the cell surface. They have no known ligand binding activity [43] although the related B_{SDR} repeats of SdrF from *S. epidermidis* bind weakly to collagen [44].

The N terminal B1 domain of SdrD makes contact with the adjacent N3 subdomain of region A resulting in the ligand binding groove between N2 and N3 opening further than was observed in the structure of the A domain alone. This interaction might influence binding activity [45].

The collagen binding protein Cna has 2–4 repeats of the B_{Cna} domain (Fig. 2.1). AFM studies were interpreted to infer that the B repeats are nanosprings that project the A domain and facilitate the conformational changes involved in the collagen hug [31]. The B_{SDR} repeats might have a similar function.

2.2.9 The Tandem β-Zipper Binding to Fibronectin Promotes Invasion of Mammalian Cells

S. aureus can invade many different types of mammalian cells that do not normally engage in phagocytosis, eg. epithelial cells, endothelial cells, fibroblasts, osteoblasts and keratinocytes [46–53]. This allows bacteria to evade host immune defences and antibiotics. Bacteria can escape into the cytosol by lysing the phagosomal membrane. They can multiply before eventually destroying the integrity of the cell and being released. Alternatively, the bacteria enter a semi-dormant state called the small colony variant (SCV) [54, 55]. SCVs express cytolytic toxins at low levels so the invaded cell stays intact. SCVs express FnBPs at high levels which facilitates efficient uptake if they are released [56].

Fibronectin (Fn) is a dimeric glycoprotein comprising three distinct domains FI, FII and FIII [57]. Fn normally exists in a compact state held together by intramolecular interactions between F1 modules in the N-terminal domain and FIII modules in the C-terminal cell binding domain. The binding of FnBP to FI modules in the N-terminal domain of Fn by the tandem β-zipper mechanism (Fig. 2.8) triggers a conformational change in the host protein [58]. This exposes a cryptic RGD integrin binding site in the tenth FIII module which binds to $\alpha5\beta1$ integrins on the surface of mammalian cells. Thus Fn acts as a bridge between *S. aureus* and host cells. Clustering of integrins triggers intracellular signalling by the focal adhesion kinase and Src kinase and the bacteria are internalized by endocytosis [50, 59, 60].

2.2.10 Flexible Stalks

Cell surface proteins from the Clf-Sdr family of MSCRAMMs are linked to the cell wall by long flexible unfolded regions comprising either serine-asparate dipeptide repeats (ClfA, ClfB, SdrC, SdrD, SdrE) or 10/11 tandem repeats that bind fibronectin (FnBRs) (Figs. 2.1 and 2.3). *S. aureus* variants expressing ClfA lacking SD

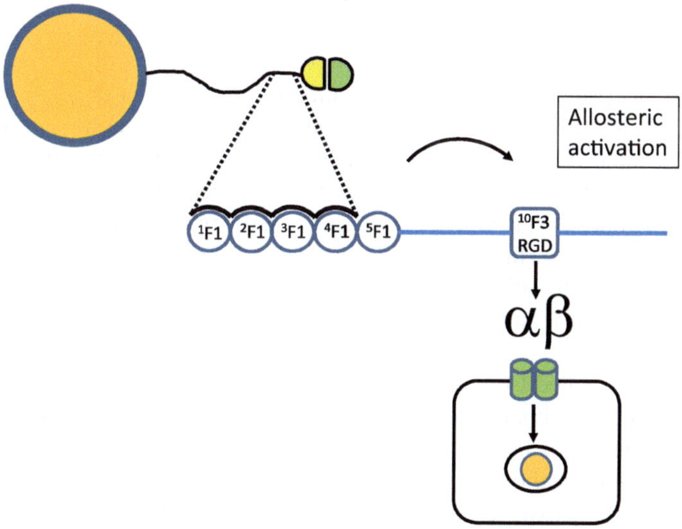

Fig. 2.8 Fibronectin binding by FnBPs. The figure also shows how one Fn binding repeat of the unstructured binding region of FnBP binds to N-terminal type I modules of Fn by the tandem β zipper mechanism. Potentially several such interactions can occur per molecule of FnBP. Intramolecular interactions between the N-terminal type I modules and C-terminal type III modules result in allosteric activation of the tenth type III module exposing an RGD motif which engages an $\alpha_5\beta_1$ integrin on the surface of a mammalian cell to promote invasion by endocytosis

repeats bound Fg weakly perhaps because the A region in close proximity to the cell surface could not undergo the conformational changes involved in DLL [61].

2.2.11 Posttranslational Modifications

The Sdr proteins are glycosylated by the action of the glysosyltransferases SdgA and SdgB (Fig. 2.3) [62]. Glycosylation creates a dominant epitope for antibodies induced in humans exposed to *S. aureus*. These antibodies can represent up to 1% of IgG. Glycosylation protects the CWA protein from degradation by host proteases such as neutrophil cathepsin G and could contribute to virulence [62]. Indeed, glycosylase defective mutants have reduced virulence in a mouse sepsis infection model [63].

The *S. aureus* protease aureolysin cleaves close to the boundary between subdomains N1 and N2 of ClfA and ClfB while the host protease thrombin can remove the N1 subdomain of FnBPA. In the case of ClfB removal of subdomain N1 prevents the truncate from binding Fg [64]. This might also reduce *S. aureus* adherence to loricrin and cytokeratin 10 since these ligands share a common binding site in ClfB. The molecular basis for loss of ligand binding is not known [20]. The biological significance of the removal of subdomain N1 from ClfA and FnBPA is unclear since it does not reduce the ability of FnBPA to promote biofilm accumulation or ClfA and FnBPA from binding to Fg [33, 65].

2.3 NEAT Motif Proteins

The Near Iron Transporter (NEAT) domain proteins (Fig. 2.1) capture haem from haemoglobin and contribute to the survival of bacteria in the human host where iron is restricted. Haem is transported by several CWA iron-regulated surface (Isd) proteins to the ATP-dependent membrane transporter (IsdEF). Once haem enters the cytoplasm, haemoxygenases (IsdG, IsdI) release free iron [10, 66]. The defining feature of surface-located Isd proteins is the presence of one or more NEAT motifs, which either bind haemoglobin or haem. The structures of NEAT domains are known and the molecular mechanisms of ligand binding elucidated [67].

The surface-located Isd proteins have additional functions. IsdA promotes adhesion to squamous epithelial cells and contributes to nasal colonization [68]. The C-terminal domain of IsdA confers resistance to bactericidal lipids [69] and supports bacterial survival on skin. IsdB binds directly to β-3 containing integrins and promotes platelet activation and invasion of mammalian cells [70, 71]. The latter is a secondary uptake mechanism compared to that promoted by FnBPs. IsdH helps *S. aureus* to avoid phagocytosis by promoting degradation of complement opsonin C3b by an as yet unknown mechanism [72].

2.4 Three Helical Bundle Repeat Proteins

The N-terminal region of protein A (Spa) comprises a tandem array of five separately folded three helical bundles, each of which can bind to several different ligands (Fig. 2.1, Table 2.1). The A1 domain of von Willebrand factor, tumour necrosis factor receptor 1 and the Fc region of IgG each bind to the interface between helices 1 and 2 while the Fab region of IgM binds to the interface between helices 2 and 3 [73–76]. The three helical bundles have a conformational plasticity that facilitates binding to structurally diverse ligands [77]. Upon binding to the Fc region of IgG conformational heterogeneity is lost and is combined with structural rearrangements at the interface.

Located between the helical bundle domains and the cell wall spanning region Xc is an octapeptide repeat region Xr which is highly variable in number. Region Xr activates interferon-β expression and is proinflammatory [78].

The *S. aureus* binder of Ig Sbi [79] contains four tandemly arrayed three helical bundles, two of which can bind to IgG. Sbi is not covalently anchored to the cell wall. It is loosely attached to the cell envelope by binding non-covalently to lipoteichoic acid [80]. This allows the most N-terminal repeats to be exposed on the cell surface and to contribute to immune evasion. The C-terminal repeats are buried but are biologically active when released into the medium by promoting fruitless consumption of complement protein C3 [81].

2.5 G5-E Repeat Proteins

Three staphylococcal CWA proteins are characterized by the presence of G5-E repeats; SasG and Pls of *S. aureus* and Aap of *S. epidermidis* (Fig. 2.1). Pls is encoded within the type I SCC*mec* cassette and is only expressed by hospital-associated MRSA strains carrying that element [82].

The proteins form highly extended fibrillar structures due to the unusual properties of the G5-E repeats. They are able to mask the ability of shorter adhesins such as ClfA and FnBPs to promote adhesion to the ECM or invasion of mammalian cells [83–86]. Expression of large long proteins is controlled by ArlS-MgrA regulators that modulate adhesion to tissue during endovascular infections [87].

The N-terminal A domains of these proteins bind to cornified squamous epithelial cells by recognizing as yet unknown ligands and thus might contribute to nasal colonization [84, 86, 88]. The A domain of Aap also promotes adhesion to hydrophobic surfaces in the initiation of biofilm formation by *S. epidermidis* [89, 90].

The ability to form very strong and stable (long and strong) extended structures is an unsual property of the intrinsically unstable repeated region [91–93]. The G5 and E domains both comprise two triple-stranded β-sheets with a central collagen-like triple helical region. When a recombinant E domain is expressed alone or in combination with a single G5 domain the proteins are intrinsically disordered and remain unfolded [91]. However, a $G5_1$-E-$G5_2$ moiety folds to form an elongated structure. $G5_2$ folds and triggers folding of the central E domain which in turn promotes folding of $G5_1$. Thus, the entire repeat region can be considered to comprise overlapping G5-E-G5 cooperative folding units.

Cell-cell adhesion and biofilm formation is promoted by G5-E repeats. Firstly, the N-terminal A domain of Aap and SasG must be removed by proteolysis [93–95]. Only then can the folded G5-E repeats engage in the homophilic interactions involving multiple contacts along the lengths of the two molecules. Molecular modelling suggested that the G5-E domains twist around each other in a twisted rope-like structure (Fig. 2.6). AFM allowed measurement of the forces involved in cell-cell interactions. Separation required the two proteins to become unwound by successively unfolding the E and G5 subdomains [96].

2.6 The Legume Lectin Domain of SraP

The serine rich adhesin of platelets (SraP) is a member of a family of glycoproteins in Gram-positive cocci [97]. The structural gene *sraP* is part of a locus comprising genes that encode glycosyltransferases (GtfA and GtfB) and accessory secretion factors (SecY2 and SecA2). A short serine-rich repeat region (SSR1) is located at the N-terminus of the CWA protein followed by the ligand binding BR domain (Fig. 2.1) which is composed of four separately folded sub-domains that form a rigid bent rod [98]. Two cadherin-like domains and a β-grasp fold domain project the legume-lectin domain outwards. This domain has structural similarity to legume lectins and binds to N-acetyl neuraminic acid (Neu5Ac) containing glycoproteins that are attached to the surface of mammalian cells. It also binds to salivary glycoprotein gp340 which is rich in a 5NeuAc containing trisaccharide [99]. SraP promotes bacterial adhesion to and invasion of mammalian cells [98]. This was studied using an experimental system where there was little or no fibronectin present and employing a strain that expressed low levels of FnBPs. Thus, there was minimal involvement of the highly efficient FnBP-mediated uptake. SraP, like IsdB, seems to act as an accessory invasin. The cadherin like domains dimerize in solution and promote cell-cell accumulation and biofilm formation [98].

2.7 Nucleotidase Motif

A nucleotidase motif was discovered in a previously uncharacterized CWA protein called SasH. It is enzymatically active and importantly contributes to the ability of *S. aureus* to evade innate immune responses [100, 101]. The ability of SasH (now called adenosine synthase, AdsA) to convert ATP to the potent immunoregulatory molecule adenosine inhibited the oxidative burst and promoted bacterial survival within neutrophils.

Neutrophils contribute to the host defences by releasing DNA which form neutrophil extracellular traps (NETs) [102]. These can entrap bacteria and facilitate killing by their attached granule contents and histones. *S. aureus* secrete deoxyribonuclease which degrades NET DNA producing phosphomononucleotides including 2′ deoxyadenosine 3′ monophosphate (dAMP) [103, 104]. AdsA converts dAMP to 2′-deoxyadenosine which triggers caspase 3-promoted apoptosis in macrophages and monocytes [103]. By excluding macrophages from a developing abscess AdsA helps to promote persistent infection.

2.8 CWA Proteins as Colonisation and Virulence Factors

2.8.1 The Contribution of CWA Proteins to the Virulence of S. aureus

One way to investigate if a CWA protein can act as a virulence factor is to isolate a mutant defective in the protein and compare its virulence to the wild type in an appropriate animal infection model. This has been called fulfilling Koch's postulates at the molecular level [105]. However functional redundancy can make it difficult to show conclusively that a mutant lacking a single factor has reduced virulence. For example, *S. aureus* expresses several CWA proteins that bind fibrinogen and most strains elaborate two fibronectin binding proteins. Until recently it was difficult to isolate mutations in clinically relevant strains. Consequently, many studies on the role of CWA proteins were performed with laboratory strains such as Newman and derivatives of NCTC8325. However, restriction barriers that prevent transfer of plasmid DNA from *Escherichia coli* cloning hosts into *S. aureus* can be circumvented rendering feasible genetic manipulation of clinical isolates [106–108].

Results of experimental infection studies must be interpreted with caution if the surface protein being analyzed has a lower affinity for ligands in the animal compared to humans. One way to overcome this is to engineer a humanized mouse expressing the human version of the ligand [109]. A transgenic mouse expressing the human version of haemoglobin has been employed in infection studies to investigate the role of IsdB [110]. It might be possible to murinize the pathogen by engineering the virulence factor to bind to the murine version of the ligand as in the case of internalin A of *Listeria monocytogenes* [111].

Table 2.2 provides a summary of studies demonstrating that bacterial mutants defective in CWA proteins have reduced virulence in different infection and colonization models. A systematic analysis of mutants defective in single surface proteins was performed with mutants of strain Newman using a mouse model of bacteraemia and kidney abscess formation [112]. This model measured both the ability of bacteria to survive in the bloodstream as well as to invade kidney tissue and establish an abscess. Several mutants had statistically significant reductions in viable counts of *S. aureus* in kidney tissue. This suggested the importance of several CWA proteins although their precise roles are not known [113]. A limitation of this study was that strain Newman does not express Cna and lacks the ability to display FnBPs on its surface [114].

The collagen binding protein promotes virulence in infections where adhesion to collagen rich tissue is important such as ocular keratitis and septic arthritis [115, 116]. The strength of adhesion of the recombinant protein to immobilized collagen in vitro correlated with the severity of the disease [117]. Also, Cna prevents complement fixation and most likely contributes to innate immune evasion in the infected host [118].

Rather than isolate null mutants that lack the ability to express a surface protein, variants expressing the protein on the cell surface with amino acid substitutions that cause loss of function can be studied. For example, strains expressing mutants of ClfA [119] and Cna [117] that bound the ligands poorly had reduced virulence in a mouse model of septic arthritis. A mutant expressing AdsA that was defective in adenosine synthase survived less well than the wild type in blood stream infection [100].

Another approach is to express the CWA protein in an avirulent surrogate host such as *Lactococcus lactis* [27, 120–122] or *S. carnosus* [123]. The ability of *L. lactis* to colonize a sterile thrombus on damaged heart valves in a rat model of endocarditis was dependent on bacteria being able to adhere to fibrinogen/fibrin on the damaged heart valves. This could be promoted either by ClfA or the A domain of FnBPA [122]. Subsequent bacterial proliferation and expansion of the lesion required FnBP-promoted invasion of cells of the adjacent heart valve endothelium.

Table 2.2 Cell wall anchored proteins as colonization and virulence factors studied using animal models

Role in colonization or infection	CWA protein	Mechanism	References
Nasal or skin colonization	ClfB	Adhesion to loricrin on squames	[27]
	IsdA	Adhesion to squames	[141]
	SasX	Adhesion to squames	[137]
Endocarditis	ClfA	Adhesion to thrombus	[187]
	FnBPA	Adhesion to thrombus. Invasion of adjacent endothelium	[122]
	ClfB	Adhesion to thrombus	[188]
	SraP	Adhesion to platelets. Colonization of thrombus	[189]
Endovascular infection	ClfA	Adhesion to endothelium via vWBFBP and vWBF	[174]
Mastitis	FnBP	Invasion of epithelial cells in mammary gland	[190]
Pneumonia	Protein A	Enhanced inflammation of lung epithelium	[191]
Foreign body infection	FnBP	MRSA. Biofilm promoted infection	[192]
		Adhesion to intra-aortic patch	[193]
Ocular keratitis	Cna	Enhanced colonization and infection	[115]
Septic death Survival in blood	ClfA	Reduced opsonophagocytosis	[112, 119, 165]
	Protein A		[194, 195]
	IsdH		[72]
	AdsA		[100]
	SasX		[137]
Kidney abscess	AdsA, ClfA, ClfB, Spa, IsdC, IsdA, IsdB, SdrD, SdrE	Increased survival in blood stream prior to kidney infection	[100, 112]
Septic arthritis	ClfA, Protein A	Enhanced survival in bloodstream prior to invasion of joint	[165, 195]
	Cna	Enhanced survival in blood stream. Adhesion to cartilage within joint	[116]
Joint infection	Fibrinogen binding MSCRAMMs	Biofilm formation in synovial fluid	[196]
Sub-cutaneous abscess	Protein A	Abscess development, bacterial load	[194]
	FnBPs, ClfA		[197]
	SasF		[185]
	ClfB		[129]
Catheter associated urinary tract infection	ClfB	Adhesion to deposited fibrinogen. Bacterial load	[128]

The *fnbA* gene of *S. aureus* strains isolated from cases of infection of cardiovascular devices were compared to those isolated from uncomplicated bacteraemia in patients with implanted cardiac devices and with those from nasal carriers [124]. It was found that the former contains one or more non-synonymous nucleotide polymorphisms compared to the reference gene from 8325-4 (Fig. 2.1). These result in amino acid changes in FnBR domains 5 and 9 that confer stronger binding of the MSCRAMM and Fn as measured by mechanical dissociation using atomic force microscopy. The same substitutions were found in strains from another continent [125]. In contrast these substitutions were not associated with strains isolated from prosthetic joint infections [126]. This suggests that strains that have stronger binding to Fn (higher affinity of binding and stronger forces required for dissociation) are more likely to colonize a cardiac device which has been coated with host proteins including Fn.

Similar polymorphisms were observed in MRSA strains causing endovascular infections [59, 127]. Some strains had an additional FnBR inserted between FnBR-9 and FnBR-10 along with a substitution in FnBR-11 which might promote cell invasion at the expense of Fn binding affinity. This is consistent with the pathogenesis of endovascular infections which involves invasion of endothelial cells during thrombus expansion on the infected heart valve.

Surprisingly the ability of FnBPA to bind to immobilized fibrinogen by DLL is influenced by the above-mentioned amino acid substitutions in the FnBRs [23]. AFM data was interpreted to indicate that initial docking of Fg in the A domain involved weak bonds and the molecules could easily be separated. The completed DLL reaction resulted in very strong separation forces and was enhanced by the amino acid substitutions in FnBR domains 5 and 9 that promoted strong binding to Fn. This implies that FnBRs fold back and interact with the A domain to promote conformational changes of DLL.

Two recent studies have shed further light on the role of ClfB in pathogenesis of *S. aureus* infections. Catheter-associated urinary tract infection (CAUTI) caused by *S. aureus* is an emerging disease that can lead to bacteraemia and sepsis. Catheterization through the urethra causes inflammation and leads to fibrinogen deposition on the implant and the bladder surface [128]. This is associated with an increased risk of bacterial infection. Infecting bacteria occur in close association with fibrinogen deposits implying that the ability to attach is important in initiating infection. In a mouse model of CAUTI a ClfB-deficient mutant colonized catheter surfaces less well than wild type and lower numbers of bacteria occurred in the bladder. No difference was seen with a ClfA mutant. Why ClfA did not feature as a virulence factor in the CAUTI model is unclear. It is possible that growth of *S. aureus* in urine might lead to differential expression of MSCRAMMS or might affect MSCRAMM function.

Community associated MRSA (CA-MRSA) strains cause particularly severe skin lesions. In a mouse model, ClfB was shown to be an important factor in the early stages of abscess development where a ClfB-deficient mutant had altered abscess structure and a reduced bacterial burden [129]. Loricrin is normally only expressed in skin corneocytes but was present in the abscess wall. Infection of loricrin knockout mice resulted in smaller abscesses with reduced pathology and a lower number of infecting bacteria implying that the ClfB-loricrin interaction is important in abscess development.

2.8.2 CWA Proteins Promote Colonisation of the Host

Approximately 20–30% of the population are colonized persistently by *S. aureus* in the nares with the remainder being intermittently colonized [130, 131]. The factors that determine whether an individual is a persistent carrier are not completely understood [132]. *S. aureus* colonizes the nares both by adhering to ciliated epithelial cells in the posterior region of the nares and to squamous epithelial cells in the anterior nares [1, 133].

Adhesion to ciliated epithelial cells is mediated in part by the zwitterionic bacterial wall teichoic acid (WTA) binding to the scavenger receptor SREC-1 [134]. This involves positively charged D-alanine residues that decorate WTA binding to negatively charged residues on the host protein [134]. Glycosylation of WTA contributes to the ability of *S. aureus* to colonize the nares of cotton rats although the receptor for the N-acetyl glucosamine residues is not known [135].

Several CWA proteins promote adhesion of *S. aureus* to the cornified envelope of squamous epithelial cells called corneocytes. ClfB and IsdA promote nasal colonization in rodent models and in the case of ClfB, in humans [27, 68, 136]. SdrC, SdrD, SasG, Pls and SasX confer bacterial adhesion to corneocytes in vitro and SasX can also promote colonization of the nares of mice [86, 121, 137]. Pls and SasX are encoded by genes located in mobile genetic elements and are only expressed by a subset of MRSA strains that carry the relevant element. The *pls* gene is carried within the type I SCC*mec* cassette [138] while the *sasX* gene is associated with a lysogenic phage in ST239 MRSA strains that are prevalent in the Far East [137]. The *sasG* gene is carried by 50% of the most prevalent lineages (clonal complexes CC1, CC5, CC8, CC15, CC22) but is absent from CC12, CC25, CC51, CC45 and CC30 strains [139].

Clumping factor B binds to cytokeratin 10 and loricrin, proteins that are thought only occur in corneocytes [18, 20, 140]. Loricrin is probably the more important ligand in vivo. Loricrin knockout mice are defective in nasal colonization by *S. aureus* [27]. ClfB binds to loricrin by DLL [27] an interaction that is strengthened by shear forces [22]. IsdA can also bind to loricin, cytokeratin 10 and involucrin but the mechanistic basis of these relatively weak interactions is not known [141]. SdrD binds to desmoglein 1, a component both of the desmosomes that hold keratinocytes together and the corneodesmosones that regulate desquamation in the stratum corneum [142, 143].

S. aureus is regarded as a minor and transitory resident of healthy skin. However, in eczema (atopic dermatitis, AD) patients suffering from a flare-up, the diversity of normal skin microbiota is reduced allowing *S. aureus* to proliferate [144]. The bacterium releases several proteins that can exacerbate inflammation in AD skin [145].

Colonization of AD skin is promoted by the ability of bacteria to adhere to corneocytes in a ClfB-dependent fashion [146]. Adhesion to loricrin and/or to other ligands that become exposed on the surface of the deformed corneocytes where corenodesmosome proteins are present on villous-like protrusions is likely to be an important first step in colonization [147].

2.9 Immune Evasion Promoted by Cell Wall Anchored Proteins

2.9.1 Evasion of Innate Immune Responses. Inhibition of Complement Activation

Protein A helps *S. aureus* evade innate responses by capturing IgG from serum. Bacterial cells become coated with IgG that is bound incorrectly. The Fc region of the antibody cannot be recognized by the Fc receptor on neutrophils [148, 149]. Furthermore, in order to initiate the classical pathway of complement fixation complement protein C1q must be able to bind to the Fc region of bound IgG [150, 151]. Thus, opsonophagocytsosis is inhibited by preventing recognition of IgG and C3b opsonins by both the Fc receptor and the complement receptor on neutrophils.

The collagen binding protein Cna blocks initiation of the classical pathway of complement fixation by binding to the collagen triple helical structures of C1q, displacing C1r-C1s and preventing complement activation [118].

2.9.2 Evasion of Innate Immune Responses. Enhanced Destruction of C3b

Complement fixation results in complement protein C3b being deposited on the bacterial cell surface in such a way that it can be recognized by the complement receptor on phagocytic cells. Accumulation of C3b triggers the amplification process that results in activation of the alternative pathway and also triggers activation of C5 convertase [151]. Three CWA proteins SdrE, ClfA and IsdH contribute to the degradation of complement C3b [72, 152–154]. Only in the case of SdrE is the mechanistic basis understood. SdrE binds the host complement regulatory protein factor H by the closed DLL mechanism which in turn recruits and activates the specific C3b-degrading protease factor I [26]. It is possible that ClfA can bind and activate factor I directly without the requirement of factor H but direct evidence for this is lacking.

As discussed above the ability of *S. aureus* CWA proteins including FnBPs to capture Plg and facilitate activation of plasmin on the cell surface leads to destruction of C3b (and IgG) opsonins [39, 155].

2.9.3 Interference with Neutrophils

The ability of CWA proteins to inhibit the function of neutrophils has been covered in several of the sections above. These actions can be summarized as follows. AdsA inhibits the activation of the oxidative burst once bacteria have been phagocytosed. Protein A attracts deposition of IgG which inhibits opsonophagocytosis. ClfA and other CWA proteins that bind soluble plasma proteins mask opsonins from neutrophils. Cell clumping and biofilm formation reduce the ability of neutrophils to engulf bacteria. The long fibrillar proteins SasG and SraP likely also contribute to opsonin masking.

2.9.4 Inhibition of Adaptive Immune Responses

A hallmark of *S. aureus* infections is the elaboration of superantigens which interfere with adaptive immunity. Most strains have the ability to secrete one or more small proteins (enterotoxins and TSST-1) that have the ability to bind the MHC-II receptor of antigen-presenting cells and promote binding to T cell receptors [156]. This triggers activation and proliferation of a plethora of T cells without appropriate guidance and recognition by bound antigen fragments. This results in uncontrolled expression of cytokines which in the worst case scenario can cause toxic shock syndrome. Depletion of T cells also results in failure to mount an appropriate adaptive immune response.

Protein A is also important in disrupting adaptive immune responses during nasal colonization and infection [157, 158]. It is released from the cell surface by the action of autolysins resulting in soluble proteins with fragments of peptiodoglycan bound to the C-terminus [159, 160]. Also Spa has a faulty sorting signal which results in a proportion of the newly translated and secreted protein being released without being anchored to the cell wall [161]. The Ig binding repeats of released Spa bind to and cross-link the variant V_H3 chain of receptors located on the surface of B cells in the lymphoid tissue [75]. This results in B cell proliferation and ultimately to their depletion. During infection there is a rise in the level of V_H3 clonal IgM and IgG in serum but these antibodies lack the ability to bind to *S. aureus* antigens [160, 162]. This inappropriate expression of Ig requires the cooperation of CD4+ T cells [160] although it is not clear if antigen presentation is involved. The C-terminal peptidoglycan fragment is implicated in activation of NOD1/NOD2 signalling which triggers maturation of the B cells into antibody secreting plasma cells [75, 160, 163]. Depletion of B cells from lymphoid tissues causes anergy and a lack of immunological memory which results in *S. aureus* being able to cause successive infections. This may also contribute to bacteria being able to proliferate in the nares of persistent carriers.

2.10 Concluding Remarks

Cell wall anchored proteins provide opportunities for *S. aureus* to interact with the host, both in the commensal state and during infection. Given that the repertoire of CWA proteins is restricted it is perhaps not surprising that many CWA proteins are multifunctional.

It has been suggested that CWA proteins be categorized based on structural motifs and functional attributes [15, 164]. The acronym MSCRAMM has been used inappropriately in referring to CWA proteins regardless of their function and whether or not they bind to the extracellular matrix.

The functions of several CWA proteins are not yet established and it is likely that new ligands will be discovered for those with known functions. Far Western blotting, phage display and two-hybrid analysis have identified potential ligands. Initial findings must be validated by detailed molecular analysis and by studying their biological relevance [34, 37, 39, 142].

Until recently most studies on CWA proteins have been performed with laboratory strains. It is now possible to manipulate genetically strains from all genetic lineages including clinically relevant bacteria [106–108]. Variations in the repertoire of CWA proteins occur in different lineages and variations in ligand binding domains might lead to different affinities for ligands. Also the regulation of gene expression might vary which can result in different levels of protein abundance on the cell surface.

Several CWA proteins have potential as antigens for vaccines to protect against *S. aureus* infections by stimulating protection in murine models of infection [68, 165–167]. In addition recombinant truncates that do not bind host ligands have improved antigenicity [119, 168]. Several combinations of antigens including fragments of CWA proteins and other surface components have been formulated [169–171] but have not progressed through clinical trials. The only phase III trial performed to date was with a single antigen (IsdB) which failed for reasons that are not entirely clear [172, 173]. Future clinical trials will require the use of an adjuvant that triggers a cellular response involving T_H1 cells and/or T_H17 cells as well as IL-17 and IFNγ-mediated recruitment of neutrophils and must suppress immune dysregulation caused by expression of superantigens [170, 173].

In conclusion CWA proteins carry out a range of important functions and are essential for successful colonization of the host and for infection. Structural and functional analysis has provided a rational framework for classification of CWA proteins and has helped to define the mechanistic basis of host-pathogen interactions.

References

1. Krismer B, Weidenmaier C, Zipperer A, Peschel A (2017) The commensal lifestyle of *Staphylococcus aureus* and its interactions with the nasal microbiota. Nat Rev Microbiol 15(11):675–687. https://doi.org/10.1038/nrmicro.2017.104
2. von Eiff C, Becker K, Machka K, Stammer H, Peters G (2001) Nasal carriage as a source of *Staphylococcus aureus* bacteremia. Study Group. N Engl J Med 344(1):11–16. https://doi.org/10.1056/nejm200101043440102
3. Peacock SJ, de Silva I, Lowy FD (2001) What determines nasal carriage of *Staphylococcus aureus*? Trends Microbiol 9(12):605–610. https://doi.org/10.1016/s0966-842x(01)02254-5
4. Lowy FD (1998) *Staphylococcus aureus* infections. N Engl J Med 339(8):520–532. https://doi.org/10.1056/nejm199808203390806
5. Chambers HF, Deleo FR (2009) Waves of resistance: *Staphylococcus aureus* in the antibiotic era. Nat Rev Microbiol 7(9):629–641. https://doi.org/10.1038/nrmicro2200
6. DeLeo FR, Chambers HF (2009) Reemergence of antibiotic-resistant *Staphylococcus aureus* in the genomics era. J Clin Invest 119(9):2464–2474. https://doi.org/10.1172/jci38226
7. Diep BA, Otto M (2008) The role of virulence determinants in community-associated MRSA pathogenesis. Trends Microbiol 16(8):361–369. https://doi.org/10.1016/j.tim.2008.05.002
8. Heilbronner S, Holden MT, van Tonder A, Geoghegan JA, Foster TJ, Parkhill J et al (2011) Genome sequence of *Staphylococcus lugdunensis* N920143 allows identification of putative colonization and virulence factors. FEMS Microbiol Lett 322(1):60–67. https://doi.org/10.1111/j.1574-6968.2011.02339.x
9. Bowden MG, Chen W, Singvall J, Xu Y, Peacock SJ, Valtulina V et al (2005) Identification and preliminary characterization of cell-wall-anchored proteins of *Staphylococcus epidermidis*. Microbiology 151(Pt 5):1453–1464. https://doi.org/10.1099/mic.0.27534-0
10. Hammer ND, Skaar EP (2011) Molecular mechanisms of *Staphylococcus aureus* iron acquisition. Annu Rev Microbiol 65:129–147. https://doi.org/10.1146/annurev-micro-090110-102851
11. Bischoff M, Dunman P, Kormanec J, Macapagal D, Murphy E, Mounts W et al (2004) Microarray-based analysis of the *Staphylococcus aureus* sigmaB regulon. J Bacteriol 186(13):4085–4099. https://doi.org/10.1128/jb.186.13.4085-4099.2004
12. Mazmanian SK, Skaar EP, Gaspar AH, Humayun M, Gornicki P, Jelenska J et al (2003) Passage of heme-iron across the envelope of *Staphylococcus aureus*. Science 299(5608):906–909. https://doi.org/10.1126/science.1081147
13. Schneewind O, Missiakas D (2014) Sec-secretion and sortase-mediated anchoring of proteins in gram-positive bacteria. Biochim Biophys Acta 1843(8):1687–1697. https://doi.org/10.1016/j.bbamcr.2013.11.009
14. Patti JM, Allen BL, McGavin MJ, Hook M (1994) MSCRAMM-mediated adherence of microorganisms to host tissues. Annu Rev Microbiol 48:585–617. https://doi.org/10.1146/annurev.mi.48.100194.003101
15. Foster TJ, Geoghegan JA, Ganesh VK, Hook M (2014) Adhesion, invasion and evasion: the many functions of the surface proteins of *Staphylococcus aureus*. Nat Rev Microbiol 12(1):49–62. https://doi.org/10.1038/nrmicro3161
16. Deivanayagam CC, Wann ER, Chen W, Carson M, Rajashankar KR, Hook M et al (2002) A novel variant of the immunoglobulin fold in surface adhesins of *Staphylococcus aureus*: crystal structure of the fibrinogen-binding MSCRAMM, clumping factor A. EMBO J 21(24):6660–6672. https://doi.org/10.1093/emboj/cdf619
17. Ponnuraj K, Bowden MG, Davis S, Gurusiddappa S, Moore D, Choe D et al (2003) A "dock, lock, and latch" structural model for a staphylococcal adhesin binding to fibrinogen. Cell 115(2):217–228. https://doi.org/10.1016/s0092-8674(03)00809-2
18. Ganesh VK, Barbu EM, Deivanayagam CC, Le B, Anderson AS, Matsuka YV et al (2011) Structural and biochemical characterization of *Staphylococcus aureus* clumping factor

B/ligand interactions. J Biol Chem 286(29):25963–25972. https://doi.org/10.1074/jbc.M110.217414
19. Bingham RJ, Rudino-Pinera E, Meenan NA, Schwarz-Linek U, Turkenburg JP, Hook M et al (2008) Crystal structures of fibronectin-binding sites from *Staphylococcus aureus* FnBPA in complex with fibronectin domains. Proc Natl Acad Sci U S A 105(34):12254–12258. https://doi.org/10.1073/pnas.0803556105
20. Xiang H, Feng Y, Wang J, Liu B, Chen Y, Liu L et al (2012) Crystal structures reveal the multi-ligand binding mechanism of *Staphylococcus aureus* ClfB. PLoS Pathog 8(6):e1002751. https://doi.org/10.1371/journal.ppat.1002751
21. Herman P, El-Kirat-Chatel S, Beaussart A, Geoghegan JA, Foster TJ, Dufrene YF (2014) The binding force of the staphylococcal adhesin SdrG is remarkably strong. Mol Microbiol 93(2):356–368. https://doi.org/10.1111/mmi.12663
22. Vitry P, Valotteau C, Feuillie C, Bernard S, Alsteens D, Geoghegan JA et al (2017) Force-induced strengthening of the interaction between *Staphylococcus aureus* clumping factor B and loricrin. MBio 8(6):e01748–e01717. https://doi.org/10.1128/mbio.01748-17
23. Casillas-Ituarte NN, DiBartola AC, Broughton MJ, Perez-Guzman L, Wheeler RM, Ibaraki M et al (2019) Fibrinogen binding is affected by amino acid substitutions in C-terminal repeat region of fibronectin binding protein a. Sci Rep 9(1):11619. https://doi.org/10.1038/s41598-019-48031-5
24. Ganesh VK, Rivera JJ, Smeds E, Ko YP, Bowden MG, Wann ER et al (2008) A structural model of the *Staphylococcus aureus* ClfA-fibrinogen interaction opens new avenues for the design of anti-staphylococcal therapeutics. PLoS Pathog 4(11):e1000226. https://doi.org/10.1371/journal.ppat.1000226
25. Ganesh VK, Liang X, Geoghegan JA, Cohen AL, Venugopalan N, Foster TJ et al (2016) Lessons from the crystal structure of the *S. aureus* surface protein clumping factor a in complex with tefibazumab, an inhibiting monoclonal antibody. EBioMedicine 13:328–338. https://doi.org/10.1016/j.ebiom.2016.09.027
26. Zhang Y, Wu M, Hang T, Wang C, Yang Y, Pan W et al (2017) *Staphylococcus aureus* SdrE captures complement factor H's C-terminus via a novel 'close, dock, lock and latch' mechanism for complement evasion. Biochem J 474(10):1619–1631. https://doi.org/10.1042/bcj20170085
27. Mulcahy ME, Geoghegan JA, Monk IR, O'Keeffe KM, Walsh EJ, Foster TJ et al (2012) Nasal colonisation by *Staphylococcus aureus* depends upon clumping factor B binding to the squamous epithelial cell envelope protein loricrin. PLoS Pathog 8(12):e1003092. https://doi.org/10.1371/journal.ppat.1003092
28. Burke FM, McCormack N, Rindi S, Speziale P, Foster TJ (2010) Fibronectin-binding protein B variation in *Staphylococcus aureus*. BMC Microbiol 10:160. https://doi.org/10.1186/1471-2180-10-160
29. Keane FM, Loughman A, Valtulina V, Brennan M, Speziale P, Foster TJ (2007) Fibrinogen and elastin bind to the same region within the a domain of fibronectin binding protein A, an MSCRAMM of *Staphylococcus aureus*. Mol Microbiol 63(3):711–723. https://doi.org/10.1111/j.1365-2958.2006.05552.x
30. Zong Y, Xu Y, Liang X, Keene DR, Hook A, Gurusiddappa S et al (2005) A 'Collagen Hug' model for *Staphylococcus aureus* CNA binding to collagen. EMBO J 24(24):4224–4236. https://doi.org/10.1038/sj.emboj.7600888
31. Herman-Bausier P, Valotteau C, Pietrocola G, Rindi S, Alsteens D, Foster TJ et al (2016) Mechanical strength and inhibition of the *Staphylococcus aureus* collagen-binding protein Cna. MBio 7(5):e01529–e01516. https://doi.org/10.1128/mbio.01529-16
32. Valotteau C, Prystopiuk V, Pietrocola G, Rindi S, Peterle D, De Filippis V et al (2017) Single-cell and single-molecule analysis unravels the multifunctionality of the *Staphylococcus aureus* collagen-binding protein Cna. ACS Nano 11(2):2160–2170. https://doi.org/10.1021/acsnano.6b08404

33. Geoghegan JA, Monk IR, O'Gara JP, Foster TJ (2013) Subdomains N2N3 of fibronectin binding protein a mediate *Staphylococcus aureus* biofilm formation and adherence to fibrinogen using distinct mechanisms. J Bacteriol 195(11):2675–2683. https://doi.org/10.1128/JB.02128-12
34. Barbu EM, Mackenzie C, Foster TJ, Hook M (2014) SdrC induces staphylococcal biofilm formation through a homophilic interaction. Mol Microbiol 94(1):172–185. https://doi.org/10.1111/mmi.12750
35. O'Neill E, Pozzi C, Houston P, Humphreys H, Robinson DA, Loughman A et al (2008) A novel *Staphylococcus aureus* biofilm phenotype mediated by the fibronectin-binding proteins, FnBPA and FnBPB. J Bacteriol 190(11):3835–3850. https://doi.org/10.1128/jb.00167-08
36. Feuillie C, Formosa-Dague C, Hays LM, Vervaeck O, Derclaye S, Brennan MP et al (2017) Molecular interactions and inhibition of the staphylococcal biofilm-forming protein SdrC. Proc Natl Acad Sci U S A 114(14):3738–3743. https://doi.org/10.1073/pnas.1616805114
37. Barbu EM, Ganesh VK, Gurusiddappa S, Mackenzie RC, Foster TJ, Sudhof TC et al (2010) β-Neurexin is a ligand for the *Staphylococcus aureus* MSCRAMM SdrC. PLoS Pathog 6(1):e1000726. https://doi.org/10.1371/journal.ppat.1000726
38. Herman-Bausier P, El-Kirat-Chatel S, Foster TJ, Geoghegan JA, Dufrene YF (2015) *Staphylococcus aureus* fibronectin-binding protein a mediates cell-cell adhesion through low-affinity homophilic bonds. MBio 6(3):e00413–e00415. https://doi.org/10.1128/mbio.00413-15
39. Pietrocola G, Nobile G, Gianotti V, Zapotoczna M, Foster TJ, Geoghegan JA et al (2016) Molecular interactions of human plasminogen with fibronectin-binding protein B (FnBPB), a fibrinogen/fibronectin-binding protein from *Staphylococcus aureus*. J Biol Chem 291(35):18148–18162. https://doi.org/10.1074/jbc.M116.731125
40. Herman-Bausier P, Pietrocola G, Foster TJ, Speziale P, Dufrene YF (2017) Fibrinogen activates the capture of human plasminogen by staphylococcal fibronectin-binding proteins. MBio 8(5):e01067–e01017. https://doi.org/10.1128/mbio.01067-17
41. Pietrocola G, Nobile G, Alfeo MJ, Foster TJ, Geoghegan JA, De Filippis V et al (2019) Fibronectin-binding protein B (FnBPB) from *Staphylococcus aureus* protects against the antimicrobial activity of histones. J Biol Chem 294(10):3588–3602. https://doi.org/10.1074/jbc.ra118.005707
42. Josefsson E, McCrea KW, Ni Eidhin D, O'Connell D, Cox J, Hook M et al (1998) Three new members of the serine-aspartate repeat protein multigene family of *Staphylococcus aureus*. Microbiology 144(Pt 12):3387–3395. https://doi.org/10.1099/00221287-144-12-3387
43. Josefsson E, O'Connell D, Foster TJ, Durussel I, Cox JA (1998) The binding of calcium to the B-repeat segment of SdrD, a cell surface protein of *Staphylococcus aureus*. J Biol Chem 273(47):31145–31152. https://doi.org/10.1074/jbc.273.47.31145
44. Herman-Bausier P, Dufrene YF (2016) Atomic force microscopy reveals a dual collagen-binding activity for the staphylococcal surface protein SdrF. Mol Microbiol 99(3):611–621. https://doi.org/10.1111/mmi.13254
45. Wang X, Ge J, Liu B, Hu Y, Yang M (2013) Structures of SdrD from *Staphylococcus aureus* reveal the molecular mechanism of how the cell surface receptors recognize their ligands. Protein Cell 4(4):277–285. https://doi.org/10.1007/s13238-013-3009-x
46. Peacock SJ, Foster TJ, Cameron BJ, Berendt AR (1999) Bacterial fibronectin-binding proteins and endothelial cell surface fibronectin mediate adherence of *Staphylococcus aureus* to resting human endothelial cells. Microbiology 145(Pt 12):3477–3486. https://doi.org/10.1099/00221287-145-12-3477
47. Sinha B, Francois PP, Nusse O, Foti M, Hartford OM, Vaudaux P et al (1999) Fibronectin-binding protein acts as *Staphylococcus aureus* invasin via fibronectin bridging to integrin $\alpha_5\beta_1$. Cell Microbiol 1(2):101–117. https://doi.org/10.1046/j.1462-5822.1999.00011.x
48. Dziewanowska K, Patti JM, Deobald CF, Bayles KW, Trumble WR, Bohach GA (1999) Fibronectin binding protein and host cell tyrosine kinase are required for internalization

of *Staphylococcus aureus* by epithelial cells. Infect Immun 67(9):4673–4678. https://doi.org/10.1128/iai.67.9.4673-4678.1999

49. Fowler T, Wann ER, Joh D, Johansson S, Foster TJ, Hook M (2000) Cellular invasion by *Staphylococcus aureus* involves a fibronectin bridge between the bacterial fibronectin-binding MSCRAMMs and host cell β1 integrins. Eur J Cell Biol 79(10):672–679. https://doi.org/10.1078/0171-9335-00104
50. Agerer F, Lux S, Michel A, Rohde M, Ohlsen K, Hauck CR (2005) Cellular invasion by *Staphylococcus aureus* reveals a functional link between focal adhesion kinase and cortactin in integrin-mediated internalisation. J Cell Sci 118(Pt 10):2189–2200. https://doi.org/10.1242/jcs.02328
51. Mempel M, Schnopp C, Hojka M, Fesq H, Weidinger S, Schaller M et al (2002) Invasion of human keratinocytes by *Staphylococcus aureus* and intracellular bacterial persistence represent haemolysin-independent virulence mechanisms that are followed by features of necrotic and apoptotic keratinocyte cell death. Br J Dermatol 146(6):943–951. https://doi.org/10.1046/j.1365-2133.2002.04752.x
52. Jett BD, Gilmore MS (2002) Internalization of *Staphylococcus aureus* by human corneal epithelial cells: role of bacterial fibronectin-binding protein and host cell factors. Infect Immun 70(8):4697–4700. https://doi.org/10.1128/IAI.70.8.4697-4700.2002
53. Ahmed S, Meghji S, Williams RJ, Henderson B, Brock JH, Nair SP (2001) *Staphylococcus aureus* fibronectin binding proteins are essential for internalization by osteoblasts but do not account for differences in intracellular levels of bacteria. Infect Immun 69(5):2872–2877. https://doi.org/10.1128/IAI.69.5.2872-2877.2001
54. Sendi P, Proctor RA (2009) *Staphylococcus aureus* as an intracellular pathogen: the role of small colony variants. Trends Microbiol 17(2):54–58. https://doi.org/10.1016/j.tim.2008.11.004
55. Guerillot R, Kostoulias X, Donovan L, Li L, Carter GP, Hachani A et al (2019) Unstable chromosome rearrangements in *Staphylococcus aureus* cause phenotype switching associated with persistent infections. Proc Natl Acad Sci U S A 16(40):20135–20140. https://doi.org/10.1073/pnas.1904861116
56. Vaudaux P, Francois P, Bisognano C, Kelley WL, Lew DP, Schrenzel J et al (2002) Increased expression of clumping factor and fibronectin-binding proteins by hemB mutants of *Staphylococcus aureus* expressing small colony variant phenotypes. Infect Immun 70(10):5428–5437. https://doi.org/10.1128/iai.70.10.5428-5437.2002
57. Henderson B, Nair S, Pallas J, Williams MA (2011) Fibronectin: a multidomain host adhesin targeted by bacterial fibronectin-binding proteins. FEMS Microbiol Rev 35(1):147–200. https://doi.org/10.1111/j.1574-6976.2010.00243.x
58. Liang X, Garcia BL, Visai L, Prabhakaran S, Meenan NA, Potts JR et al (2016) Allosteric regulation of fibronectin/$\alpha_5\beta_1$ interaction by fibronectin-binding MSCRAMMs. PLoS One 11(7):e0159118. https://doi.org/10.1371/journal.pone.0159118
59. Hauck CR, Ohlsen K (2006) Sticky connections: extracellular matrix protein recognition and integrin-mediated cellular invasion by *Staphylococcus aureus*. Curr Opin Microbiol 9(1):5–11. https://doi.org/10.1016/j.mib.2005.12.002
60. Schwarz-Linek U, Hook M, Potts JR (2004) The molecular basis of fibronectin-mediated bacterial adherence to host cells. Mol Microbiol 52(3):631–641. https://doi.org/10.1111/j.1365-2958.2004.04027.x
61. Hartford O, Francois P, Vaudaux P, Foster TJ (1997) The dipeptide repeat region of the fibrinogen-binding protein (clumping factor) is required for functional expression of the fibrinogen-binding domain on the *Staphylococcus aureus* cell surface. Mol Microbiol 25(6):1065–1076. https://doi.org/10.1046/j.1365-2958.1997.5291896.x
62. Hazenbos WL, Kajihara KK, Vandlen R, Morisaki JH, Lehar SM, Kwakkenbos MJ et al (2013) Novel staphylococcal glycosyltransferases SdgA and SdgB mediate immunogenicity and protection of virulence-associated cell wall proteins. PLoS Pathog 9(10):e1003653. https://doi.org/10.1371/journal.ppat.1003653

63. Thomer L, Becker S, Emolo C, Quach A, Kim HK, Rauch S et al (2014) N-acetylglucosaminylation of serine-aspartate repeat proteins promotes *Staphylococcus aureus* bloodstream infection. J Biol Chem 289(6):3478–3486. https://doi.org/10.1074/jbc.m113.532655
64. McAleese FM, Walsh EJ, Sieprawska M, Potempa J, Foster TJ (2001) Loss of clumping factor B fibrinogen binding activity by *Staphylococcus aureus* involves cessation of transcription, shedding and cleavage by metalloprotease. J Biol Chem 276(32):29969–29978. https://doi.org/10.1074/jbc.M102389200
65. McCormack N, Foster TJ, Geoghegan JA (2014) A short sequence within subdomain N1 of region a of the *Staphylococcus aureus* MSCRAMM clumping factor a is required for export and surface display. Microbiology 160(Pt 4):659–670. https://doi.org/10.1099/mic.0.074724-0
66. Cassat JE, Skaar EP (2012) Metal ion acquisition in *Staphylococcus aureus*: overcoming nutritional immunity. Semin Immunopathol 34(2):215–235. https://doi.org/10.1007/s00281-011-0294-4
67. Grigg JC, Ukpabi G, Gaudin CF, Murphy ME (2010) Structural biology of heme binding in the *Staphylococcus aureus* Isd system. J Inorg Biochem 104(3):341–348. https://doi.org/10.1016/j.jinorgbio.2009.09.012
68. Clarke SR, Brummell KJ, Horsburgh MJ, McDowell PW, Mohamad SA, Stapleton MR et al (2006) Identification of in vivo-expressed antigens of *Staphylococcus aureus* and their use in vaccinations for protection against nasal carriage. J Infect Dis 193(8):1098–1108. https://doi.org/10.1086/501471
69. Clarke SR, Mohamed R, Bian L, Routh AF, Kokai-Kun JF, Mond JJ et al (2007) The *Staphylococcus aureus* surface protein IsdA mediates resistance to innate defenses of human skin. Cell Host Microbe 1(3):199–212. https://doi.org/10.1016/j.chom.2007.04.005
70. Miajlovic H, Zapotoczna M, Geoghegan JA, Kerrigan SW, Speziale P, Foster TJ (2010) Direct interaction of iron-regulated surface determinant IsdB of *Staphylococcus aureus* with the GPIIb/IIIa receptor on platelets. Microbiology 156(Pt 3):920–928. https://doi.org/10.1099/mic.0.036673-0
71. Zapotoczna M, Jevnikar Z, Miajlovic H, Kos J, Foster TJ (2013) Iron-regulated surface determinant B (IsdB) promotes *Staphylococcus aureus* adherence to and internalization by non-phagocytic human cells. Cell Microbiol 15(6):1026–1041. https://doi.org/10.1111/cmi.12097
72. Visai L, Yanagisawa N, Josefsson E, Tarkowski A, Pezzali I, Rooijakkers SH et al (2009) Immune evasion by *Staphylococcus aureus* conferred by iron-regulated surface determinant protein IsdH. Microbiology 155(Pt 3):667–679. https://doi.org/10.1099/mic.0.025684-0
73. Graille M, Stura EA, Corper AL, Sutton BJ, Taussig MJ, Charbonnier JB et al (2000) Crystal structure of a *Staphylococcus aureus* protein a domain complexed with the fab fragment of a human IgM antibody: structural basis for recognition of B-cell receptors and superantigen activity. Proc Natl Acad Sci U S A 97(10):5399–5404. https://doi.org/10.1073/pnas.97.10.5399
74. Gomez MI, O'Seaghdha M, Magargee M, Foster TJ, Prince AS (2006) *Staphylococcus aureus* protein a activates TNFR1 signaling through conserved IgG binding domains. J Biol Chem 281(29):20190–20196. https://doi.org/10.1074/jbc.m601956200
75. Silverman GJ, Goodyear CS (2006) Confounding B-cell defences: lessons from a staphylococcal superantigen. Nat Rev Immunol 6(6):465–475. https://doi.org/10.1038/nri1853
76. O'Seaghdha M, van Schooten CJ, Kerrigan SW, Emsley J, Silverman GJ, Cox D et al (2006) *Staphylococcus aureus* protein a binding to von Willebrand factor A1 domain is mediated by conserved IgG binding regions. FEBS J 273(21):4831–4841. https://doi.org/10.1111/j.1742-4658.2006.05482.x
77. Deis LN, Wu Q, Wang Y, Qi Y, Daniels KG, Zhou P et al (2015) Suppression of conformational heterogeneity at a protein-protein interface. Proc Natl Acad Sci U S A 112(29):9028–9033. https://doi.org/10.1073/pnas.1424724112

78. Martin FJ, Gomez MI, Wetzel DM, Memmi G, O'Seaghdha M, Soong G et al (2009) *Staphylococcus aureus* activates type I IFN signaling in mice and humans through the Xr repeated sequences of protein A. J Clin Invest 119(7):1931–1939. https://doi.org/10.1172/jci35879
79. Burman JD, Leung E, Atkins KL, O'Seaghdha MN, Lango L, Bernado P et al (2008) Interaction of human complement with Sbi, a staphylococcal immunoglobulin-binding protein: indications of a novel mechanism of complement evasion by *Staphylococcus aureus*. J Biol Chem 283(25):17579–17593. https://doi.org/10.1074/jbc.m800265200
80. Smith EJ, Corrigan RM, van der Sluis T, Grundling A, Speziale P, Geoghegan JA et al (2012) The immune evasion protein Sbi of *Staphylococcus aureus* occurs both extracellularly and anchored to the cell envelope by binding lipoteichoic acid. Mol Microbiol 83(4):789–804. https://doi.org/10.1111/j.1365-2958.2011.07966.x
81. Upadhyay A, Burman JD, Clark EA, Leung E, Isenman DE, van den Elsen JM et al (2008) Structure-function analysis of the C3 binding region of *Staphylococcus aureus* immune subversion protein Sbi. J Biol Chem 283(32):22113–22120. https://doi.org/10.1074/jbc.M802636200
82. Werbick C, Becker K, Mellmann A, Juuti KM, von Eiff C, Peters G et al (2007) Staphylococcal chromosomal cassette mec type I, spa type, and expression of Pls are determinants of reduced cellular invasiveness of methicillin-resistant *Staphylococcus aureus* isolates. J Infect Dis 195(11):1678–1685. https://doi.org/10.1086/517517
83. Banner MA, Cunniffe JG, Macintosh RL, Foster TJ, Rohde H, Mack D et al (2007) Localized tufts of fibrils on *Staphylococcus epidermidis* NCTC 11047 are comprised of the accumulation-associated protein. J Bacteriol 189(7):2793–2804. https://doi.org/10.1128/jb.00952-06
84. Corrigan RM, Rigby D, Handley P, Foster TJ (2007) The role of *Staphylococcus aureus* surface protein SasG in adherence and biofilm formation. Microbiology 153(Pt 8):2435–2446. https://doi.org/10.1099/mic.0.2007/006676-0
85. Juuti KM, Sinha B, Werbick C, Peters G, Kuusela PI (2004) Reduced adherence and host cell invasion by methicillin-resistant *Staphylococcus aureus* expressing the surface protein Pls. J Infect Dis 189(9):1574–1584. https://doi.org/10.1086/383348
86. Roche FM, Meehan M, Foster TJ (2003) The *Staphylococcus aureus* surface protein SasG and its homologues promote bacterial adherence to human desquamated nasal epithelial cells. Microbiology 149(Pt 10):2759–2767. https://doi.org/10.1099/mic.0.26412-0
87. Kwiecinski JM, Crosby HA, Valotteau C, Hippensteel JA, Nayak MK, Chauhan AK et al (2019) *Staphylococcus aureus* adhesion in endovascular infections is controlled by the ArlRS-MgrA signaling cascade. PLoS Pathog 15(5):e1007800. https://doi.org/10.1371/journal.ppat.1007800
88. Huesca M, Peralta R, Sauder DN, Simor AE, McGavin MJ (2002) Adhesion and virulence properties of epidemic Canadian methicillin-resistant *Staphylococcus aureus* strain 1: identification of novel adhesion functions associated with plasmin-sensitive surface protein. J Infect Dis 185(9):1285–1296. https://doi.org/10.1086/340123
89. Schaeffer CR, Woods KM, Longo GM, Kiedrowski MR, Paharik AE, Buttner H et al (2015) Accumulation-associated protein enhances *Staphylococcus epidermidis* biofilm formation under dynamic conditions and is required for infection in a rat catheter model. Infect Immun 83(1):214–226. https://doi.org/10.1128/iai.02177-14
90. Conlon BP, Geoghegan JA, Waters EM, McCarthy H, Rowe SE, Davies JR et al (2014) Role for the a domain of unprocessed accumulation-associated protein (Aap) in the attachment phase of the *Staphylococcus epidermidis* biofilm phenotype. J Bacteriol 196(24):4268–4275. https://doi.org/10.1128/jb.01946-14
91. Gruszka DT, Whelan F, Farrance OE, Fung HK, Paci E, Jeffries CM et al (2015) Cooperative folding of intrinsically disordered domains drives assembly of a strong elongated protein. Nat Commun 6:7271. https://doi.org/10.1038/ncomms8271

92. Gruszka DT, Wojdyla JA, Bingham RJ, Turkenburg JP, Manfield IW, Steward A et al (2012) Staphylococcal biofilm-forming protein has a contiguous rod-like structure. Proc Natl Acad Sci U S A 109(17):E1011–E1018. https://doi.org/10.1073/pnas.1119456109
93. Conrady DG, Wilson JJ, Herr AB (2013) Structural basis for Zn^{2+}-dependent intercellular adhesion in staphylococcal biofilms. Proc Natl Acad Sci U S A 110(3):E202–E211. https://doi.org/10.1073/pnas.1208134110
94. Rohde H, Burdelski C, Bartscht K, Hussain M, Buck F, Horstkotte MA et al (2005) Induction of *Staphylococcus epidermidis* biofilm formation via proteolytic processing of the accumulation-associated protein by staphylococcal and host proteases. Mol Microbiol 55(6):1883–1895. https://doi.org/10.1111/j.1365-2958.2005.04515.x
95. Geoghegan JA, Corrigan RM, Gruszka DT, Speziale P, O'Gara JP, Potts JR et al (2010) Role of surface protein SasG in biofilm formation by *Staphylococcus aureus*. J Bacteriol 192(21):5663–5673. https://doi.org/10.1128/jb.00628-10
96. Formosa-Dague C, Speziale P, Foster TJ, Geoghegan JA, Dufrene YF (2016) Zinc-dependent mechanical properties of *Staphylococcus aureus* biofilm-forming surface protein SasG. Proc Natl Acad Sci U S A 113(2):410–415. https://doi.org/10.1073/pnas.1519265113
97. Lizcano A, Sanchez CJ, Orihuela CJ (2012) A role for glycosylated serine-rich repeat proteins in gram-positive bacterial pathogenesis. Mol Oral Microbiol 27(4):257–269. https://doi.org/10.1111/j.2041-1014.2012.00653.x
98. Yang YH, Jiang YL, Zhang J, Wang L, Bai XH, Zhang SJ et al (2014) Structural insights into SraP-mediated *Staphylococcus aureus* adhesion to host cells. PLoS Pathog 10(6):e1004169. https://doi.org/10.1371/journal.ppat.1004169
99. Kukita K, Kawada-Matsuo M, Oho T, Nagatomo M, Oogai Y, Hashimoto M et al (2013) *Staphylococcus aureus* SasA is responsible for binding to the salivary agglutinin gp340, derived from human saliva. Infect Immun 81(6):1870–1879. https://doi.org/10.1128/iai.00011-13
100. Thammavongsa V, Kern JW, Missiakas DM, Schneewind O (2009) *Staphylococcus aureus* synthesizes adenosine to escape host immune responses. J Exp Med 206(11):2417–2427. https://doi.org/10.1084/jem.20090097
101. Thammavongsa V, Schneewind O, Missiakas DM (2011) Enzymatic properties of *Staphylococcus aureus* adenosine synthase (AdsA). BMC Biochem 12:56. https://doi.org/10.1186/1471-2091-12-56
102. Brinkmann V, Reichard U, Goosmann C, Fauler B, Uhlemann Y, Weiss DS et al (2004) Neutrophil extracellular traps kill bacteria. Science 303(5663):1532–1535. https://doi.org/10.1126/science.1092385
103. Thammavongsa V, Missiakas DM, Schneewind O (2013) *Staphylococcus aureus* degrades neutrophil extracellular traps to promote immune cell death. Science 342(6160):863–866. https://doi.org/10.1126/science.1242255
104. Berends ET, Horswill AR, Haste NM, Monestier M, Nizet V, von Kockritz-Blickwede M (2010) Nuclease expression by *Staphylococcus aureus* facilitates escape from neutrophil extracellular traps. J Innate Immun 2(6):576–586. https://doi.org/10.1159/000319909
105. Falkow S (1988) Molecular Koch's postulates applied to microbial pathogenicity. Rev Infect Dis 10(Suppl 2):S274–S276. https://doi.org/10.1093/cid/10.supplement_2.s274
106. Monk IR, Foster TJ (2012) Genetic manipulation of staphylococci-breaking through the barrier. Front Cell Infect Microbiol 2:49. https://doi.org/10.3389/fcimb.2012.00049
107. Monk IR, Shah IM, Xu M, Tan MW, Foster TJ (2012) Transforming the untransformable: application of direct transformation to manipulate genetically *Staphylococcus aureus* and *Staphylococcus epidermidis*. MBio 3(2):e00277–e00211. https://doi.org/10.1128/mbio.00277-11
108. Monk IR, Tree JJ, Howden BP, Stinear TP, Foster TJ (2015) Complete bypass of restriction systems for major *Staphylococcus aureus* lineages. MBio 6(3):e00308–e00315. https://doi.org/10.1128/mbio.00308-15

109. Parker D (2017) Humanized mouse models of *Staphylococcus aureus* infection. Front Immunol 8:512. https://doi.org/10.3389/fimmu.2017.00512
110. Pishchany G, McCoy AL, Torres VJ, Krause JC, Crowe JE Jr, Fabry ME et al (2010) Specificity for human hemoglobin enhances *Staphylococcus aureus* infection. Cell Host Microbe 8(6):544–550. https://doi.org/10.1016/j.chom.2010.11.002
111. Tsai YH, Disson O, Bierne H, Lecuit M (2013) Murinization of internalin extends its receptor repertoire, altering *listeria monocytogenes* cell tropism and host responses. PLoS Pathog 9(5):e1003381. https://doi.org/10.1371/journal.ppat.1003381
112. Cheng AG, Kim HK, Burts ML, Krausz T, Schneewind O, Missiakas DM (2009) Genetic requirements for *Staphylococcus aureus* abscess formation and persistence in host tissues. FASEB J 23(10):3393–4404. https://doi.org/10.1096/fj.09-135467
113. Cheng AG, DeDent AC, Schneewind O, Missiakas D (2011) A play in four acts: *Staphylococcus aureus* abscess formation. Trends Microbiol 19(5):225–232. https://doi.org/10.1016/j.tim.2011.01.007
114. Grundmeier M, Hussain M, Becker P, Heilmann C, Peters G, Sinha B (2004) Truncation of fibronectin-binding proteins in *Staphylococcus aureus* strain Newman leads to deficient adherence and host cell invasion due to loss of the cell wall anchor function. Infect Immun 72(12):7155–7163. https://doi.org/10.1128/IAI.72.12.7155-7163.2004
115. Rhem MN, Lech EM, Patti JM, McDevitt D, Hook M, Jones DB et al (2000) The collagen-binding adhesin is a virulence factor in *Staphylococcus aureus* keratitis. Infect Immun 68(6):3776–3779. https://doi.org/10.1128/IAI.68.6.3776-3779.2000
116. Patti JM, Bremell T, Krajewska-Pietrasik D, Abdelnour A, Tarkowski A, Ryden C et al (1994) The *Staphylococcus aureus* collagen adhesin is a virulence determinant in experimental septic arthritis. Infect Immun 62(1):152–161. https://doi.org/10.1128/iai.62.1.152-161.1994
117. Xu Y, Rivas JM, Brown EL, Liang X, Hook M (2004) Virulence potential of the staphylococcal adhesin CNA in experimental arthritis is determined by its affinity for collagen. J Infect Dis 189(12):2323–2333. https://doi.org/10.1086/420851
118. Kang M, Ko YP, Liang X, Ross CL, Liu Q, Murray BE et al (2013) Collagen-binding microbial surface components recognizing adhesive matrix molecule (MSCRAMM) of gram-positive bacteria inhibit complement activation via the classical pathway. J Biol Chem 288(28):20520–20531. https://doi.org/10.1074/jbc.m113.454462
119. Josefsson E, Higgins J, Foster TJ, Tarkowski A (2008) Fibrinogen binding sites P336 and Y338 of clumping factor a are crucial for *Staphylococcus aureus* virulence. PLoS One 3(5):e2206. https://doi.org/10.1371/journal.pone.0002206
120. O'Brien LM, Walsh EJ, Massey RC, Peacock SJ, Foster TJ (2002) *Staphylococcus aureus* clumping factor B (ClfB) promotes adherence to human type I cytokeratin 10: implications for nasal colonization. Cell Microbiol 4(11):759–770. https://doi.org/10.1046/j.1462-5822.2002.00231.x
121. Corrigan RM, Miajlovic H, Foster TJ (2009) Surface proteins that promote adherence of *Staphylococcus aureus* to human desquamated nasal epithelial cells. BMC Microbiol 9:22. https://doi.org/10.1186/1471-2180-9-22
122. Que YA, Haefliger JA, Piroth L, Francois P, Widmer E, Entenza JM et al (2005) Fibrinogen and fibronectin binding cooperate for valve infection and invasion in *Staphylococcus aureus* experimental endocarditis. J Exp Med 201(10):1627–1635. https://doi.org/10.1084/jem.20050125
123. Sinha B, Francois P, Que YA, Hussain M, Heilmann C, Moreillon P et al (2000) Heterologously expressed *Staphylococcus aureus* fibronectin-binding proteins are sufficient for invasion of host cells. Infect Immun 68(12):6871–6878. https://doi.org/10.1128/iai.68.12.6871-6878.2000
124. Lower SK, Lamlertthon S, Casillas-Ituarte NN, Lins RD, Yongsunthon R, Taylor ES et al (2011) Polymorphisms in fibronectin binding protein a of *Staphylococcus aureus* are associated with infection of cardiovascular devices. Proc Natl Acad Sci U S A 108(45):18372–18377. https://doi.org/10.1073/pnas.1109071108

125. Hos NJ, Rieg S, Kern WV, Jonas D, Fowler VG, Higgins PG et al (2015) Amino acid alterations in fibronectin binding protein a (FnBPA) and bacterial genotype are associated with cardiac device related infection in *Staphylococcus aureus* bacteraemia. J Infect 70(2):153–159. https://doi.org/10.1016/j.jinf.2014.09.005
126. Eichenberger EM, Thaden JT, Sharma-Kuinkel B, Park LP, Rude TH, Ruffin F et al (2015) Polymorphisms in fibronectin binding proteins a and B among *Staphylococcus aureus* bloodstream isolates are not associated with arthroplasty infection. PLoS One 10(11):e0141436. https://doi.org/10.1371/journal.pone.0141436
127. Xiong YQ, Sharma-Kuinkel BK, Casillas-Ituarte NN, Fowler VG Jr, Rude T, DiBartola AC et al (2015) Endovascular infections caused by methicillin-resistant *Staphylococcus aureus* are linked to clonal complex-specific alterations in binding and invasion domains of fibronectin-binding protein a as well as the occurrence of fnbB. Infect Immun 83(12):4772–4780. https://doi.org/10.1128/iai.01074-15
128. Walker JN, Flores-Mireles AL, Pinkner CL, Schreiber HL, Joens MS, Park AM et al (2017) Catheterization alters bladder ecology to potentiate *Staphylococcus aureus* infection of the urinary tract. Proc Natl Acad Sci U S A 114(41):E8721–E8730. https://doi.org/10.1073/pnas.1707572114
129. Lacey KA, Mulcahy ME, Towell AM, Geoghegan JA, McLoughlin RM (2019) Clumping factor B is an important virulence factor during *Staphylococcus aureus* skin infection and a promising vaccine target. PLoS Pathog 15(4):e1007713. https://doi.org/10.1371/journal.ppat.1007713
130. Wertheim HF, Vos MC, Ott A, van Belkum A, Voss A, Kluytmans JA et al (2004) Risk and outcome of nosocomial *Staphylococcus aureus* bacteraemia in nasal carriers versus non-carriers. Lancet 364(9435):703–705. https://doi.org/10.1016/s0140-6736(04)16897-9
131. van Belkum A, Verkaik NJ, de Vogel CP, Boelens HA, Verveer J, Nouwen JL et al (2009) Reclassification of *Staphylococcus aureus* nasal carriage types. J Infect Dis 199(12):1820–1826. https://doi.org/10.1086/599119
132. Sivaraman K, Venkataraman N, Cole AM (2009) *Staphylococcus aureus* nasal carriage and its contributing factors. Future Microbiol 4(8):999–1008. https://doi.org/10.2217/fmb.09.79
133. Weidenmaier C, Goerke C, Wolz C (2012) *Staphylococcus aureus* determinants for nasal colonization. Trends Microbiol 20(5):243–250. https://doi.org/10.1016/j.tim.2012.03.004
134. Baur S, Rautenberg M, Faulstich M, Grau T, Severin Y, Unger C et al (2014) A nasal epithelial receptor for *Staphylococcus aureus* WTA governs adhesion to epithelial cells and modulates nasal colonization. PLoS Pathog 10(5):e1004089. https://doi.org/10.1371/journal.ppat.1004089
135. Winstel V, Kuhner P, Salomon F, Larsen J, Skov R, Hoffmann W et al (2015) Wall teichoic acid glycosylation governs *Staphylococcus aureus* nasal colonization. MBio 6(4):e00632. https://doi.org/10.1128/mbio.00632-15
136. Wertheim HF, Walsh E, Choudhurry R, Melles DC, Boelens HA, Miajlovic H et al (2008) Key role for clumping factor B in *Staphylococcus aureus* nasal colonization of humans. PLoS Med 5(1):e17. https://doi.org/10.1371/journal.pmed.0050017
137. Li M, Du X, Villaruz AE, Diep BA, Wang D, Song Y et al (2012) MRSA epidemic linked to a quickly spreading colonization and virulence determinant. Nat Med 18(5):816–819. https://doi.org/10.1038/nm.2692
138. Savolainen K, Paulin L, Westerlund-Wikstrom B, Foster TJ, Korhonen TK, Kuusela P (2001) Expression of pls, a gene closely associated with the mecA gene of methicillin-resistant *Staphylococcus aureus*, prevents bacterial adhesion in vitro. Infect Immun 69(5):3013–3020. https://doi.org/10.1128/IAI.69.5.3013-3020.2001
139. Lindsay JA, Moore CE, Day NP, Peacock SJ, Witney AA, Stabler RA et al (2006) Microarrays reveal that each of the ten dominant lineages of *Staphylococcus aureus* has a unique combination of surface-associated and regulatory genes. J Bacteriol 188(2):669–676. https://doi.org/10.1128/jb.188.2.669-676.2006

140. Walsh EJ, O'Brien LM, Liang X, Hook M, Foster TJ (2004) Clumping factor B, a fibrinogen-binding MSCRAMM (microbial surface components recognizing adhesive matrix molecules) adhesin of *Staphylococcus aureus*, also binds to the tail region of type I cytokeratin 10. J Biol Chem 279(49):50691–50699. https://doi.org/10.1074/jbc.m408713200
141. Clarke SR, Andre G, Walsh EJ, Dufrene YF, Foster TJ, Foster SJ (2009) Iron-regulated surface determinant protein a mediates adhesion of *Staphylococcus aureus* to human corneocyte envelope proteins. Infect Immun 77(6):2408–2416. https://doi.org/10.1128/iai.01304-08
142. Askarian F, Uchiyama S, Valderrama JA, Ajayi C, Sollid JU, van Sorge NM et al (2017) Serine-aspartate repeat protein D increases *Staphylococcus aureus* virulence and survival in blood. Infect Immun 85(1):e00559–e00516. https://doi.org/10.1128/iai.00559-16
143. Ishida-Yamamoto A, Igawa S (2015) The biology and regulation of corneodesmosomes. Cell Tissue Res 360(3):477–482. https://doi.org/10.1007/s00441-014-2037-z
144. Kong HH, Oh J, Deming C, Conlan S, Grice EA, Beatson MA et al (2012) Temporal shifts in the skin microbiome associated with disease flares and treatment in children with atopic dermatitis. Genome Res 22(5):850–859. https://doi.org/10.1101/gr.131029.111
145. Ong PY, Leung DY (2016) Bacterial and viral infections in atopic dermatitis: a comprehensive review. Clin Rev Allergy Immunol 51(3):329–337. https://doi.org/10.1007/s12016-016-8548-5
146. Fleury OM, McAleer MA, Feuillie C, Formosa-Dague C, Sansevere E, Bennett DE et al (2017) Clumping factor B promotes adherence of *Staphylococcus aureus* to corneocytes in atopic dermatitis. Infect Immun 85(6):e00994–e00916. https://doi.org/10.1128/iai.00994-16
147. Riethmuller C, McAleer MA, Koppes SA, Abdayem R, Franz J, Haftek M et al (2015) Filaggrin breakdown products determine corneocyte conformation in patients with atopic dermatitis. J Allergy Clin Immunol 136(6):1573–1580.e2. https://doi.org/10.1016/j.jaci.2015.04.042
148. Forsgren A, Quie PG (1974) Effects of staphylococcal protein a on heat labile opsonins. J Immunol 112(3):1177–1180
149. Cedergren L, Andersson R, Jansson B, Uhlen M, Nilsson B (1993) Mutational analysis of the interaction between staphylococcal protein a and human IgG1. Protein Eng 6(4):441–448. https://doi.org/10.1093/protein/6.4.441
150. Thielens NM, Tedesco F, Bohlson SS, Gaboriaud C, Tenner AJ (2017) C1q: a fresh look upon an old molecule. Mol Immunol 89:73–83. https://doi.org/10.1016/j.molimm.2017.05.025
151. Tosi MF (2005) Innate immune responses to infection. J Allergy Clin Immunol 116(2):241–249. https://doi.org/10.1016/j.jaci.2005.05.036
152. Hair PS, Echague CG, Sholl AM, Watkins JA, Geoghegan JA, Foster TJ et al (2010) Clumping factor a interaction with complement factor I increases C3b cleavage on the bacterial surface of *Staphylococcus aureus* and decreases complement-mediated phagocytosis. Infect Immun 78(4):1717–1727. https://doi.org/10.1128/iai.01065-09
153. Hair PS, Ward MD, Semmes OJ, Foster TJ, Cunnion KM (2008) *Staphylococcus aureus* clumping factor a binds to complement regulator factor I and increases factor I cleavage of C3b. J Infect Dis 198(1):125–133. https://doi.org/10.1086/588825
154. Sharp JA, Echague CG, Hair PS, Ward MD, Nyalwidhe JO, Geoghegan JA et al (2012) *Staphylococcus aureus* surface protein SdrE binds complement regulator factor H as an immune evasion tactic. PLoS One 7(5):e38407. https://doi.org/10.1371/journal.pone.0038407
155. Rooijakkers SH, van Wamel WJ, Ruyken M, van Kessel KP, van Strijp JA (2005) Anti-opsonic properties of staphylokinase. Microbes Infect 7(3):476–484. https://doi.org/10.1016/j.micinf.2004.12.014
156. Spaulding AR, Salgado-Pabon W, Kohler PL, Horswill AR, Leung DY, Schlievert PM (2013) Staphylococcal and streptococcal superantigen exotoxins. Clin Microbiol Rev 26(3):422–447. https://doi.org/10.1128/cmr.00104-12
157. Votintseva AA, Fung R, Miller RR, Knox K, Godwin H, Wyllie DH et al (2014) Prevalence of *Staphylococcus aureus* protein a (spa) mutants in the community and hospitals in Oxfordshire. BMC Microbiol 14:63. https://doi.org/10.1186/1471-2180-14-63

158. Cole AL, Muthukrishnan G, Chong C, Beavis A, Eade CR, Wood MP et al (2016) Host innate inflammatory factors and staphylococcal protein a influence the duration of human *Staphylococcus aureus* nasal carriage. Mucosal Immunol 9(6):1537–1548. https://doi.org/10.1038/mi.2016.2
159. Becker S, Frankel MB, Schneewind O, Missiakas D (2014) Release of protein a from the cell wall of *Staphylococcus aureus*. Proc Natl Acad Sci U S A 111(4):1574–1579. https://doi.org/10.1073/pnas.1317181111
160. Kim HK, Falugi F, Missiakas DM, Schneewind O (2016) Peptidoglycan-linked protein a promotes T cell-dependent antibody expansion during *Staphylococcus aureus* infection. Proc Natl Acad Sci U S A 113(20):5718–5723. https://doi.org/10.1073/pnas.1524267113
161. O'Halloran DP, Wynne K, Geoghegan JA (2015) Protein a is released into the *Staphylococcus aureus* culture supernatant with an unprocessed sorting signal. Infect Immun 83(4):1598–1609. https://doi.org/10.1128/iai.03122-14
162. Pauli NT, Kim HK, Falugi F, Huang M, Dulac J, Henry Dunand C et al (2014) *Staphylococcus aureus* infection induces protein A-mediated immune evasion in humans. J Exp Med 211(12):2331–2339. https://doi.org/10.1084/jem.20141404
163. Missiakas D, Schneewind O (2016) *Staphylococcus aureus* vaccines: deviating from the carol. J Exp Med 213(9):1645–1653. https://doi.org/10.1084/jem.20160569
164. Geoghegan JA, Foster TJ (2015) Cell wall-anchored surface proteins of *Staphylococcus aureus*: many proteins, multiple functions. Curr Top Microbiol Immunol 409:95–120. https://doi.org/10.1007/82_2015_5002
165. Josefsson E, Hartford O, O'Brien L, Patti JM, Foster T (2001) Protection against experimental *Staphylococcus aureus* arthritis by vaccination with clumping factor a, a novel virulence determinant. J Infect Dis 184(12):1572–1580. https://doi.org/10.1086/324430
166. Nilsson IM, Patti JM, Bremell T, Hook M, Tarkowski A (1998) Vaccination with a recombinant fragment of collagen adhesin provides protection against *Staphylococcus aureus*-mediated septic death. J Clin Invest 101(12):2640–2649. https://doi.org/10.1172/jci1823
167. Stranger-Jones YK, Bae T, Schneewind O (2006) Vaccine assembly from surface proteins of *Staphylococcus aureus*. Proc Natl Acad Sci U S A 103(45):16942–16947. https://doi.org/10.1073/pnas.0606863103
168. Kim HK, Cheng AG, Kim HY, Missiakas DM, Schneewind O (2010) Nontoxigenic protein a vaccine for methicillin-resistant *Staphylococcus aureus* infections in mice. J Exp Med 207(9):1863–1870. https://doi.org/10.1084/jem.20092514
169. Anderson AS, Miller AA, Donald RG, Scully IL, Nanra JS, Cooper D et al (2012) Development of a multicomponent *Staphylococcus aureus* vaccine designed to counter multiple bacterial virulence factors. Hum Vaccin Immunother 8(11):1585–1594. https://doi.org/10.4161/hv.21872
170. Jansen KU, Girgenti DQ, Scully IL, Anderson AS (2013) Vaccine review: "*Staphyloccocus aureus* vaccines: problems and prospects". Vaccine 31(25):2723–2730. https://doi.org/10.1016/j.vaccine.2013.04.002
171. Levy J, Licini L, Haelterman E, Moris P, Lestrate P, Damaso S et al (2015) Safety and immunogenicity of an investigational 4-component *Staphylococcus aureus* vaccine with or without AS03B adjuvant: results of a randomized phase I trial. Hum Vaccin Immunother 11(3):620–631. https://doi.org/10.1080/21645515.2015.1011021
172. Fowler VG, Allen KB, Moreira ED, Moustafa M, Isgro F, Boucher HW et al (2013) Effect of an investigational vaccine for preventing *Staphylococcus aureus* infections after cardiothoracic surgery: a randomized trial. JAMA 309(13):1368–1378. https://doi.org/10.1001/jama.2013.3010
173. Bagnoli F, Bertholet S, Grandi G (2012) Inferring reasons for the failure of *Staphylococcus aureus* vaccines in clinical trials. Front Cell Infect Microbiol 2:16. https://doi.org/10.3389/fcimb.2012.00016
174. Claes J, Liesenborghs L, Peetermans M, Veloso TR, Missiakas D, Schneewind O et al (2017) Clumping factor A, von Willebrand factor-binding protein and von Willebrand factor anchor

Staphylococcus aureus to the vessel wall. J Thromb Haemost 15(5):1009–1019. https://doi.org/10.1111/jth.13653
175. Walsh EJ, Miajlovic H, Gorkun OV, Foster TJ (2008) Identification of the *Staphylococcus aureus* MSCRAMM clumping factor B (ClfB) binding site in the αC-domain of human fibrinogen. Microbiology 154(Pt 2):550–558. https://doi.org/10.1099/mic.0.2007/010868-0
176. Vazquez V, Liang X, Horndahl JK, Ganesh VK, Smeds E, Foster TJ et al (2011) Fibrinogen is a ligand for the *Staphylococcus aureus* microbial surface components recognizing adhesive matrix molecules (MSCRAMM) bone sialoprotein-binding protein (Bbp). J Biol Chem 286(34):29797–29805. https://doi.org/10.1074/jbc.m110.214981
177. Burke FM, Di Poto A, Speziale P, Foster TJ (2011) The a domain of fibronectin-binding protein B of *Staphylococcus aureus* contains a novel fibronectin binding site. FEBS J 278(13):2359–2371. https://doi.org/10.1111/j.1742-4658.2011.08159.x
178. Schwarz-Linek U, Werner JM, Pickford AR, Gurusiddappa S, Kim JH, Pilka ES et al (2003) Pathogenic bacteria attach to human fibronectin through a tandem β-zipper. Nature 423(6936):177–181. https://doi.org/10.1038/nature01589
179. Clarke SR, Wiltshire MD, Foster SJ (2004) IsdA of *Staphylococcus aureus* is a broad spectrum, iron-regulated adhesin. Mol Microbiol 51(5):1509–1519. https://doi.org/10.1111/j.1365-2958.2003.03938.x
180. Clarke SR, Foster SJ (2008) IsdA protects *Staphylococcus aureus* against the bactericidal protease activity of apolactoferrin. Infect Immun 76(4):1518–1526. https://doi.org/10.1128/IAI.01530-07
181. Pilpa RM, Robson SA, Villareal VA, Wong ML, Phillips M, Clubb RT (2009) Functionally distinct NEAT (NEAr transporter) domains within the *Staphylococcus aureus* IsdH/HarA protein extract heme from methemoglobin. J Biol Chem 284(2):1166–1176. https://doi.org/10.1074/jbc.M806007200
182. Lasa I, Penades JR (2006) Bap: a family of surface proteins involved in biofilm formation. Res Microbiol 157(2):99–107. https://doi.org/10.1016/j.resmic.2005.11.003
183. Valle J, Latasa C, Gil C, Toledo-Arana A, Solano C, Penades JR et al (2012) Bap, a biofilm matrix protein of *Staphylococcus aureus* prevents cellular internalization through binding to GP96 host receptor. PLoS Pathog 8(8):e1002843. https://doi.org/10.1371/journal.ppat.1002843
184. Schroeder K, Jularic M, Horsburgh SM, Hirschhausen N, Neumann C, Bertling A et al (2009) Molecular characterization of a novel *Staphylococcus aureus* surface protein (SasC) involved in cell aggregation and biofilm accumulation. PLoS One 4(10):e7567. https://doi.org/10.1371/journal.pone.0007567
185. Kenny JG, Ward D, Josefsson E, Jonsson IM, Hinds J, Rees HH et al (2009) The *Staphylococcus aureus* response to unsaturated long chain free fatty acids: survival mechanisms and virulence implications. PLoS One 4(2):e4344. https://doi.org/10.1371/journal.pone.0004344
186. Roche FM, Massey R, Peacock SJ, Day NP, Visai L, Speziale P et al (2003) Characterization of novel LPXTG-containing proteins of *Staphylococcus aureus* identified from genome sequences. Microbiology 149(Pt 3):643–654. https://doi.org/10.1099/mic.0.25996-0
187. Moreillon P, Entenza JM, Francioli P, McDevitt D, Foster TJ, Francois P et al (1995) Role of *Staphylococcus aureus* coagulase and clumping factor in pathogenesis of experimental endocarditis. Infect Immun 63(12):4738–4743. https://doi.org/10.1128/iai.63.12.4738-4743.1995
188. Entenza JM, Foster TJ, Ni Eidhin D, Vaudaux P, Francioli P, Moreillon P (2000) Contribution of clumping factor B to pathogenesis of experimental endocarditis due to *Staphylococcus aureus*. Infect Immun 68(9):5443–5446. https://doi.org/10.1128/IAI.68.9.5443-5446.2000
189. Siboo IR, Chambers HF, Sullam PM (2005) Role of SraP, a serine-rich surface protein of *Staphylococcus aureus*, in binding to human platelets. Infect Immun 73(4):2273–2280. https://doi.org/10.1128/IAI.73.4.2273-2280.2005
190. Brouillette E, Grondin G, Shkreta L, Lacasse P, Talbot BG (2003) In vivo and in vitro demonstration that *Staphylococcus aureus* is an intracellular pathogen in the presence or absence

of fibronectin-binding proteins. Microb Pathog 35(4):159–168. https://doi.org/10.1016/S0882-4010(03)00112-8
191. Gomez MI, Lee A, Reddy B, Muir A, Soong G, Pitt A et al (2004) *Staphylococcus aureus* protein a induces airway epithelial inflammatory responses by activating TNFR1. Nat Med 10(8):842–848. https://doi.org/10.1038/nm1079
192. Vergara-Irigaray M, Valle J, Merino N, Latasa C, Garcia B, de Los R, Mozos I et al (2009) Relevant role of fibronectin-binding proteins in *Staphylococcus aureus* biofilm-associated foreign-body infections. Infect Immun 77(9):3978–3991. https://doi.org/10.1128/iai.00616-09
193. Arrecubieta C, Asai T, Bayern M, Loughman A, Fitzgerald JR, Shelton CE et al (2006) The role of *Staphylococcus aureus* adhesins in the pathogenesis of ventricular assist device-related infections. J Infect Dis 193(8):1109–1119. https://doi.org/10.1086/501366
194. Patel AH, Nowlan P, Weavers ED, Foster T (1987) Virulence of protein A-deficient and alpha-toxin-deficient mutants of *Staphylococcus aureus* isolated by allele replacement. Infect Immun 55(12):3103–3110. https://doi.org/10.1128/iai.55.12.3103-3110.1987
195. Palmqvist N, Foster T, Tarkowski A, Josefsson E (2002) Protein a is a virulence factor in *Staphylococcus aureus* arthritis and septic death. Microb Pathog 33(5):239–249. https://doi.org/10.1006/mpat.2002.0533
196. Dastgheyb S, Parvizi J, Shapiro IM, Hickok NJ, Otto M (2015) Effect of biofilms on recalcitrance of staphylococcal joint infection to antibiotic treatment. J Infect Dis 211(4):641–650. https://doi.org/10.1093/infdis/jiu514
197. Kwiecinski J, Jin T, Josefsson E (2014) Surface proteins of *Staphylococcus aureus* play an important role in experimental skin infection. APMIS 122(12):1240–1250. https://doi.org/10.1111/apm.12295

Chapter 3
Exotoxins of *Staphylococcus aureus*

Dong-Liang Hu, Hisaya K. Ono, Shaowen Li, and Rendong Fang

Abstract *Staphylococcus aureus* produces a variety of exotoxins, which are important virulence factors for *S. aureus*-related diseases. Representatively, staphylococcal enterotoxins induce staphylococcal food-born poisoning; toxic shock syndrome toxin-1, which is a typical superantigen, induces toxic shock syndrome; hemolysins induce cell damages of erythrocyte and leukocyte; and exfoliative toxin induces staphylococcal skin scalded syndrome. Recently, a cytotoxin, Panton-Valentine leukocidin which is produced by community-associated MRSA is noticeable. This chapter addresses novel information in activities and stability of the exotoxins produced by *S. aureus*.

Keywords Staphylococcal enterotoxin · Superantigen · Membrane-damaging toxin · Hemolysin · Panton-valentine leukocidin · Exfoliative toxin

Abbreviations

5-HT	Serotonin
ADAM10	A disintegrin and metalloproteinase 10
agr	Accessory gene regulator
APC	Antigen-presenting cell
BCR	B-cell receptor
CDR	Complementarity-determining region
DARC	DUFFY antigen receptor for chemokines

D.-L. Hu (✉) · H. K. Ono
Department of Zoonoses, Kitasato University School of Veterinary Medicine, Towada, Aomori, Japan
e-mail: hudl@vmas.kitasato-u.ac.jp

S. Li
College of Veterinary Medicine, Huazhong Agricultural University, Wuhan, Hubei, China

R. Fang
College of Veterinary Medicine, Southwest University, Congqing, China

A. Nakane, K. Asano (eds.), *Staphylococcus aureus*,
https://doi.org/10.1007/978-981-99-9428-1_3

Dsg1	Desmoglein 1
egc	Enterotoxin gene cluster
ET	Exfoliative toxin
FPR2	Formyl peptide receptor 2
IFN-γ	Gamma interferon
Ig	Immunoglobulin
IL-1	Interleukin 1
INCSS	International Nomenclature Committee for Staphylococcal Superantigens
Luk	Leukocidin
MDT	Membrane-damaged toxin
MHC	Major histocompatibility complex
PAMP	Pathogen recognition molecular pattern
PC	Phosphatidylcholine
PE	Phosphatidyletanolamine
PFT	Pore-forming toxin
PG	Phosphatidylglycerol
PI	Phosphatidylinositol
Pmt	Phenol-soluble modulin transporter
PS	Phosphatidylserine
PSM	Phenol-soluble modulin
PVL	Panton-Valentine leucocidin
RBC	Red blood cell
SAg	Superantigen
SaPI	*Staphylococcus aureus* pathogenicity island
SE	Staphylococcal enterotoxin
SEA	Staphylococcal enterotoxin A
SEl	Staphylococcal enterotoxin-like toxin
SElU	Staphylococcal enterotoxin-like toxin U
SMase	Sphingomyelinase
SpA	Staphylococcal protein A
TCR	T-cell receptor
TNF-α	Tumor necrosis factor α
TSS	Toxic shock syndrome
TSST-1	Toxic shock syndrome toxin-1

3.1 Introduction

Staphylococcus aureus produces a variety of exotoxins and proteases that contribute to their ability to colonize and cause disease in humans and animals [1, 2]. The main exotoxins of *S. aureus* can be divided into three major groups: superantigen (SAg) toxins, membrane-damaged toxins (MDTs), and exfoliative toxins (ETs). The SAg toxins include staphylococcal enterotoxins (SEs), which have emetic activity, induce staphylococcal food poisoning; Toxic shock syndrome toxin-1 (TSST-1), which is a typical superantigen, induces toxic shock syndrome (TSS). MDTs can be further divided into hemolysins, leukocidins including Panton-Valentine leukocidin (PVL), and phenol-soluble modulins (PSMs) [3, 4]. The main function of the toxins may be to convert local host tissues into nutrients needed for bacterial growth, and/or have an effective effect on cells of the immune system, such as modulating and disturbing the host's immune response to *S. aureus*. Representatively, SEs, TSST-1, PVL, and exfoliative toxin are associated with the increased virulence of certain strains of *S. aureus*. In this chapter, we focus on the knowledge and new information in the characteristics, activities and pathogenic mechanisms of these exotoxins.

3.2 Superantigen (SAg) Toxins

SAg toxins are a large family of structurally related toxins, including SEs, staphylococcal enterotoxins-like toxins (SEls), and TSST-1. SEs, SEls and TSST-1 are the most potent SAgs due to their ability to polyclonally activate T-cells at picomolar concentrations. They activate macrophages and T-cells without normal antigen presentation process. Staphylococcal superantigen-like (SSL) proteins exhibit various immunomodulatory activities. Whereas, staphylococcal protein A (SpA) produced by *S. aureus* is a B cell superantigen capable of interacting with both antibody Fc and Fab regions of VH3+ immunoglublins (Igs).

3.2.1 Staphylococcal Enterotoxins (SEs) and Staphylococcal Enterotoxin-Like Toxins (SEls)

S. aureus produces a large variety of SEs that are important causive toxins of food poisoning and representative SAg toxins. SEs and SEls share common phylogenetic relationships, structure, function, and sequence homology. These toxins are simple proteins composed of approximately 168 to 261 amino acids and demonstrate molecular size of 19 to 30 kD (Table 3.1). Since the first characterization of staphylococcal enterotoxin A (SEA) in 1959 by Casman and Bergdoll, five SEs, SEA to SEE, have been known from the differences in the antigenicity of SEs [5–7]. The SEC was further divided into three antigenic subtypes, SEC1, SEC2 and SEC3 [7, 8]. Todd et al. had reported a toxin that induce TSS caused by *S. aureus* infection [9], and the toxin had been named SEF. However, SEF does not have emetic activity, later it was renamed TSST-1. From 1990s, the presence of additional SEs began to be revealed, and after 2001, many new types of SEs and SEls were discovered, one after another, by our research group [10–14] and other researchers [15, 16]. To date, the presence of total 29 different SEs and SEls have been reported and described (Table 3.1). These toxins include SEA to SEE, SEG to SET, and staphylococcal

Table 3.1 Characteristics and pathogenicity of staphylococcal superantigen toxins[a]

	Molecular weight (kDa)	Superantigen activity	Genetic elements	Emetic activity[b]		First reported
				Monkey[c]	Shrew[d]	
SEA	27.1	+	Prophage	25	0.3	1963
SEB	28.3	+	Chromosome, SaPI3, plasmid (pZA10)	100	10	1963
SEC1	27.5	+	SaPI	5	NE	1965
SEC2	27.6	+	SaPI	NE	1000	1965
SEC3	27.6	+	SaPI	<50	NE	1965
SED	26.9	+	Plasmid(pIB485)	NE	40	1967
SEE	26.4	+	Prophage (hypothetical location)	NE	10	1971
SEG	27	+	*egc*1, *egc*2, *egc*3, *egc*4	160–320	200	1998
SEH	25.1	+	Transposon (MGEmw2/mssa476 *seh*/Δ*seo*)	30	1000	1994
SEI	24.9	+	*egc1*, *egc*2, *egc*3	300–600	1	1998
SElJ	28.6	+	Plasmid(pIB485, pF5)	NE	NE	1998
SEK	25.3	+	Prophages, SaPI1, SaPI3, SaPI5, SaPIbov1	100 (2/6)	1000	1998
SEL	24.7	+	SaPIn1, SaPIm1, SaPImw2, SaPIbov1	100 (1/6)	500	2001
SEM	24.8	+	*egc*1, *egc*2,	100 (1/7)	100	2001
SEN	26.1	+	*egc*1, *egc*2, *egc*3, *egc*4	100 (1/8)	1000	2001
SEO	26.8	+	*egc*1, *egc*2, *egc*3, *egc*4, transposon	100 (2/6)	500	2001
SEP	26.7	+	Prophage(Sa3n)	100 (3/6)	50	2001
SEQ	25.2	+	SaPI1, SaPI3, SaPI5, prophage	100 (2/6)	4	2002
SER	27	+	Plasmid(pIB485, pF5)	<100	<1000	2003
SES	26.2	+	Plasmid(pF5)	<100	20	2008
SET	22.6	+	Plasmid(pF5)	<100	1000	2008

SElU	27.2	+	*egc*2, *egc*3	NE	NE	2003
SElV	26.7	+	*egc*4	NE	NE	2006
SElW	27.3	+	*egc*4	NE	NE	2006
SElX	19.5	+	Chromosome	NE	NE	2011
SEY	23.3	+/–	Chromosome	250	500	2015
SElZ	27.4	+	Chromosome	NE	NE	2015
SE01	26.4	+	Plasmid	NE	NE	2017
SE02	27.2	+	Chromosome	250	NE	2020
SEl26	25.1	+	SaPI	NE	NE	2018
SEl27	27.1	+	SaPI	NE	NE	2018
TSST-1	22.1	+	SaPIs	–	–	1981

+ positive reaction, *NE* not examined
[a]This table is modified from Hu, D.-L et al. Animal Diseases. 1:7, 2021. https://doi.org/10.1186/s44149-021-00007-7
[b]μg/animal
[c]Oral administration
[d]Intraperitoneal administration

enterotoxin-like toxin U (SElU) to SElZ, SE01, SE02, SE26 and SE27 [2, 17]. The International Nomenclature Committee for Staphylococcal Superantigens (INCSS) proposed the standard nomenclature for the toxins that have been newly discovered [18]. INCSS naming convention is to emphasize the relevance of the food poisoning (emetic activity). To name the SE, it is obliged to prove emetic activity by oral administration to a primate model. If it shows negative emetic potential in the vomiting experiments in a primate model or the vomiting experiments have not yet been carried out, this toxin should be named "SEl" toxin, even the superantigen is considered closely related to SE structure.

Omoe et al. [14] assessed the emetic potentials of some newly discovered SEls, including SElK, SElL, SElM, SElN, SElO, SElP and SElQ, using a monkey-feeding assay. The results showed all of the SEls tested induced emetic responses in the monkeys at a dose of 100 μg/kg, although the numbers of affected monkeys were smaller than the numbers that were affected after consuming SEA or SEB. These results indicated that the newly identified SEls are emetic toxins and the names should be re-designated as SEK, SEL, SEM, SEN, SEO, SEP and SEQ, respectively, according to the INCSS naming convention [14, 19, 20]. More recently, Ono et al. reported a novel staphylococcal emetic toxin, SElY, that had strong emetic activity in a small emetic animal model, house musk shrew [12]. SEs and SEls have become superfamily toxins.

3.2.1.1 Genes of SEs and Sels

The genes of SEs and SEls are encoded by diverse accessory genetic elements, and many of them were harbored by numerous mobile elements, including *S. aureus* pathogenicity islands (SaPIs), genomic islands (υSa), prophages, and plasmids [6, 17, 21, 22]. SaPIs with the exceptions of SaPIbov2 (27 kb) and a highly degenerated SaPI (3.14 kb) present in some sequenced genomes [22, 23]. Some of SaPIs carry genes encoding one or more SEs. *sek* and *seq* are found together with *tst* in SaPI1; *sel* and *sec* are found in SaPIbov1; *seb, seq* and *sek* have been reported in SaPI3 [17, 24]. Strains carrying different SaPIs were found to have significantly different production of superantigen toxins [17, 22]. Genomic islands, υSaα and υSaβ, contain clusters of genes encoding virulence factors. υSaβ carries enterotoxin gene cluster (*egc*), which includes a variable number of *se* or *sel* genes forming an operon. *egc*1 consists of five *se* genes (*seg, sei, sem, sen and seo*) [1, 25, 26] . The *egc*2 contains an additional *sel* gene (*selu*) [24]. Allelic variants of each of the *egc*2 genes compose the *egc*3 cluster [27, 28], and two new *sel* genes (*selv, selw*) are present in *egc*4 [15]. SEs genes can also be carried by prophage [29–32]. Three *se/sel* genes (*sea, selk* and *selq*) are present together in ϕSa3ms and ϕSa3mw, while a single *se/sel* gene (*sea* or *sep*) is carried by ϕMu3A, ϕSa3a and other prophages [31]. Two kinds of plasmids, pIB485 and pF5, carrying *se/sel* genes have been characterized [5, 10, 11, 33]. The pIB485, a 27.6 kb plasmid, in which first *sed*, *selj*, and latter *ser* were identified.

3.2.1.2 Molecular Structures of SEs and SEls

SEs and SEls are single-chain proteins with molecular weights ranging from 19 to 30 kDa (Fig. 3.1). According to the homology of nucleotide and amino acid sequences of the toxins, they can be classified into several groups. The SEA group includes SEA, SED, SEE, SEH, SElJ, SEN and SEP, which contain a cystine loop with nine amino acids [1, 13, 34]. These toxins have a low-affinity α-chain major histocompatibility complex (MHC) II binding site and a high-affinity site, referred to as a Zn^{2+}-dependent β-chains MHC II binding site [35, 36]. The presence of the Zn^{2+}-dependent high-affinity site on the toxins make them 10- to 100-fold more active overall in causing cytokine production from T cells and antigen-presenting cells (APCs) than other superantigen toxins. SEB group includes SEB, SEC, SEG, SElU and SElW (SElU2) which contain a core superantigen structure plus a cystine loop that has a varying 10- to 19-amino-acid sequence separating the cysteine residues [37–40]. These toxins contain one α-chain MHC II binding site, and this interaction does not depend on interaction with the antigenic peptide within the MHC II peptide-binding groove [41, 42]. SEI group includes SEI, SEK, SEL, SEM, SEQ, SET and SElV, which contain both low- and high-affinity MHC II binding sites, but lack the cystine loop [43, 44]. The three-dimensional structures of SEs and SEls show very similar conformation that the canonical structure consists of one A domain and one B domain, and one α-helix that spans the center of the structure and connects the A and B domains [45–47]. The interfaces between the A and B domains are marked by a set of α-helices, which form a long groove in the back side of the molecule and a shallow cavity at the top.

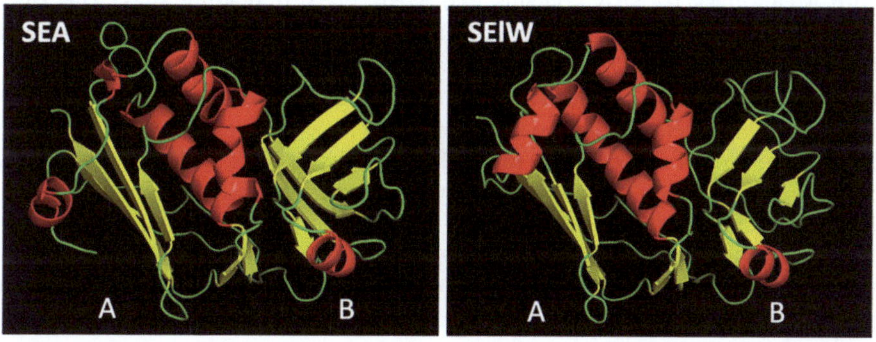

Fig. 3.1 Molecular structures of SEs and SEls. SEs and SEls are single-chain proteins with molecular weights ranging from 19 to 30 kDa. The three-dimensional structures of SEs and SEls show very similar conformation that the canonical structure consists of one A domain and one B domain, and one α-helix that spans the center of the structure and connects the A and B domains

3.2.1.3 Heat- and Enzyme-Resistance of SEs and SEls

SEs and SEls are remarkably resistant to heat and enzymatic degradation. The potency of these toxins can only be gradually decreased by prolonged boiling or autoclaving. They are highly stable to most proteolytic enzymes, and thus retain their activity in the digestive tract after ingestion. Several groups have compared the respective integrity and toxicity of SEA and TSST-1 after treatment with heat, pepsin and trypsin in relation to the condition of food cooking, or luminal location in stomach and intestine. SEA retained significant superantigenic and emetic activity, although it was degraded to smaller fragments after treatment with heat or pepsin, indicating that cooked foods which contaminated with the toxins may also induce food poisoning [44, 48, 49].

3.2.1.4 Superantigenic Activity of SEs and SEls

SEs and SEls, unlike conventional antigens, bypass normal processing by antigen-presenting cells (APCs) and induce a large proportion (5%–30%) of T-cells to proliferate, and subsequently stimulate massive cytokine release (Fig. 3.2a). The excessive release of cytokines, such as tumor necrosis factor α (TNF-α), interleukin 1 (IL-1), IL-2 and gamma interferon (IFN-γ), that mediate the toxic effects of these toxins [35, 50–52]. Each SE or SEl binds to a distinct repertoire of Vβ-bearing T-cells, revealing a unique biological "fingerprint" [53–55]. SEs have evolved several distinct modes of interaction with the MHC class II molecules. The information on the SE-MHC interaction was provided by the crystal structure of SE complexed with MHC class II. SEA contains two MHC class II binding sites. The zinc-dependent site located in domain A is the major interaction region, and several important residues (H187, H225, and D227) for binding to MHC class II were identified by mutagenesis [1, 53, 56, 57]. The second (minor) binding site on SEA is F47 located in domain B, which is not zinc dependent. The cooperation between the two binding sites may be responsible for the high affinity of SEA for MHC class II [58, 59]. However, no such site is predicted by the structure of SEB and SEC. The binding of SEB to the MHC-peptide complex was examined by using soluble HLA-DR1 loaded with hemagglutinin peptide HA 306–318. SEB had a considerably higher affinity for the MHC class II molecule, which may well explain the difference in activity [60, 61].

3.2.1.5 Emetic Activity of SEs

SEs are the leading causes of foodborne bacterial intoxications worldwide [20, 49, 62–64]. SEA is most common enterotoxin recovered from food poisoning outbreaks in many countries [65–67]. The symptoms of food poisoning are vomiting, abdominal cramps, nausea, sometimes followed by diarrhea after a short period of incubation [68, 69].

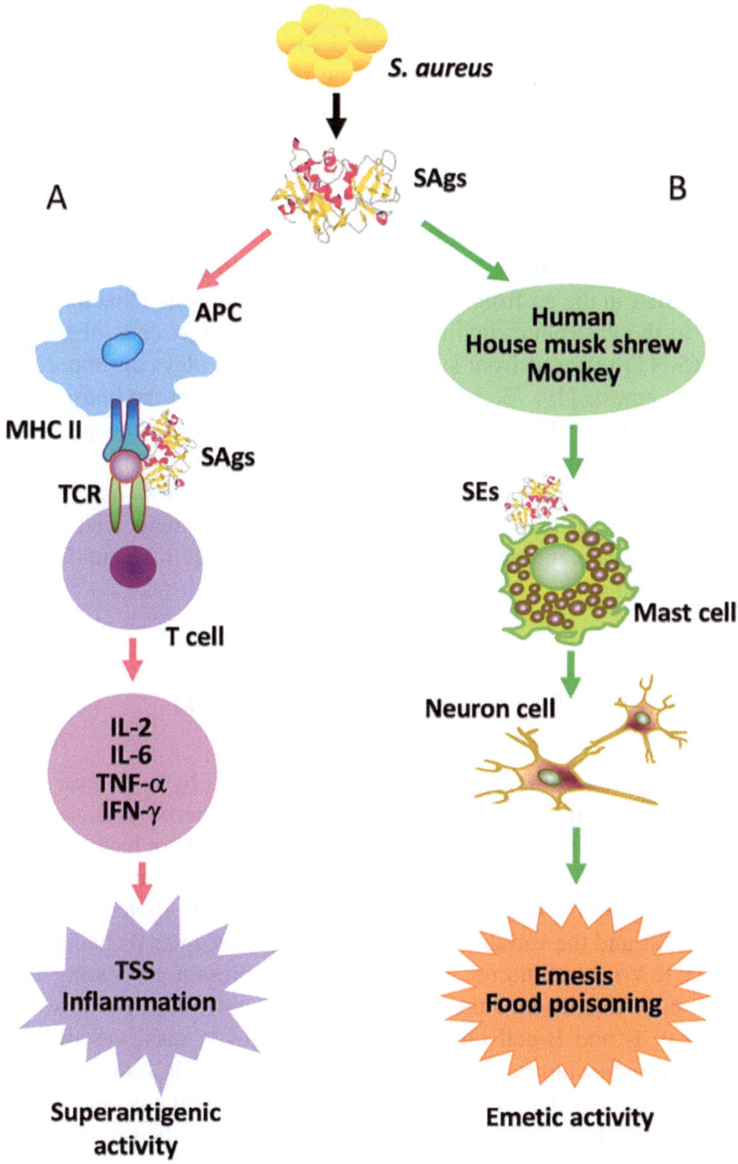

Fig. 3.2 Biological activities of SEs and SEls. (**a**) Superantigenic activity of SEs and SEls. Superantigens (SAgs), unlike conventional antigens, bypass normal processing by APC and induce a large proportion of T-cells to proliferate, and subsequently stimulate massive cytokines release, that mediate the toxic effects of the toxins. (**b**) Emetic activity of SEs. The submucosal mast cells in the gastrointestinal tract are one of the target cells of SEs and the serotonin released from mast cells and/or neuron cells plays an important role in SE-induced emesis and food poisoning

Although the clinical manifestations are well-known, the mechanisms of SEs-induced emesis are only partially understood. Monkeys have been considered to be the primary animal models, but the use in investigating SEs is severely restricted by the high cost, the availability of the animals, and ethical considerations. Other animals like mouse, rat, rabbit and cat are less susceptible to SEs or their responses to SEs are not specific [11, 28, 70, 71]. Hu et al. investigated the emetic response of house musk shrews to the classic and several newly discoverd SEls, and showed that all of the tested SEs and SEls caused vomiting responses in the animals, although they have different emetic activities [10, 44, 69, 72]. Ono et al. investigated the behavior of SEA in the gastrointestinal tract in vivo, suggested that the submucosal mast cells in the gastrointestinal tract are one of the target cells of SEA and that serotonin (5-HT) released from submucosal mast cells plays an important role in SEA-induced emesis [19, 73, 74] (Fig. 3.2b). The emesis in the house musk shrew is inhibited by the 5-HT synthesis inhibitor and the 5-HT_3 receptor antagonist, showing the important role of 5-HT in SEA-induced emesis [56]. SEA-induced emesis is blocked by surgical vagotomy in house musk shrews and primates, suggesting that 5-HT released from submucosal mast cells may bind to the 5-HT_3 receptor expressed on enteric nerves in the gastrointestinal tract and thereby induce the depolarization of these nerves [56].

3.2.2 Toxic Shock Syndrome Toxin-1 (TSST-1)

TSST-1 is a 22 kDa extracellular protein toxin secreted by *S. aureus*. Its physical and chemical properties are very similar to those of enterotoxins, so it had been once called SEF, but it was renamed TSST-1 because it did not have vomiting activity [48, 75, 76]. TSST-1 is a superantigenic toxin that stimulate polyclonal T-cell proliferation through co-ligation between MHC class II molecules on antigen-presenting cells and the variable portion of the T-cell receptor β chain (TCR Vβ) or α chain (TCR Vα), with no need for prior antigen-present processing. The toxin-activated T-cell/APC leads to the release of various cytokines, enhances endotoxic shock, causes T- and B-cell immunosuppression, fever, rash, vascular disorders, TSS, multiple organ failure and decrease blood pressure in humans and animals. TSST-1 has unique primary amino acid sequences and contains only a low-affinity MHC II binding site in their O/B folds that interacts with the α-chains of MHC II molecules [76–78].

3.2.3 Staphylococcal Protein A (SpA)

SpA produced by *S. aureus* is a B cell superantigen capable of interacting with both antibody Fc and Fab regions of VH3+ Igs. The gene *spa* is encoded in the core genome of majority clinical isolates [79] and the SpA proteins can be released from the bacterial cell wall by the hydrolase LytM in the cell wall [80] . The SpA molecule contains 4 to 5 highly conserved Ig binding domains and X hypervariable regions composed of sub-regions Xr and Xc [81, 82]. *S. aureus* isolates can be classified by the highly variable and repetitive octapeptides in Xr of SpA [79]. The Ig-binding domain on the SpA molecule confers the ability of SpA to bind to the Fcγ portion of the Ig and prevents the opsonization of the host's cells [83]. SpA has two distinct Ig-binding sites; one for IgG Fc domains, while a separate site binds an evolutionarily conserved surface on Fab encoded by VH3 clan related genes.

The Ig-binding domains can also mediate the binding of SpA to B cells by cross-linking VH3- expressing B-cell receptors (BCRs) and thereby activate the B cells. SpA exerts its mitogenic activity by binding to the variable regions of the heavy chain, rather than the complementarity-determining regions (CDRs) required for common antigens, thereby bypassing the antigen specificity required for B cell activation [84, 85]. SpA may also represent a paradigm relevant to other microbial toxins with unconventional V region targeted activity in *S. aureus* and other microbial commensal/opportunistic pathogens [86]. Such toxins with superantigen properties may be highly effective at subverting host defenses. SpA is required for persistence of *S. aureus* in the nasopharynx. Compared to animals colonized by wild-type *S. aureus*, mice colonized with the *Δspa* variant mount increased IgG responses against staphylococcal colonization determinants. Immunization of mice with a nontoxigenic SpA variant, which cannot cross-link B cell receptors and divert antibody responses, elicits protein A-neutralizing antibodies that promote IgG responses against colonizing *S. aureus* and diminish pathogen persistence [87].

3.3 Membrane-Damaging Toxins

3.3.1 α-Hemolysin (α-toxin or Hla)

α-Hemolysin, also termed hemolysin-α, α-toxin or Hla, is a small β-pore-forming toxin (PFT) and considered to be an important virulence factor of *S. aureus* in human and animal diseases. α-Hemolysin is encoded by a single genetic locus, *hla*, and is secreted as a 292 residue, 33.2 kDa water soluble monomer. α-Hemolysin is not heat-resistant and can be destroyed at 65 °C for 30 min. The regulation of α-toxin expression is mainly controlled by the quorum-sensing accessory gene regulator (*agr*) system, which is activated from the late logarithmic growth phase to the stationary phase of growth, thereby inducing the transcription and translation of *hla* [88, 89]. In response to environmental conditions (such as the *SaeR/S* two-component system) or other systems (such as the *Sar* family), the Agr system can also be adjusted to fine-tune *hla* expression [90]. After binding to its target cells, α-toxin oligomerizes to form a pre-porous structure, and then attacks the cell membrane by extruding the β-barrel through the lipid bilayer, forming a hydrophilic transmembrane channel (Fig. 3.3). α-Toxin assembles into a heptameric pore

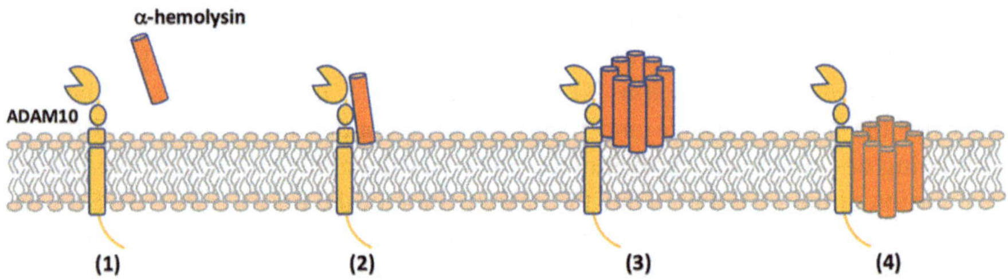

Fig. 3.3 The hemolytic mechanism of α-hemolysin. (1) α-Hemolysin is a water-soluble monomer. (2) α-Hemolysin binds to the transmembrane protein ADAM10 that is a hemolysin receptor. (3) The toxin then oligomerizes at the plasma membrane to form heptamers and form pre-pores, and (4) finally forms transmembrane channels

composed of three major domains, including an extracellular-facing cap domain; a marginal domain that interacts with the extracellular leaflets of the host cell membrane; and a stem domain that forms a β-barrel transmembrane channel. The fully assembled toxin is 100 Å wide and 100 Å high with a minimum pore diameter of 14 Å [91].

The hemolytic effect of α-toxin is cell-type specific because it selectively lyses rabbit red blood cells (RBCs), but does not lyse human RBCs [92]. However, α-toxin has been shown to lyse human platelets, endothelial cells, epithelial cells, and leukocytes [93]. *S. aureus* PFTs receptors provide a powerful explanation for their cellular tropism, laid the foundation for further study of the physiological effects of toxins. Earlier studies showed that α-toxins are able to bind and cleave protein-free membranes, suggesting that the α-toxin receptor is the lipid moiety and the binding is specific because the pore-formation on the liposomes depends on their lipid composition [94]. Liposomes composed of phosphatidylcholine (PC) or a combination of sphingomyelin and cholesterol were easily cleaved, while phosphatidylethanolamine (PE), phosphatidylserine (PS), phosphatidylglycerol (PG), or phosphatidylinositol (PI) composition of liposomes were not [95, 96]. Among them, PC may become a receptor for α-toxin because exogenous PC competes for α-toxin binding to prevent hemolysis [96]. Later, several studies have shown that liposomes formed from lipids extracted from human or rabbit red blood cells have similar sensitivity to α-toxin, and that treatment with amylase can reduce the sensitivity of rabbit RBC to the toxins, suggesting that the main recepter of α-toxin is a protein, but not a lipid.

Wilke and Bubeck Wardenburg [97] used a biochemical method to identify cell surface metalloproteinases, a disintegrin and metalloproteinase 10 (ADAM10), as the specific protein receptors for α-toxin. Loss of ADAM10 expression reduces α-toxin binding and cytotoxicity. Overexpression of ADAM10 results in increased α-toxin binding and toxicity [98–100]. ADAM10 is expressed on cell types known to be susceptible to α-toxin poisoning, such as epithelial cells [101], keratinocytes [102], endothelial cells [103], and platelets [104]. The surface expression levels of ADAM10 on various cell lines correlates with Hla binding levels [97]. ADAM10 is a cell surface dehydrogenase involved in development and homeostasis processes

[102, 105]. Factors that regulate the surface expression of ADAM10 also regulate α-toxin sensitivity [99]. ADAM10 not only promotes pore formation, but its enzyme activity is also up-regulated by α-toxin-pore [8].

α-Hemolysin is a β-barrel toxin secreted by 95% of clinical *S. aureus* strains as a water-soluble monomer [106, 107]. This toxin has been shown to affect multiple human cell types such as epithelial cells, endothelial cells, T cells, monocytes and macrophages [108–112]. For many years, pore formation and cytolysis were considered to be the most prominent results of α-toxin action. However, recent studies have demonstrated the importance of this toxin in altering cellular signaling pathways, including cell proliferation, inflammatory response, cytokine secretion, and intercellular interactions [93, 108, 113]. When the toxin approaches the cell, the pores formed allow ATP and K^+ ions to be released quickly, while also limiting the movement of macromolecules on the cell membrane [114, 115]. After the pores are formed, extracellular Ca^{2+} floods into the cells. Increased intracellular Ca^{2+} stimulates the hydrolysis of membrane phospholipids and the metabolism of arachidonic acid to leukotrienes, prostaglandins, and thromboxane A2 [116, 117]. The toxin then also activates protein kinase K and induces NF-κB nuclear translocation [116, 117]. All these events and the production of IL-1β, IL-6 and IL-8 indicate pro-inflammatory stimuli. After the α-toxin bond disrupts the barrier function of the epithelial tissue, the enzyme that degrades E-cadherin is activated, which allows *S. aureus* to invade [118, 119]. This toxin has been shown to play an important role in the pathogenesis of *S. aureus*, and the expression of α-toxin may be necessary for the pathogenesis of invasive diseases in healthy individuals [120, 121].

3.3.2 β-Hemolysin (β-Toxin or Sphingomyelinase C)

The gene *hlb* encoding β-hemolysin is part of the core genome of *S. aureus*. β-Hemolysin is a Mg^{2+}-dependent neutral sphingomyelinase (SMase) that specifically cleaves sphingomyelin to produce ceramide and phosphocholine [122]. β-Hemolysin is a single-domain protein consisting of four layers: two β-sheet layers in the center and two α-helix and β-chains in the outer layer [123]. The crystal structure of β-hemolysin reveals structural homology with members of the DNase I superfamily [123]. Based on the structural homology to the DNase I superfamily, a secondary function of β-hemolysin was speculated. In addition, β-hemolysin enhances biofilm formation by catalyzing the formation of a nucleoprotein matrix in biofilms, and thus β-hemolysin is also a biofilm ligase [124].

In the presence of α-toxin-neutralizing serum, sheep erythrocytes were lysed, but not rabbit erythrocytes, and the induced hemolysis would be enhanced when the temperature was changed from 37 °C to a lower temperature. In the presence of α-toxin neutralizing serum, hemolysis of erythrocytes in the presence of α-toxin neutralizing serum, lysis of sheep but not rabbit erythrocytes, and enhanced hemolysis caused by temperature shifting from 37 °C to a lower temperature. Therefore, β-hemolysin is also called as a hot-cold hemolysin. This unique phenomenon is the result of membrane aggregation of ceramide hydrolyzates at 37 °C. When the temperature is lowered (i.e., 4 °C), phase separation occurs, resulting in accumulation of ceramides in the pool, disrupting the lipid bilayer and sinking the erythrocyte membrane [125].

β-Hemolysin exhibits species-related hemolytic activity. Sheep, cattle, and goat red blood cells are highly sensitive to the toxins, rabbits and humans are moderately sensitive, and mouse and dog red blood cells are resistant. The difference in sensitivity of different red blood cells may be due to different sphingomyelin content of these cells. The activity of β-hemolysin SMase also causes the lysis of human keratinocytes, monocytes, T cells and bovine epithelial cells [123, 126, 127]. β-Hemolysin stimulates the production of proinflammatory cytokines in human monocytes, but inhibits the production of IL-8 and the expression of cell adhesion molecules in human endothelial cells, so the toxin can prevent the migration of leukocytes across endothelial cells [128].

β-Hemolysin is produced in large quantities in strains isolated from bovine mastitis [89] and chronic skin infections [90]. Several studies have demonstrated the importance of this toxin for the pathogenicity of *S. aureus*. Infecting β-hemolysin producing *S. aureus* can cause larger lesions in organs [129]. The presence of β-hemolysin enhances the colonization of *S. aureus* on the skin during keratitis in mice [127] and causes damage to scleral epithelial cells [90]. Intranasal injection of β-hemolysin induced shedding of lung epithelial cells, syndecan-1 and infiltration of neutrophils into the lungs of mice [130]. In pneumonia and mouse ear skin infection models, the toxicity of mutant *S. aureus* strains lacking HLB is reduced [127, 130].

3.3.3 Leukotoxin

Leukotoxins lyse cells of the leukotocytic lineage and also known to kill neutrophils, of which γ-hemolysin and LukED also have lytic activity against red blood cells [131–133]. Leukotoxins are bicomponent PFTs, which is composed of two different protein components that are assembled together to form β-barrel pores [134, 135], which has structural homology with α-toxin and has a similar pore formation mechanism (Fig. 3.4). Bicomponent PFTs require two subunits: the fast-eluting subunit, F-subunit, and the slow-eluting subunit, S-subunit [136]. S-subunit recognizes and binds to a surface receptor on the target cells, and then recruits the F-subunit for dimerization [106, 137]. Before inserting of the β-barrel transmembrane channel into the target cell membrane, oligomerization occurs to form the pre-pore prior to resulting in cell lysis [134, 138]. Similar to the α-toxin heptamer, the bicomponent PFT octamer also resembles a mushroom, consisting of the cap, rim, and stem domains.

Leukotoxins mainly target leukocytes, so they are also called leukocidins (Luk). In additional to mediating cytolysis, many of the leukocidins also have sublytic effects, causing extracellular Ca^{2+} influx on host cells and production of

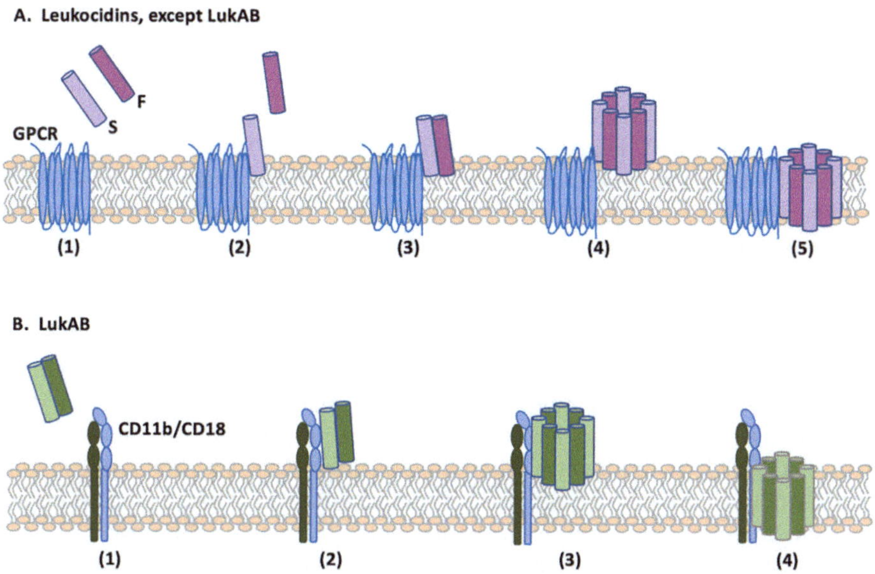

Fig. 3.4 Mechanism of action of leukotoxin. (**a**) Bicomponent leukocidins (except LukAB). (1) The leukotoxins are secreted as monomers. (2) The S-subunit recognizes the target cell by binding to cell surface receptors that are typically GPCRs. (3) The S-subunit dimerizes with the F-subunit. (4) These dimers oligomerize at the plasma membrane and a pre-pore appears. (5) Finally, transmembrane channels are formed, thus disrupting the target cell membrane. (**b**) The bicomponent leukotoxin, LukAB. (1) LukAB is secreted as dimers. (2) LukAB recognizes the target cell by binding to cell surface receptors which are the integrin, CD11b. (3) Upon receptor binding, they dimerize with additional leukocidin dimers to form an octameric pre-pore. (4) The prestem domains of the pre-pore extend to form a β-barrel pore, which eventually destroys the target cell membrane

proinflammatory cytokines [139–142]. Several of the leukocidins, PVL, HlgAB and LukAB, stimulate K^+ efflux, production of the NLRP3-inflammasome, and activation of caspase-1, resulting in a form of inflammatory cell death, known as pyroptosis. At present, four of leukotoxins have been identified from *S. aureus* strains associated to human infections: (1) γ-hemolysins AB and CB (HlgAB, HlgCB), (2) LukSF-PV (Panton-Valentine leukocidin, PVL), (3) Leukotoxin ED (LukE, LukD) and (4) Leukotoxin AB/GH (LukAB/LukGH) [131], and two of the leukotoxins, LukMF' and LukPQ, are associated with animal infections [143, 144].

3.3.3.1 γ-Hemolysins (HlgAB, HlgCB)

The γ-hemolysin locus is part of the core genome of *S. aureus* and accounts for approximately 99% of the sequenced *S. aureus* genome [145, 146]. The γ-hemolysin locus encodes three genes *hlg*ACB: two S subunits HlgA and HlgC, and one F subunit HlgB. *hlg*CB forms one transcription unit and hlgA forms another transcription unit, each with an independent promoter [147]. In the γ-hemolysin group, HlgAB, HlgCB and HlgACB are three proteins with different combinations of subunits. HlgAB and HlgCB share the same F subunit (HlgB), but the S subunit composition is different (HlgA or HlgC) [131]. HlgA and HlgB combine to form the classic γ-hemolysin (HlgAB). HlgAB is particularly effective in lysing human red blood cells and shows cytolytic activity on human and rabbit leukocytes [147–149]. Although HlgCB is also known as γ-hemolysin, it is mainly leukocytotoxic and has limited activity towards red blood cells. Early studies referred to it simply as leukocidin (LukSF) or leukocidin R (LukR) [149, 150].

HlgAB exhibits tropism towards human red blood cells, neutrophils and macrophages and lyses these cells. Human CXCR1, CXCR2, CCR2, and CXCR4 (at very high concentrations), were identified as receptors for HlgA [149, 151]. HlgAB can also target and lyse murine monocytes and macrophages through CCR2, but does not lyse murine neutrophils. Here are two reasons: (1) mouse neutrophils mainly express CXCR2, unlike human neutrophils mainly express CXCR1; (2) murine CXCR2 does not bind the toxins [151]. Thus, host tropism of the toxin depends not only on structural compatibility of the receptors, but also on differences in gene expression of the cells. Duffy antigen receptor for chemokines (DARC) is also a receptor for HlgAB [149]. HlgA and LukE have different binding sites on DARC. Many polymorphisms of the DARC gene, coding region of the receptor responsible for antigenic differences, exist in humans. Antigen differences have no effect on HlgA or LukE binding, but DARC expression levels are related to hemolysis [149]. HlgCB, on the other hand, is a human-specific toxin that targets cells expressing the receptors C5aR1 and C5aR2 of human cells [151].

The role of HlgACB in the virulence of *S. aureus* remains unclear. Some studies suggest that HlgAB is necessary for the survival and proliferation of *S. aureus* during a blood infection, which is most likely caused by macrophage escape and nutrients (Fe2 +) released by red blood cells [149, 151]. Approximately 89–100% of the strains associated with human or bovine colonization contain the hlgACB locus, but have not evidence associated with any particular type of infection [152, 153]. γ-Hemolysin causes acute tissue damage and inflammation in different animal models, and causes *S. aureus* disease. Post-orbital administration of micrograms of HlgAB is fatal to mice. Intravitreal injection of γ-hemolysin into rabbits is highly toxic, causing eye and tissue damage [154]. Tissue damage may be the result of a combined effect of toxin-mediated cell lysis and apoptosis due to excessive concentration of toxins [155]. The contribution of HlgAB to the disease has been further demonstrated in several infection models with the strains which do not produce HlgAB [151, 156, 157].

3.3.3.2 Panton-Valentine Leukocidin (LukF-PV and LukS-PV)

The PVL genes (lukS-PV and lukF-PV) are encoded within the genomes of at least six different prophages (ex. ϕPVL, ϕPVL108, ϕSa2mw, ϕSa2USA300, ϕSLT) expressing the Sa2 integrase [158]. PVL was firstly purified from the culture supernatants of *S. aureus* V8 [136]. PVL-producing isolates produce class S (slow-eluted) and class F (fast-eluted) proteins specific for PVL (LukS-PV and LukF-PV) [158, 159]. PVL is species-specific and only kills rabbit and human leukocytes. The species specificity is due to the targeting human and rabbit G-protein-coupled receptors (GPCRs), C5aR1 and C5aR2, but not the murine counterparts [160, 161] (Fig. 3.4a). Therefore, rabbit models and human *ex vivo* models have provided insight into the complexity of PVL-mediated pathology, but regular mice are not suitable for studying this toxin.

PVL is a cytotoxin that affects leukocytes and causes tissue necrosis and has been associated to furuncles, cutaneous abscesses and severe necrotic skin infections [162, 163]. PVL is a critical factor leading to invasive diseases such as osteomyelitis and pneumonia in rabbits. Deletion of *pvl* can reduce inflammation, tissue damage and bacterial burden, and promote host survival [164, 165]. Sublytic levels of PVL can enhance phagocytosis and killing of the bacteria by primary human neutrophils [166]. However, the role of PVL during skin and soft tissue infections is unclear. Some studies seem to suggest that the pathogenic effect of PVL could be dependent on the site of infection [167, 168].

The contribution of PVL to the virulence of *S. aureus* has not yet been confirmed [98]. Molecular epidemiological studies have found that PVL accounts for a small percentage (5%) of clinical *S. aureus* strains, but is closely related to CA-MRSA strains (85%), especially in the strains causing pneumonia, skin and soft tissue Infections [131]. Phylogenetic analysis of CA-MRSA strains shows rather than a single virulent clone, multiple strains containing methicillin resistance gene (SCC*mec*IV) and PVL had arisen around the world, corroborating the possibility that the virulence associated with CA-MRSA could be horizontally transferred [169]. Clinical studies have shown that pneumonitis involving PVL-positive *S. aureus* progresses faster and is more lethal than PVL-negative pneumonia [170]. Autopsy revealed ulcers and bleeding in the lungs of these patients, indicating severe inflammation, which was attributed to PVL-induced lysis of macrophages and neutrophils [170].

3.3.3.3 LukAB/GH

LukAB, also known as LukGH, is the most recently discovered member of the bicomponent leukotoxin family [86]. It is also the most distantly related member of the bicomponent leukocidin family, with LukA exhibiting only 30% amino acid identity with other S-subunits and LukB only 40% identity to other F-subunits [86, 106]. LukAB/GH heterodimerize before binding its cell-surface receptor, suggesting that LukAB/GH dimerization occurs prior to secretion, as the co-expression of subunits appear to be more toxic than the mixture of subunits individually purified [171, 172] (Fig. 3.4b). This is different for other leukotoxins which are secreted as monomers and oligomerize after binding to the cell surface.

The lukAB locus is part of the core genome of *S. aureus* and is found in 99% of *S. aureus* strains. A large number of LukAB is found in the secreted proteome during the late exponential growth phase of *S. aureus* [86, 173]. The C-terminal region of LukA is critical for toxin activity because deletion or mutation within this region render the toxin inactive [171]. LukAB has species specificity – it has the greatest effect on human and primate cells, followed by rabbit, and it is ~1000-fold less active in mice [174]. LukAB mediates cytotoxicity by targeting the I-domain of the CD11b receptor present on leukocytes, including neutrophils, monocytes, macrophages, dendritic cells and NK cells [171, 175] (Fig. 3.4b).

The role of LukAB during infection remains to be fully elucidated. This toxin appears to be encoded in most sequenced *S. aureus* strains [176]. Higher LukAB neutralizing antibody titers were observed in the sera of patients with invasive *S. aureus* disease, indicating that LukAB is produced during infection [177, 178]. Studies *in vitro* and *ex vivo* have provided insights into the role of LukAB in diseases. Primary human neutrophils infected with ΔlukAB strains exhibit enhanced survival compared to wildtype, suggesting that the virulence of lukAB-deficient mutants is greatly reduced [86]. LukAB can promote survival of *S. aureus* in leukocytes, which may be related to the role of the toxins in escaping from phagocytes and neutrophils [86, 179].

3.3.3.4 LukED

LukED was originally reported more than two decades ago. The *luk*ED locus is located in the νSaβ gene cluster, and the two genes *luk*E and *luk*D in this locus are co-transcribed at the late exponential phase [180]. Epidemiological data indicate that *luk*ED is widespread, but is more common in the prevalent CA-MRSA strains (~99%) [133, 181]. The *luk*ED locus is also present in approximately 90% of *S. aureus* strains isolated from cows with subclinical to severe mastitis, suggesting that LukED may play a role in virulence or infectivity in different host species [182].

Early studies have shown that LukED has cytolytic activity against rabbit and human erythrocytes and leukocytes [133]. Subsequently, LukED has been demonstrated to mediate to the lysis of many different human and murine bone-marrow derived cells, and is the only toxin that can efficiently kill mouse phagocytes [183]. LukED has a wide range of leukocyte-killing activities, and its pathogenic effects is receptor-dependent [184]. LukED can target G protein-coupled receptors on neutrophils, monocytes, macrophages, dendritic cells, natural killer cells, T-cells, and red blood cells [149, 184]. A panel of human cell lines were treated with purified LukED and cytotoxicity was found in T-cell lines expressing CCR5. Knockdown and overexpression studies have confirmed that CCR5 is a LukED receptor [185]. The S-component LukE, but not LukD, directly interacts with the receptor [185]. Furthermore, CXCR1 and CXCR2 were identified as two other LukE receptors expressed in PMNs and monocytes [184].

LukED is an important virulence factor for *S. aureus* infection. Studies have shown that LukED causes dermonecrosis of rabbit skin [180]. Retroorbital administration of LukED can cause acute lethality in mice [186]. In addition, the virulence of *ΔlukED* strain is greatly attenuated, resulting in lowered inflammation, reduced bacterial burden, and improved host survival in murine infections [184, 185].

3.3.3.5 LukMF'

LukMF' expression is restricted to *S. aureus* strains that are transmitted in non-human hosts, such as cattle, sheep, and goats [134]. LukMF' can be isolated from bovine mastitis tissue samples, suggesting that this leukocidin is commonly associated with bovine mastitis, which is a major cause of milk loss and also global economic problem. This data indicated a role of LukMF' in disease progression of bovine mastitis [134, 135]. The *luk*MF' locus is encoded in the temperate prophage ΦSa1, which can spread horizontally across strains, much like PVL [144]. However, unlike the other leukocidins, purified LukMF' does not cause a strong proinflammatory response when incubated with primary bovine macrophages [187].

LukMF' is well adapted to bovine host and shows cytolytic activity towards bovine neutrophils and macrophages [188]. Compared with other species, LukMF' is particularly toxic to PMNs of bovine, indicating a possible host-specific adaptation [134]. LukM is able to bind surface receptor CCR1, CCR2 and CCR5 of the cells, and enable intoxication of bovine macrophages and neutrophils [188]. However, LukM binding is restricted to CCR1 and macrophages in humans. The tropism of LukM primarily comes from the DR4 loop of the fringe domain, so that it binds to bCCR1 and bCCR2, although it does not affect the binding of bCCR5 [188].

3.3.3.6 LukPQ

LukPQ is a novel equid-adapted leukocidin of *S. aureus*. LukPQ is encoded on a 45 kb prophage (ΦSaeq1), almost exclusively in strains isolated from horses. Like other bicomponent PFTs, LukPQ preferentially targets leucocytes and consists of an S- and an F-subunit. The S-subunit recognizes the receptor on the host cells, enabling high-affinity binding to the cell surface, and then the toxins form a pore that penetrates the lipid bilayer of the cells.

LukPQ is a effective and specific killer of equine neutrophils. Equine neutrophils are most sensitive to the lysis of LukPQ, followed by bovine neutrophils, human neutrophils are relatively insensitive to the toxin [189] (Fig. 3.3). Although the S-subunit (LukP) is highly similar to that of LukED, the species specificities of LukPQ and LukED are different. Uniquely, the F-subunit of LukQ is responsible for conferring species specificity of LukPQ, whereas, species specificity is conferred by the S-subunit in other leukocidins [189]. Equine-GPCRs, CXCRA and CXCR2, are the target receptors for LukPQ. At high concentration, the toxin also targets equine CCR5. LukPQ binding can be inhibited by cytokines binding to these receptors, suggesting that LukP and the receptor ligands may share a common binding site [189].

3.3.3.7 ε-Toxin

ε-Toxin is a newly identified cytotoxin secreted by a surgical site isolate of *S. aureus*, since its lack of homology to known exotoxins [190]. The gene encoding ε-toxin, cytE, is conserved in the core genome of *S. aureus*. The toxin has a molecular weight of 15.7 kDa and an isoelectric point of 8.9. The cloned and purified protein has cytotoxic and proinflammatory properties, as shown *in vitro* and *in vivo*. ε-toxin lyses rabbit erythrocytes and human keratinocytes.

Lytic concentration of ε-toxin in keratinocytes promotes the secretion of proinflammatory cytokine, IL-8 [190]. In contrast, lower concentration of ε-toxin slows the proliferation of keratinocytes. When administered subcutaneously, microgram amount of ε-toxin can cause neutrophil recruitment to the injection site in rabbits, and also cause rabbits to develop fever after intravenous administration of the toxin [190]. The potent biological effects on keratinocytes and rabbit skin suggest that the toxin may play an important role in impairing normal wound healing and preventing re-epithelialization.

3.3.3.8 Phenol Soluble Modulins (PSMs)

PSMs are amphipathic peptides uniquely found in *Staphylococci*. Three peptides, PSMα, PSMβ and PSMγ were identified by hot phenol extraction from the culture filtrate [191, 192]. PSMγ is the same as previously described *S. epidermidis* δ-toxin. In *S. aureus*, PSMs are encoded in three loci of the core genome [192]. PSMα peptides (PSMα1-PSMα4) are encoded in the *psma* operon; PSMβ (PSMβ1, PSMβ2) are encoded in *psmβ* operon; and PSMγ (δ-toxin) is encoded in *hld*, which is also part of the coding sequence for RNAIII [193]. PSMs are secreted with an N-terminal N-formylmethionine, but without a signal peptide [192]. Lack of a signal peptide indicates a dedicated mechanism of secretion, which was identified to be a four-component ABC transporter, phenol-soluble modulin transporter (Pmt).

According to the length of amino acid, PSMs are divided into 2 types [193]. PSMα and δ-toxin are α-type PSMs. They are usually 20–25 amino acids long, having neutral or positive net charge, and forming one α-helix [194]. In contrast, PSMβ2 is the β-type PSMs. they are longer, typically 43–45 amino acids in length, having negative net charge, and forming 3 α-helices that fold to a "v"-like shape [194]. PSMs peptides form an α-helix amphipathic structure, which stretches over virtually the entire length of the peptide in the shorter α-type PSMs, and are located in the carboxy-terminal region in the longer β-type PSMs. PSMs have different charge characteristics: PSMα are positively charged and PSMβ are negatively charged, while δ-toxin is neutral [193]. PSMs can attach to the cytoplasmic membrane in a non-specific manner and cause membrane to disintegrate [195] (Fig. 3.5). Phospholipid composition and membrane charge are important for cell susceptibility to PSMs [195]. The tendency of PSMs to aggregate in oligomers, forming short-lived pores.

PSMs have multiple roles in the pathogenesis of *S. aureus*, including cell lysis, biofilm formation, and immune modulation. The ability to promote surface diffusion or form biofilms appears to be a major aspect of the pathogenesis [193]. PSMs have a key impact on the capacity of virulent *S. aureus* to cause skin infections and bacteremia in animal infection models. α-type PSM peptides have high potency in lysing eukaryotic cells in a receptor independent manner by targeting cell

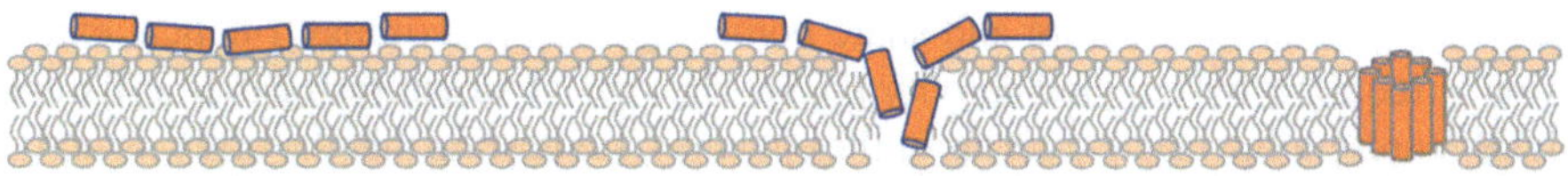

Fig. 3.5 Pore-formation mechanism of phenol-soluble modulins (PSMs). (1) PSMs attach the cytoplasmic membrane in a non-specific way. (2) The attachment can lead to membrane disintegration. (3) PSMs aggregate in oligomers and form a short-lived pore

membranes [192, 196]. Phagocytosed *S. aureus* produces PSMs to lyse neutrophils and osteoblasts intracellularly, indicating that PSMs may mediate the intracellular escape of *S. aureus* [196, 197].

PSMs have immune modulatory effects on host cells. At nanomolar concentrations, they stimulate leukocytes and initiate pro-inflammatory responses including neutrophil chemoattraction, activation, and IL-8 release. The neutrophil-attracting properties of PSMs are important in local *S. aureus* infections and contribute to inflammation (Fig. 3.4). In humans, the pattern recognition receptor formyl peptide receptor 2 (FPR2) can bind to PSM [198]. FPR2 is a member of the G-protein-coupled receptor family that specializes in recognizing pathogen-associated molecular patterns (PAMPs) produced by bacteria. FPR2 is predominately expressed on innate immune cell types, including neutrophils, monocytes, macrophages, and immature dendritic cells. FPR2 not only detects PAMPs, but also monitors the invader's pathogenicity to properly regulate the immune response. After activation by PSMs, FPR2 induces a series of pro-inflammatory responses, including leukocyte activation, neutrophil chemotaxis, and cytokine production [198].

PSMs have also biofilm-structuring activities, and influence biofilm development via their shared physico-chemical properties. PSM expression can also lead to biofilm dispersal, i.e. the detachment of cells or cellular clusters from biofilms, which is a key mechanism leading to the systemic dissemination of biofilm infection [199].

3.4 Exfoliative Toxins (ETs)

ETs are extremely specific serine proteases secreted by *S. aureus*. They are also known as epidermolytic toxins. The principal ETs are known as ETA, ETB, ETC and ETD according the antigenically distinct forms. The gene of each ET is encoded on a different mobile genetic element, *eta* is encoded in the genome of a temperate phage [200], *etb* is found on the plasmid pETB, *etc* has not been described, although the ETC was purified from a *S. aureus* isolate associated horse infection, and *etd* is encoded as part of a pathogenicity island [201].

ETs are glutamate-specific serine proteases of the chymotrypsin family. The catalytic triad (histidine, aspartate, serine) is conserved in all ETs [202]. ETA and

ETB are similar in structure and share homology. The N-terminal α-helical extension is required for enzyme activity [203]. The crystal structures of both ETA and ETB represent the inactive forms of the enzymes, suggesting that the protease activity of ET may require a specific cellular target and occur under specific condition [203]. ETs recognize and hydrolyse desmosome cadherins in the superficial layers of the skin [202]. ETs can cleavage the keratinocytes junctions and cell-cell adhesion in the epidermis of the host, which induce skin peeling and blister formation [201, 204]. ETA and ETB are the most implicated in human skin damage, while ETC was only isolated from a horse infection and no association with human disease [202]. ETD was only identified in a clinical sample of *S. aureus*.

Approximately 5% of *S. aureus* strains produce ETs [205]. The production of ETs in certain strains is related to localized epidermal infections and generalized diseases, such as SSSS including Ritter's disease, toxic epidermal necrosis, bullous impetigo and certain erythema cases [202]. ETs were shown to interact with human and mouse desmoglein 1 (Dsg1), causing blistering of the superficial skin [204, 206]. ETs can affect a large part of the body, predominantly in neonates, infants, and immunocompromised adult patients [202].

The syndromes of SSSS are characterized by formation of blisters and superficial desquamation and skin exfoliation, but its early manifestations include fever, skin hypersensitivity, and erythema, followed by superficial fluid-filled blister formation and skin separation [202, 207]. The lesions characteristic of SSSS are often sterile, because the ETs can be distributed through the bloodstream from a distant site to cause symptoms [208].

3.5 Exoenzymes

3.5.1 Coagulase

Plasma coagulase is a protease with a molecular weight of 64 kDa, which forms fibrin and coagulates plasma, and can coagulate human or rabbit plasma containing sodium citrate or heparin anticoagulant. Coagulases are heat-resistant proteins, but is easily destroyed by proteolytic enzymes. They have been divided into ten (I to X) serotypes according to their antigen differences [209]. The crude product remains partially active at 100 °C for 30 min or after autoclaving. Plasma coagulase binds to prothrombin in plasma and changes its conformation to form a thrombin-like substance. There are two types of coagulase. One is secreted outside the bacteria, called free coagulase. Its effect is similar to prothrombin, which can be activated by the cofactor in human or rabbit plasma. After it becomes a thrombin-like substance, the liquid fibrinogen becomes solid fibrin, thereby coagulating the plasma. Another type of coagulase is bound to the surface of bacteria, called bound coagulase or clumping factor. It acts as a specific receptor for fibrinogen on the surface of bacteria. When bacteria are suspended in human or rabbit plasma, fibrinogen crosslinks with bacterial receptors and causes bacteria to aggregate [210]. Free coagulase was tested using the test tube method, while combined coagulase was tested using the slide method.

Coagulase is closely related to the virulence of *S. aureus*, especially pathogenic strains, which is often used as an important marker to identify *S. aureus* [211]. After the coagulase-positive strain enters the body, fibrin in blood or plasma is deposited on the surface of the bacterial body, hindering the phagocytosis of phagocytes, and even after being phagocytosed, it is not easy to be killed. The clot formed by plasma coagulation can be used as a place for bacterial growth and reproduction, and can resist the phagocytosis of white blood cells and the neutralization of antibodies in plasma. At the same time, the coagulation enzyme is gathered around the bacteria, which can also protect the bacteria from the sterilizing substances in the serum. Staphylococcal infections are prone to localization and thrombosis, and are related to the production of plasma coagulase. Coagulase is immunogenic, and the antibodies produced by stimulating the body have a certain protective effect on coagulase-positive bacterial infection [212, 213].

3.5.2 *Staphylokinase*

Staphylokinase is a protease with a molecular weight of 15 kDa. Also known as staphylococcal fibrinolysin [214]. Plasminogen is activated in the serum to form plasmin, which causes fibrinolysis. Staphylokinase can dissolve the plasma agglomerates caused by plasma coagulase, and cause the bacterial clots to break down, thereby causing the bacteria to spread infection [215].

3.5.3 *Other Pathogenic Factors*

S. aureus can produce heat-resistant nucleic acid degrading enzymes (DNase and RNase) [216]. Among them, the DNase can withstand high temperatures, which can be used as one of the basis for identifying *S. aureus* [217]. In addition, *S. aureus* also produces some other phosphatases, proteases and lipases, which are involved in the breakdown of host proteins, nucleic acids, or lipids, destroying surrounding tissues, and making bacterial infections spread in the body [218, 219].

References

1. Hu DL, Nakane A (2014) Mechanisms of staphylococcal enterotoxin-induced emesis. Eur J Pharmacol 722:95–107. https://doi.org/10.1016/j.ejphar.2013.08.050
2. Hu DL, Li S, Fang R, Ono HK (2021) Update on molecular diversity and multipathogenicity of staphylococcal superantigen toxins. Anim Dis 1(2):1–15. https://doi.org/10.1186/s44149-021-00007-7

3. Tam K, Torres VJ (2019) *Staphylococcus aureus* secreted toxins and extracellular enzymes. Microbiol Spectr 7(2). https://doi.org/10.1128/microbiolspec.GPP3-0039-2018
4. Oliveira D, Borges A, Simoes M (2018) *Staphylococcus aureus* toxins and their molecular activity in infectious diseases. Toxins (Basel) 10(6):252. https://doi.org/10.3390/toxins10060252
5. Sato'o Y, Omoe K, Aikawa Y, Kano M, Ono HK, Hu DL et al (2021) Investigation of *Staphylococcus aureus* positive for staphylococcal enterotoxin S and T genes. J Vet Med Sci 83(7):1120–1127. https://doi.org/10.1292/jvms.20-0662
6. Ono HK, Suzuki Y, Kubota H, Asano K, Takai S, Nakane A et al (2021) Complete genome sequence of *Staphylococcus aureus* strain 834, isolated from a septic patient in Japan. Microbiol Resour Announc 10(9). https://doi.org/10.1128/MRA.01477-20
7. Balaban N, Rasooly A (2000) Staphylococcal enterotoxins. Int J Food Microbiol 61(1):1–10. https://doi.org/10.1016/s0168-1605(00)00377-9
8. Swaminathan S, Furey W, Pletcher J, Sax M (1995) Preliminary X-ray studies on two new crystal forms of staphylococcal enterotoxin C2. Acta Crystallogr D Biol Crystallogr 51(Pt 6):1080–1081. https://doi.org/10.1107/S0907444995003180
9. Todd J, Fishaut M, Kapral F, Welch T (1978) Toxic-shock syndrome associated with phage-group-I staphylococci. Lancet 2(8100):1116–1118. https://doi.org/10.1016/s0140-6736(78)92274-2
10. Omoe K, Hu DL, Takahashi-Omoe H, Nakane A, Shinagawa K (2003) Identification and characterization of a new staphylococcal enterotoxin-related putative toxin encoded by two kinds of plasmids. Infect Immun 71(10):6088–6094. https://doi.org/10.1128/IAI.71.10.6088-6094.2003
11. Ono HK, Omoe K, Imanishi K, Iwakabe Y, Hu DL, Kato H et al (2008) Identification and characterization of two novel staphylococcal enterotoxins, types S and T. Infect Immun 76(11):4999–5005. https://doi.org/10.1128/IAI.00045-08
12. Ono HK, Sato'o Y, Narita K, Naito K, Hirose S, Hisatsune J et al (2015) Identification and characterization of a novel staphylococcal emetic toxin. Appl Environ Microbiol 81(20):7034–7040. https://doi.org/10.1128/AEM.01873-15
13. Suzuki Y, Ono HK, Shimojima Y, Kubota H, Kato R, Kakuda T et al (2020) A novel staphylococcal enterotoxin SE02 involved in a staphylococcal food poisoning outbreak that occurred in Tokyo in 2004. Food Microbiol 92:103588. https://doi.org/10.1016/j.fm.2020.103588
14. Omoe K, Hu DL, Ono HK, Shimizu S, Takahashi-Omoe H, Nakane A et al (2013) Emetic potentials of newly identified staphylococcal enterotoxin-like toxins. Infect Immun 81(10):3627–3631. https://doi.org/10.1128/IAI.00550-13
15. Thomas DY, Jarraud S, Lemercier B, Cozon G, Echasserieau K, Etinne J et al (2006) Staphylococcal enterotoxin-like toxins U2 and V, two new staphylococcal superantigens arising from recombination within the enterotoxin gene cluster. Infect Immun 74(8):4724–4734. https://doi.org/10.1128/IAI.00132-06
16. Wilson GJ, Seo KS, Cartwright RA, Connelley T, Chuang-Smith ON, Merriman JA et al (2011) A novel core genome-encoded superantigen contributes to lethality of community-associated MRSA necrotizing pneumonia. PLoS Pathog 7(10):e1002271. https://doi.org/10.1371/journal.ppat.1002271
17. Sato'o Y, Omoe K, Naito I, Ono HK, Nakane A, Sugai M et al (2014) Molecular epidemiology and identification of a *Staphylococcus aureus* clone causing food poisoning outbreaks in Japan. J Clin Microbiol 52(7):2637–2640. https://doi.org/10.1128/JCM.00661-14
18. Lina G, Bohach GA, Nair SP, HIramatsu K, Jouvin-Marche E, Mariuzza R. (2004) Standard nomenclature for the superantigens expressed by *Staphylococcus*. J Infect Dis 189(12):2334–2336. https://doi.org/10.1086/420852
19. Ono HK, Hirose S, Narita K, Sugiyama M, Asano K, Hu DL et al (2019) Histamine release from intestinal mast cells induced by staphylococcal enterotoxin a (SEA) evokes vomiting reflex in common marmoset. PLoS Pathog 15(5):e1007803. https://doi.org/10.1371/journal.ppat.1007803

20. Umeda K, Ono HK, Wada T, Motooka D, Nakamura S, Nakamura H et al (2021) High production of egc2-related staphylococcal enterotoxins caused a food poisoning outbreak. Int J Food Microbiol 357:109366. https://doi.org/10.1016/j.ijfoodmicro.2021.109366
21. Alibayov B, Zdenkova K, Sykorova H, Demnerova K (2014) Molecular analysis of *Staphylococcus aureus* pathogenicity islands (SaPI) and their superantigens combination of food samples. J Microbiol Methods 107:197–204. https://doi.org/10.1016/j.mimet.2014.10.014
22. Sato'o Y, Omoe K, Ono HK, Nakane A, Hu DL (2013) A novel comprehensive analysis method for *Staphylococcus aureus* pathogenicity islands. Microbiol Immunol 57(2):91–99. https://doi.org/10.1111/1348-0421.12007
23. Tallent SM, Langston TB, Moran RG, Christie GE (2007) Transducing particles of *Staphylococcus aureus* pathogenicity Island SaPI1 are comprised of helper phage-encoded proteins. J Bacteriol 1899(20):7520–7524. https://doi.org/10.1128/JB.00738-07
24. Novick RP, Subedi A (2007) The SaPIs: mobile pathogenicity islands of *Staphylococcus*. Chem Immunol Allergy 93:42–57. https://doi.org/10.1159/000100857
25. Letertre C, Perelle S, Dilasser F, Fach P (2003) Identification of a new putative enterotoxin SEU encoded by the *egc* cluster of *Staphylococcus aureus*. J Appl Microbiol 95(1):38–43. https://doi.org/10.1046/j.1365-2672.2003.01957.x
26. Monday SR, Bohach GA (2001) Genes encoding staphylococcal enterotoxins G and I are linked and separated by DNA related to other staphylococcal enterotoxins. J Nat Toxins 10(1):1–8
27. Collery MM, Smyth DS, Tumilty JJG, Twohig JM, Smyth CJ (2009) Associations between enterotoxin gene cluster types *egc1*, *egc2* and *egc3*, *agr* types, enterotoxin and enterotoxin-like gene profiles, and molecular typing characteristics of human nasal carriage and animal isolates of *Staphylococcus aureus*. J Med Microbiol 58(Pt 1):13–25. https://doi.org/10.1099/jmm.0.005215-0
28. Omoe K, Hu DL, Takahashi-Omoe H, Nakane A, Shinagawa K (2005) Comprehensive analysis of classical and newly described staphylococcal superantigenic toxin genes in Staphylococcus aureus isolates. FEMS Microbiol Lett 246(2):191–198. https://doi.org/10.1016/j.femsle.2005.04.007
29. Betley MJ, Mekalanos JJ (1985) Staphylococcal enterotoxin a is encoded by phage. Science 229(4709):185–187. https://doi.org/10.1126/science.3160112
30. Soltis MT, Mekalanos JJ, Betley MJ (1990) Identification of a bacteriophage containing a silent staphylococcal variant enterotoxin gene (sezA+). Infect Immun 58(6):1614–1619. https://doi.org/10.1128/iai.58.6.1614-1619.1990
31. Coleman DC, Sullivan DJ, Russell RJ, Arbuthnott JP, Carey BF, Pomeroy HM (1989) *Staphylococcus aureus* bacteriophages mediating the simultaneous lysogenic conversion of beta-lysin, staphylokinase and enterotoxin a: molecular mechanism of triple conversion. J Gen Microbiol 135(6):1679–1697. https://doi.org/10.1099/00221287-135-6-1679
32. Zeaki N, Susilo YB, Pregiel A, Radstrom P, Schelin J (2015) Prophage-encoded staphylococcal enterotoxin a: regulation of production in *Staphylococcus aureus* strains representing different sea regions. Toxins (Basel) 7(12):5359–5376. https://doi.org/10.3390/toxins7124889
33. Couch JL, Soltis MT, Betley MJ (1988) Cloning and nucleotide sequence of the type E staphylococcal enterotoxin gene. J Bacteriol 170(7):2954–2960. https://doi.org/10.1128/jb.170.7.2954-2960.1988
34. Spaulding AR, Salgado-Pabon W, Kohler PL, Horswill AR, Leung DY, Schlievert PM (2013) Staphylococcal and streptococcal superantigen exotoxins. Clin Microbiol Rev 26(3):422–447. https://doi.org/10.1128/CMR.00104-12
35. Günther S, Varma AK, Moza B, Kasper KJ, Wyatt AW, Zhu P et al (2007) A novel loop domain in superantigens extends their T cell receptor recognition site. J Mol Biol 371(1):210–221. https://doi.org/10.1016/j.jmb.2007.05.038

36. Petersson K, Thunnissen M, Forsberg G, Walse B (2002) Crystal structure of a SEA variant in complex with MHC class II reveals the ability of SEA to crosslink MHC molecules. Structure 10(12):1619–1626. https://doi.org/10.1016/s0969-2126(02)00895-x
37. Mollick JA, Chintagumpala M, Cook RG, Rich RR (1991) Staphylococcal exotoxin activation of T cells. Role of exotoxin-MHC class II binding affinity and class II isotype. J Immunol 146(2):463–468
38. Herman A, Croteau G, Sekaly RP, Kappler J, Marrack P (1990) HLA-DR alleles differ in their ability to present staphylococcal enterotoxins to T cells. J Exp Med 172(3):709–717. https://doi.org/10.1084/jem.172.3.709
39. Ferry T, Thomas D, Perpoint T, Lina G, Monneret G, Mohammedi I et al (2008) Analysis of superantigenic toxin Vβ T-cell signatures produced during cases of staphylococcal toxic shock syndrome and septic shock. Clin Microbiol Infect 14(6):546–554. https://doi.org/10.1111/j.1469-0691.2008.01975.x
40. Marrack P, Kappler J (1990) The staphylococcal enterotoxins and their relatives. Science 248(4956):705–711. https://doi.org/10.1126/science.2185544
41. Faulkner L, Cooper A, Fantino C, Altmann DM, Sriskandan S (2005) The mechanism of superantigen-mediated toxic shock: not a simple Th1 cytokine storm. J Immunol 175(10):6870–6877. https://doi.org/10.4049/jimmunol.175.10.6870
42. Leder L, Llera A, Lavoie PM, Lebedeva MI, Li H, Sékaly RP et al (1998) A mutational analysis of the binding of staphylococcal enterotoxins B and C3 to the T cell receptor β chain and major histocompatibility complex class II. J Exp Med 187(6):823–833. https://doi.org/10.1084/jem.187.6.823
43. Li H, Llera A, Tsuchiya D, Leder L, Ysern X, Schlievert PM et al (1998) Three-dimensional structure of the complex between a T cell receptor β chain and the superantigen staphylococcal enterotoxin B. Immunity 9(6):807–816. https://doi.org/10.1016/s1074-7613(00)80646-9
44. Ono HK, Hirose S, Naito I, Sato'o Y, Asano K, Hu DL et al (2017) The emetic activity of staphylococcal enterotoxins, SEK, SEL, SEM, SEN and SEO in a small emetic animal model, the house musk shrew. Microbiol Immunol 61(1):12–16. https://doi.org/10.1111/1348-0421.12460
45. Swaminathan S, Furey W, Pletcher J, Sax M (1992) Crystal structure of staphylococcal enterotoxin B, a superantigen. Nature 359(6398):801–806. https://doi.org/10.1038/359801a0
46. Swaminathan S, Furey W, Pletcher J, Sax M (1995) Residues defining Vβ specificity in staphylococcal enterotoxins. Nat Struct Biol 2(8):680–686. https://doi.org/10.1038/nsb0895-680
47. Swaminathan S, Yang DS, Furey W, Abrams L, Pletcher J, Sax M (1988) Crystallization and preliminary X-ray study of staphylococcal enterotoxin B. J Mol Biol 199(2):397. https://doi.org/10.1016/0022-2836(88)90326-9
48. Li SJ, Hu DL, Maina EK, Shinagawa K, Omoe K, Nakane A (2011) Superantigenic activity of toxic shock syndrome toxin-1 is resistant to heating and digestive enzymes. J Appl Microbiol 110(3):729–736. https://doi.org/10.1111/j.1365-2672.2010.04927.x
49. Hu DL, Ono HK, Isayama S, Okada R, Okamura M, Lei LC et al (2017) Biological characteristics of staphylococcal enterotoxin Q and its potential risk for food poisoning. J Appl Microbiol 122(6):1672–1679. https://doi.org/10.1111/jam.13462
50. Jupin C, Anderson S, Damais C, Alouf JE, Parant M (1988) Toxic shock syndrome toxin 1 as an inducer of human tumor necrosis factors and gamma interferon. J Exp Med 167(3):752–761. https://doi.org/10.1084/jem.167.3.752
51. Trede NS, Geha RS, Chatila T (1991) Transcriptional activation of IL-1 beta and tumor necrosis factor-alpha genes by MHC class II ligands. J Immunol 146(7):2310–2315
52. Tessier PA, Naccache PH, Diener KR, Gladue RP, Neote KS, Clark-Lewis I et al (1998) Induction of acute inflammation in vivo by staphylococcal superantigens. II. Critical role for chemokines, ICAM-1, and TNF-α. J Immunol 161(3):1204–1211. https://doi.org/10.4049/jimmunol.161.3.1204

53. Irwin MJ, Hudson KR, Fraser JD, Gascoigne NR (1992) Enterotoxin residues determining T-cell receptor Vβ binding specificity. Nature 359(6398):841–843. https://doi.org/10.1038/359841a0
54. Irwin MJ, Hudson KR, Ames KT, Fraser JD, Gascoigne NR (1993) T-cell receptor β-chain binding to enterotoxin superantigens. Immunol Rev 131:61–78. https://doi.org/10.1111/j.1600-065x.1993.tb01530.x
55. Hu DL, Cui JC, Omoe K, Sashinami H, Yokomizo Y, Shinagawa K et al (2005) A mutant of staphylococcal enterotoxin C devoid of bacterial superantigenic activity elicits a Th2 immune response for protection against *Staphylococcus aureus* infection. Infect Immun 73(1):174–180. https://doi.org/10.1128/IAI.73.1.174-180.2005
56. Hu DL, Zhu G, Mori F, Omoe K, Okada M, Wakabayashi K et al (2007) Staphylococcal enterotoxin induces emesis through increasing serotonin release in intestine and it is downregulated by cannabinoid receptor 1. Cell Microbiol 9(9):2267–2277. https://doi.org/10.1111/j.1462-5822.2007.00957.x
57. Hu DL, Omoe K, Sashinami H, Shinagawa K, Nakane A (2009) Immunization with a nontoxic mutant of staphylococcal enterotoxin a, SEAD227A, protects against enterotoxin-induced emesis in house musk shrews. J Infect Dis 199(3):302–310. https://doi.org/10.1086/596065
58. Schad EM, Zaitseva I, Zaitsev VN, Dohlsten M, Kalland T, Schlievert PM et al (1995) Crystal structure of the superantigen staphylococcal enterotoxin type A. EMBO J 14(4):3292–3301. https://doi.org/10.1002/j.1460-2075.1995.tb07336.x
59. Hu DL, Omoe K, Nakane A, Sugii S, Ono K, Sasaki S et al (1999) Studies on the functional site on staphylococcal enterotoxin a responsible for production of murine gamma interferon. FEMS Immunol Med Microbiol 25(3):237–244. https://doi.org/10.1111/j.1574-695X.1999.tb01348.x
60. Krakauer T (2019) Staphylococcal superantigens: pyrogenic toxins induce toxic shock. Toxins (Basel) 11(3):178. https://doi.org/10.3390/toxins11030178
61. Omoe K, Nunomura W, Kato H, Li ZJ, Igarashi O, Araake M et al (2010) High affinity of interaction between superantigen and T cell receptor Vβ molecules induces a high level and prolonged expansion of superantigen-reactive CD4+ T cells. J Biol Chem 285(40):30427–30435. https://doi.org/10.1074/jbc.M110.140871
62. Suzuki Y, Omoe K, Hu DL, Sato'o Y, Ono HK, Monm C et al (2014) Molecular epidemiological characterization of *Staphylococcus aureus* isolates originating from food poisoning outbreaks that occurred in Tokyo, Japan. Microbiol Immunol 58(10):570–580. https://doi.org/10.1111/1348-0421.12188
63. Chai SJ, Gu W, O'Connor KA, Richardson LC, Tauxe RV (2019) Incubation periods of enteric illnesses in foodborne outbreaks, United States, 1998-2013. Epidemiol Infect 147:e285. https://doi.org/10.1017/S0950268819001651
64. Tauxe RV (2002) Emerging foodborne pathogens. Int J Food Microbiol 78(1–2):31–41. https://doi.org/10.1016/s0168-1605(02)00232-5
65. Wieneke AA, Roberts D, Gilbert RJ (1993) Staphylococcal food poisoning in the United Kingdom, 1969-90. Epidemiol Infect 110(3):519–531. https://doi.org/10.1017/s0950268800050949
66. Veras JF, do Carmo LS, Tong LC, Shupp JW, Cummings C, Dos Santos DA et al (2008) A study of the enterotoxigenicity of coagulase-negative and coagulase-positive staphylococcal isolates from food poisoning outbreaks in Minas Gerais, Brazil. Int J Infect Dis 12(4):410–415. https://doi.org/10.1016/j.ijid.2007.09.018
67. Kitamoto M, Kito K, Niimi Y, Shoda S, Takamura A, Hiramatsu T et al (2009) Food poisoning by *Staphylococcus aureus* at a university festival. Jpn J Infect Dis 62(3):242–243. https://doi.org/10.7883/yoken.JJID.2009.242
68. Le Loir Y, Baron F, Gautier M (2003) *Staphylococcus aureus* and food poisoning. Genet Mol Res 2(1):63–76

69. Hu DL, Omoe K, Shimoda Y, Nakane A, Shinagawa K (2003) Induction of emetic response to staphylococcal enterotoxins in the house musk shrew (*Suncus murinus*). Infect Immun 71(1):567–570. https://doi.org/10.1128/IAI.71.1.567-570.2003
70. Hu DL, Omoe K, Saleh MH, Sugii S, Nakane A, Shinagawa K (2001) Analysis of the epitopes on staphylococcal enterotoxin a responsible for emetic activity. J Vet Med Sci 63(3):237–241. https://doi.org/10.1292/jvms.63.237
71. Omoe K, Imanishi K, Hu DL, Kato H, Fugane Y, Abe Y et al (2005) Characterization of novel staphylococcal enterotoxin-like toxin type P. Infect Immun 73(9):5540–5546. https://doi.org/10.1128/IAI.73.9.5540-5546.2005
72. Omoe K, Imanishi K, Hu DL, Kato H, Takahashi-Omoe H, Nakane A et al (2004) Biological properties of staphylococcal enterotoxin-like toxin type R. Infect Immun 72(6):3664–3667. https://doi.org/10.1128/IAI.72.6.3664-3667.2004
73. Ono HK, Nishizawa M, Yamamoto Y, Hu DL, Nakane A, Shinagawa K et al (2012) Submucosal mast cells in the gastrointestinal tract are a target of staphylococcal enterotoxin type A. FEMS Immunol Med Microbiol 64(3):392–402. https://doi.org/10.1111/j.1574-695X.2011.00924.x
74. Hirose S, Ono HK, Omoe K, Hu DL, Asano K, Yamamoto Y et al (2016) Goblet cells are involved in translocation of staphylococcal enterotoxin a in the intestinal tissue of house musk shrew (*Suncus murinus*). J Appl Microbiol 120(3):781–789. https://doi.org/10.1111/jam.13029
75. Lindsay JA, Ruzin A, Ross HF, Kurepina N, Novick RP (1998) The gene for toxic shock toxin is carried by a family of mobile pathogenicity islands in *Staphylococcus aureus*. Mol Microbiol 29(2):527–543. https://doi.org/10.1046/j.1365-2958.1998.00947.x
76. McCormick JK, Yarwood JM, Schlievert PM (2001) Toxic shock syndrome and bacterial superantigens: an update. Annu Rev Microbiol 55:77–104. https://doi.org/10.1146/annurev.micro.55.1.77
77. Kim J, Urban RG, Strominger JL, Wiley DC (1994) Toxic shock syndrome toxin-1 complexed with a class II major histocompatibility molecule HLA-DR1. Science 266(5192):1870–1874. https://doi.org/10.1126/science.7997880
78. Cui JC, Zhang BJ, Lin YC, Wang QK, Qian AD, Nakane A et al (2010) Protective effect of glutathione S-transferase-fused mutant staphylococcal enterotoxin C against *Staphylococcus aureus*-induced bovine mastitis. Vet Immunol Immunopathol 135(1–2):64–70. https://doi.org/10.1016/j.vetimm.2009.11.001
79. Shopsin B, Gomez M, Montgomery SO, Smith DH, Waddington M, Dodge DE et al (1999) Evaluation of protein a gene polymorphic region DNA sequencing for typing of *Staphylococcus aureus* strains. J Clin Microbiol 37(11):3556–3563. https://doi.org/10.1128/JCM.37.11.3556-3563.1999
80. Becker S, Frankel MB, Schneewind O, Missiakas D (2014) Release of protein a from the cell wall of *Staphylococcus aureus*. Proc Natl Acad Sci U S A 111(4):1574–1579. https://doi.org/10.1073/pnas.1317181111
81. Kim HK, Thammavongsa V, Schneewind O, Missiakas D (2012) Recurrent infections and immune evasion strategies of *Staphylococcus aureus*. Curr Opin Microbiol 15(1):92–99. https://doi.org/10.1016/j.mib.2011.10.012
82. Guss B, Uhlen M, Nilsson B, Lindberg M, Sjoquist J, Sjodahl J et al (1984) The cell-wall-attachment part of staphylococcal protein A. Eur J Biochem 138(2):413–420. https://doi.org/10.1111/j.1432-1033.1984.tb07931.x
83. Peterson PK, Verhoef J, Sabath LD, Quie PG (1977) Effect of protein A on staphylococcal opsonization. Infect Immun 15(3):760–764. https://doi.org/10.1128/iai.15.3.760-764.1977
84. Graille M, Stura EA, Corper AL, Sutton BJ, Taussig MJ, Charbonnjer J-B et al (2000) Crystal structure of a *Staphylococcus aureus* protein a domain complexed with the fab fragment of a human IgM antibody: structural basis for recognition of B-cell receptors and superantigen activity. Proc Natl Acad Sci U S A 97(10):5399–5404. https://doi.org/10.1073/pnas.97.10.5399

85. Silverman GJ, Goodyear CS (2006) Confounding B-cell defences: lessons from a staphylococcal superantigen. Nat Rev Immunol 6(6):465–475. https://doi.org/10.1038/nri1853
86. Dumont AL, Nygaard TK, Watkins RL, Smith A, Kuzhaya L, Kreiswirth BN et al (2011) Characterization of a new cytotoxin that contributes to *Staphylococcus aureus* pathogenesis. Mol Microbiol 79(3):814–825. https://doi.org/10.1111/j.1365-2958.2010.07490.x
87. Sun Y, Emolo C, Holtfreter S, Wiles S, Kreiswirth B, Missiakas D et al (2018) Staphylococcal protein a contributes to persistent colonization of mice with *Staphylococcus aureus*. J Bacteriol 200(9):e00735–e00717. https://doi.org/10.1128/JB.00735-17
88. Kennedy AD, Wardenburg JB, Gardner DJ, Long D, Whitney AR, Braughton KR et al (2010) Targeting of alpha-hemolysin by active or passive immunization decreases severity of USA300 skin infection in a mouse model. J Infect Dis 202(7):1050–1058. https://doi.org/10.1086/656043
89. Kielian T, Cheung A, Hickey WF (2001) Diminished virulence of an alpha-toxin mutant of *Staphylococcus aureus* in experimental brain abscesses. Infect Immun 69(11):6902–6911. https://doi.org/10.1128/IAI.69.11.6902-6911.2001
90. O'Callaghan RJ, Callegan MC, Moreau JM, Green LC, Foster TJ, Hartford OM et al (1997) Specific roles of alpha-toxin and beta-toxin during *Staphylococcus aureus* corneal infection. Infect Immun 65(5):1571–1578. https://doi.org/10.1128/iai.65.5.1571-1578.1997
91. Suttorp N, Seeger W, Dewein E, Bhakdi S, Roka L (1985) Staphylococcal alpha-toxin-induced PGI2 production in endothelial cells: role of calcium. Am J Physiol 248(1 Pt1):C127–C134. https://doi.org/10.1152/ajpcell.1985.248.1.C127
92. Bhakdi S, Muhly M, Korom S, Hugo F (1989) Release of interleukin-1 beta associated with potent cytocidal action of staphylococcal alpha-toxin on human monocytes. Infect Immun 57(11):3512–3519. https://doi.org/10.1128/iai.57.11.3512-3519.1989
93. Song L, Hobaugh MR, Shustak C, Cheley S, Bayley H, Gouaux JE (1996) Structure of staphylococcal α-hemolysin, a heptameric transmembrane pore. Science 274(5294):1859–1866. https://doi.org/10.1126/science.274.5294.1859
94. Freer JH, Arbuthnott JP, Bernheimer AW (1968) Interaction of staphylococcal alpha-toxin with artificial and natural membranes. J Bacteriol 95(3):1153–1168. https://doi.org/10.1128/jb.95.3.1153-1168.1968
95. Watanabe M, Tomita T, Yasuda T (1987) Membrane-damaging action of staphylococcal alpha-toxin on phospholipid-cholesterol liposomes. Biochim Biophys Acta 898(3):257–265. https://doi.org/10.1016/0005-2736(87)90065-4
96. Valeva A, Hellmann N, Walev I, Strand D, Plate M, Boukhallouk F et al (2006) Evidence that clustered phosphocholine head groups serve as sites for binding and assembly of an oligomeric protein pore. J Biol Chem 281(36):26014–26021. https://doi.org/10.1074/jbc.M601960200
97. Wilke GA, Wardenburg JB (2010) Role of a disintegrin and metalloprotease 10 in *Staphylococcus aureus* α-hemolysin-mediated cellular injury. Proc Natl Acad Sci U S A 107(30):13473–13478. https://doi.org/10.1073/pnas.1001815107
98. Popov LM, Marceau CD, Starkl PM, Lumb JH, Sheh J, Guerrera D et al (2015) The adherens junctions control susceptibility to *Staphylococcus aureus* α-toxin. Proc Natl Acad Sci U S A 112(46):14337–14342. https://doi.org/10.1073/pnas.1510265112
99. Virreira Winter S, Zychlinsky A, Bardoel BW (2016) Genome-wide CRISPR screen reveals novel host factors required for *Staphylococcus aureus* α-hemolysin-mediated toxicity. Sci Rep 6:24242. https://doi.org/10.1038/srep24242
100. von Hoven G, Rivas AJ, Neukirch C, Klein S, Hamm C, Qin Q et al (2016) Dissecting the role of ADAM10 as a mediator of *Staphylococcus aureus* α-toxin action. Biochem J 473(13):1929–1940. https://doi.org/10.1042/BCJ20160062
101. Lemjabbar H, Basbaum C (2002) Platelet-activating factor receptor and ADAM10 mediate responses to *Staphylococcus aureus* in epithelial cells. Nat Med 8(1):41–46. https://doi.org/10.1038/nm0102-41

102. Maretzky T, Reiss K, Ludwig A, Buchholz J, Scholz F, Proksch E et al (2005) ADAM10 mediates E-cadherin shedding and regulates epithelial cell-cell adhesion, migration, and β-catenin translocation. Proc Natl Acad Sci U S A 102(26):9182–9187. https://doi.org/10.1073/pnas.0500918102
103. Schulz B, Pruessmeyer J, Maretzky T, Ludwig A, Blobel CP, Safting P et al (2008) ADAM10 regulates endothelial permeability and T-cell transmigration by proteolysis of vascular endothelial cadherin. Circ Res 102(10):1192–1201. https://doi.org/10.1161/CIRCRESAHA.107.169805
104. Colciaghi F, Borroni B, Pastorino L, Marcello E, Zimmermann CF et al (2002) αSecretase ADAM10 as well as αAPPs is reduced in platelets and CSF of Alzheimer disease patients. Mol Med 8(2):67–74. https://doi.org/10.1007/BF03402076
105. Hattori M, Osterfield M, Flanagan JG (2000) Regulated cleavage of a contact-mediated axon repellent. Science 289(5483):1360–1365. https://doi.org/10.1126/science.289.5483.1360
106. Alonzo F 3rd, Torres VJ (2014) The bicomponent pore-forming leucocidins of *Staphylococcus aureus*. Microbiol Mol Biol Rev 78(2):199–230. https://doi.org/10.1128/MMBR.00055-13
107. Sugawara-Tomita N, Tomita T, Kamio Y (2002) Stochastic assembly of two-component staphylococcal gamma-hemolysin into heteroheptameric transmembrane pores with alternate subunit arrangements in ratios of 3:4 and 4:3. J Bacteriol 184(17):4747–4756. https://doi.org/10.1128/JB.184.17.4747-4756.2002
108. Bhakdi S, Tranum-Jensen J (1991) Alpha-toxin of *Staphylococcus aureus*. Microbiol Rev 55(4):733–751. https://doi.org/10.1128/mr.55.4.733-751.1991
109. Powers ME, Kim HK, Wang Y, Wardenburg JB (2012) ADAM10 mediates vascular injury induced by Staphylococcus aureus α-hemolysin. J Infect Dis 206(3):352–356. https://doi.org/10.1093/infdis/jis192
110. Bhakdi S, Muhly M, Mannhardt U, Hugo F, Klapettek K, Mueller-Eckhardt C et al (1988) Staphylococcal α toxin promotes blood coagulation via attack on human platelets. J Exp Med 168(2):527–542. https://doi.org/10.1084/jem.168.2.527
111. Nygaard TK, Pallister KB, DuMont AL, DeWald M, Watkins RL, Pallister EQ et al (2012) Alpha-toxin induces programmed cell death of human T cells, B cells, and monocytes during USA300 infection. PloS One 7(5):e36532. https://doi.org/10.1371/journal.pone.0036532
112. Bubeck Wardenburg J, Patel RJ, Schneewind O (2007) Surface proteins and exotoxins are required for the pathogenesis of *Staphylococcus aureus* pneumonia. Infect Immun 75(2):1040–1044. https://doi.org/10.1128/IAI.01313-06
113. Parker MW, Feil SC (2005) Pore-forming protein toxins: from structure to function. Prog Biophys Mol Biol 88(1):91–142. https://doi.org/10.1016/j.pbiomolbio.2004.01.009
114. Berube BJ, Bubeck WJ (2013) *Staphylococcus aureus* α-toxin: nearly a century of intrigue. Toxins (Basel) 5(6):1140–1166. https://doi.org/10.3390/toxins5061140
115. Lizak M, Yarovinsky TO (2012) Phospholipid scramblase 1 mediates type I interferon-induced protection against staphylococcal α-toxin. Cell Host Microbe 11(1):70–80. https://doi.org/10.1016/j.chom.2011.12.004
116. Grimminger F, Rose F, Sibelius U, Meinhardt M, Pötzsch B, Spriestersbach R et al (1997) Human endothelial cell activation and mediator release in response to the bacterial exotoxins *Escherichia coli* hemolysin and staphylococcal alpha-toxin. J Immunol 159(4):1909–1916
117. Rose F, Dahlem G, Guthmann B, Grimminger F, Maus U, Hänze J et al (2002) Mediator generation and signaling events in alveolar epithelial cells attacked by *S. aureus* α-toxin. Am J Physiol Lung Cell Mol Physiol 282(2):L207–L214. https://doi.org/10.1152/ajplung.00156.2001
118. Grumann D, Nubel U, Broker BM (2014) *Staphylococcus aureus* toxins—their functions and genetics. Infect Genet Evol 21:583–592. https://doi.org/10.1016/j.meegid.2013.03.013
119. Inoshima I, Inoshima N, Wilke GA, Wilke GA, Powers ME, Frank KM et al (2011) A *Staphylococcus aureus* pore-forming toxin subverts the activity of ADAM10 to cause lethal infection in mice. Nat Med 17(10):1310–1314. https://doi.org/10.1038/nm.2451

120. Kolata J, Bode LG, Holtfreter S, Steil L, Kusch H, Holtfreter B et al (2011) Distinctive patterns in the human antibody response to *Staphylococcus aureus* bacteremia in carriers and non-carriers. Proteomics 11(19):3914–3927. https://doi.org/10.1002/pmic.201000760
121. Fritz SA, Tiemann KM, Hogan PG, Epplin EK, Rodriguez M, Al-Zubeidi DN et al (2013) A serologic correlate of protective immunity against community-onset *Staphylococcus aureus* infection. Clin Infect Dis 56(1):1554–1561. https://doi.org/10.1093/cid/cit123
122. Doery HM, Magnusson BJ, Cheyne IM, Sulasekharam J (1963) A phospholipase in staphylococcal toxin which hydrolyses sphingomyelin. Nature 198:1091–1092. https://doi.org/10.1038/1981091a0
123. Huseby M, Shi K, Brown CK, Digre J, Mengistu F, Seo KS et al (2007) Structure and biological activities of beta toxin from *Staphylococcus aureus*. J Bacteriol 189(23):8719–8726. https://doi.org/10.1128/JB.00741-07
124. Huseby MJ, Kruse AC, Digre J, Kohler PL, Vocke JA, Mann EE et al (2010) Beta toxin catalyzes formation of nucleoprotein matrix in staphylococcal biofilms. Proc Natl Acad Sci U S A 107(32):14407–14412. https://doi.org/10.1073/pnas.0911032107
125. Low DK, Freer JH, Arbuthnott JP, Mollby R, Wadström T (1974) Consequences of spingomyelin degradation in erythrocyte ghost membranes by staphylococcal β-toxin (sphingomyelinase C). Toxicon 12(3):279–285. https://doi.org/10.1016/0041-0101(74)90070-1
126. Cifrian E, Guidry AJ, Bramley AJ, Norcross NL, Bastida-Corcuera FD, Marquardt WW (1996) Effect of staphylococcal β toxin on the cytotoxicity, proliferation and adherence of *Staphylococcus aureus* to bovine mammary epithelial cells. Vet Microbiol 48(3–4):187–198. https://doi.org/10.1016/0378-1135(95)00159-x
127. Katayama Y, Baba T, Sekine M, Fukuda M, Hiramatsu K (2013) Beta-hemolysin promotes skin colonization by *Staphylococcus aureus*. J Bacteriol 195(6):1194–1203. https://doi.org/10.1128/JB.01786-12
128. Tajima A, Iwase T, Shinji H, Seki K, Mizunoe Y (2009) Inhibition of endothelial interleukin-8 production and neutrophil transmigration by *Staphylococcus aureus* beta-hemolysin. Infect Immun 77(1):327–334. https://doi.org/10.1128/IAI.00748-08
129. Salgado-Pabon W, Herrera A, Vu BG, Stach CS, Merriman JA, Spaulding AR et al (2014) *Staphylococcus aureus* β-toxin production is common in strains with the β-toxin gene inactivated by bacteriophage. J Infect Dis 210(5):784–792. https://doi.org/10.1093/infdis/jiu146
130. Hayashida A, Bartlett AH, Foster TJ, Park PW (2009) *Staphylococcus aureus* beta-toxin induces lung injury through syndecan-1. Am J Pathol 174(2):509–518. https://doi.org/10.2353/ajpath.2009.080394
131. Yoong P, Torres VJ (2013) The effects of *Staphylococcus aureus* leukotoxins on the host: cell lysis and beyond. Curr Opin Microbiol 16(1):63–69. https://doi.org/10.1016/j.mib.2013.01.012
132. Prevost G, Cribier B, Couppie P, Petiau P, Supersac G, Finck-Barbançon V et al (1995) Panton-valentine leucocidin and gamma-hemolysin from *Staphylococcus aureus* ATCC 49775 are encoded by distinct genetic loci and have different biological activities. Infect Immun 63(10):4121–4129. https://doi.org/10.1128/iai.63.10.4121-4129.1995
133. Morinaga N, Kaihou Y, Noda M (2003) Purification, cloning and characterization of variant LukE-LukD with strong leukocidal activity of staphylococcal bi-component leukotoxin family. Microbiol Immunol 47(1):81–90. https://doi.org/10.1111/j.1348-0421.2003.tb02789.x
134. Yamashita K, Kawai Y, Tanaka Y, Hirano N, Kaneko J, Tomita N et al (2011) Crystal structure of the octameric pore of staphylococcal γ-hemolysin reveals the β-barrel pore formation mechanism by two components. Proc Natl Acad Sci U S A 108(42):17314–17319. https://doi.org/10.1073/pnas.1110402108
135. Aman MJ, Karauzum H, Bowden MG, Nguyen TL (2010) Structural model of the pre-pore ring-like structure of Panton-valentine leukocidin: providing dimensionality to biophysical and mutational data. J Biomol Struct Dyn 28(1):1–12. https://doi.org/10.1080/073911010010524952

136. Woodin AM (1960) Purification of the two components of leucocidin from *Staphylococcus aureus*. Biochem J 75:158–165. https://doi.org/10.1042/bj0750158
137. Spaan AN, van Strijp JAG, Torres VJ (2017) Leukocidins: staphylococcal bi-component pore-forming toxins find their receptors. Nat Rev Microbiol 15(7):435–447. https://doi.org/10.1038/nrmicro.2017.27
138. Yamashita D, Sugawara T, Takeshita M, Koneko J, Kamio Y, Tanaka I et al (2014) Molecular basis of transmembrane beta-barrel formation of staphylococcal pore-forming toxins. Nat Commun 5:4897. https://doi.org/10.1038/ncomms5897
139. Yanai M, Rocha MA, Matolek AZ, Chintalacharuvu A, Taira Y, Chintalacharuvu K et al (2014) Separately or combined, LukG/LukH is functionally unique compared to other staphylococcal bicomponent leukotoxins. PloS One 9(2):e89308. https://doi.org/10.1371/journal.pone.0089308
140. Perret M, Badiou C, Lina G, Burbaud S, Benito Y, Bes M et al (2012) Cross-talk between *Staphylococcus aureus* leukocidins-intoxicated macrophages and lung epithelial cells triggers chemokine secretion in an inflammasome-dependent manner. Cell Microbiol 14(7):1019–1036. https://doi.org/10.1111/j.1462-5822.2012.01772.x
141. Holzinger D, Gieldon L, Mysore V, Nippe N, Taxman DJ, Duncan JA et al (2012) *Staphylococcus aureus* Panton-valentine leukocidin induces an inflammatory response in human phagocytes via the NLRP3 inflammasome. J Leukoc Biol 92(5):1069–1081. https://doi.org/10.1189/jlb.0112014
142. Melehani JH, James DB, DuMont AL, Torres VJ, Duncan JA (2015) *Staphylococcus aureus* leukocidin a/B (LukAB) kills human monocytes via host NLRP3 and ASC when extracellular, but not intracellular. PLoS Pathog 11(6):e1004970. https://doi.org/10.1371/journal.ppat.1004970
143. Vrieling M, Boerhout EM, van Wigcheren GF, Koymans KJ, Mols-Vorstermans TG, de Haas CJ et al (2016) LukMF' is the major secreted leukocidin of bovine *Staphylococcus aureus* and is produced in vivo during bovine mastitis. Sci Rep 6:37759. https://doi.org/10.1038/srep37759
144. Yamada T, Tochimaru N, Nakasuji S, Hata E, Kobayashi H, Eguchi M et al (2005) Leukotoxin family genes in *Staphylococcus aureus* isolated from domestic animals and prevalence of lukM-lukF-PV genes by bacteriophages in bovine isolates. Vet Microbiol 110(1–2):97–103. https://doi.org/10.1016/j.vetmic.2005.07.006
145. McCarthy AJ, Lindsay JA (2013) *Staphylococcus aureus* innate immune evasion is lineage-specific: a bioinfomatics study. Infect Genet Evol 19:7–14. https://doi.org/10.1016/j.meegid.2013.06.012
146. von Eiff C, Friedrich AW, Peters G, Becker K (2004) Prevalence of genes encoding for members of the staphylococcal leukotoxin family among clinical isolates of *Staphylococcus aureus*. Diagn Microbiol Infect Dis 49(3):157–162. https://doi.org/10.1016/j.diagmicrobio.2004.03.009
147. Seals DF, Courtneidge SA (2003) The ADAMs family of metalloproteases: multidomain proteins with multiple functions. Genes Dev 17(1):7–30. https://doi.org/10.1101/gad.1039703
148. Fackrell HB, Wiseman GM (1976) Properties of the gamma haemolysin of *Staphylococcus aureus* 'Smith 5R'. J Gen Microbiol 92(1):11–24. https://doi.org/10.1099/00221287-92-1-11
149. Spaan AN, Reyes-Robles T, Badiou C, Cochet S, Boguslawski KM, Yoong P et al (2015) *Staphylococcus aureus* targets the Duffy antigen receptor for chemokines (DARC) to lyse erythrocytes. Cell Host Microbe 18(3):363–370. https://doi.org/10.1016/j.chom.2015.08.001
150. Noda M, Hirayama T, Kato I, Matsuda F (1980) Crystallization and properties of staphylococcal leukocidin. Biochim Biophys Acta 633(1):33–44. https://doi.org/10.1016/0304-4165(80)90035-5
151. Spaan AN, Vrieling M, Wallet P, Badiou C, Reyes-Robles T, Ohneck EA et al (2014) The staphylococcal toxins γ-haemolysin AB and CB differentially target phagocytes by employing specific chemokine receptors. Nat Commun 5:5438. https://doi.org/10.1038/ncomms6438

152. Hildebrand A, Pohl M, Bhakdi S (1991) *Staphylococcus aureus* alpha-toxin. Dual mechanism of binding to target cells. J Biol Chem 266(26):17195–17200. https://doi.org/10.1016/S0021-9258(19)47358-4
153. DeLeo FR, Kennedy AD, Chen L, Wardenburg JB, Kobayashi SD, Mathema B et al (2011) Molecular differentiation of historic phage-type 80/81 and contemporary epidemic *Staphylococcus aureus*. Proc Natl Acad Sci U S A 108(44):18091–18096. https://doi.org/10.1073/pnas.1111084108
154. Siqueira JA, Speeg-Schatz C, Freitas FI, Sahel J, Monteil H, Prevost G (1997) Channel-forming leucotoxins from *Staphylococcus aureus* cause severe inflammatory reactions in a rabbit eye model. J Med Microbiol 46(6):486–494. https://doi.org/10.1099/00222615-46-6-486
155. Munoz-Planillo R, Franchi L, Miller LS, Nunez G (2009) A critical role for hemolysins and bacterial lipoproteins in *Staphylococcus aureus*-induced activation of the Nlrp3 inflammasome. J Immunol 183(6):3942–3948. https://doi.org/10.4049/jimmunol.0900729
156. Nilsson IM, Hartford O, Foster T, Tarkowski A (1999) Alpha-toxin and gamma-toxin jointly promote *Staphylococcus aureus* virulence in murine septic arthritis. Infect Immun 67(3):1045–1049. https://doi.org/10.1128/IAI.67.3.1045-1049.1999
157. Malachowa N, Whitney AR, Kobayashi SD, Sturdevant D, Kennedy AD, Braughton K et al (2011) Global changes in *Staphylococcus aureus* gene expression in human blood. PloS One 6(4):e18617. https://doi.org/10.1371/journal.pone.0018617
158. Goerke C, Pantucek R, Holtfreter S, Schulte B, Zink M, Grumann D et al (2009) Diversity of prophages in dominant *Staphylococcus aureus* clonal lineages. J Bacteriol 191(11):3462–3468. https://doi.org/10.1128/JB.01804-08
159. Kaneko J, Kimura T, Narita S, Tomita T, Kamio Y (1998) Complete nucleotide sequence and molecular characterization of the temperate staphylococcal bacteriophage phiPVL carrying Panton-valentine leukocidin genes. Gene 215(1):57–67. https://doi.org/10.1016/s0378-1119(98)00278-9
160. Spaan AN, Henry T, van Rooijen WJM, Perret M, Badiou C, Aerts PC et al (2013) The staphylococcal toxin Panton-valentine Leukocidin targets human C5a receptors. Cell Host Microbe 13(5):584–594. https://doi.org/10.1016/j.chom.2013.04.006
161. Spaan AN, Schiepers A, de Haas CJ et al (2015) Differential interaction of the staphylococcal toxins Panton-valentine Leukocidin and γ-hemolysin CB with human C5a receptors. J Immunol 195(3):1034–1043. https://doi.org/10.4049/jimmunol.1500604
162. Couppie P, Cribier B, Prevost G (1994) Leukocidin from *Staphylococcus aureus* and cutaneous infections: an epidemiologic study. Arch Dermatol 130(9):1208–1209. https://doi.org/10.1001/archderm.130.9.1208
163. Prevost G, Couppie P, Prevost P, Gayet S, Petiau P, Cribier B et al (1995) Epidemiological data on *Staphylococcus aureus* strains producing synergohymenotropic toxins. J Med Microbiol 42(4):237–245. https://doi.org/10.1099/00222615-42-4-237
164. Diep BA, Chan L, Tattevin P, Kajikawa O, Nartin TR, Basuino L et al (2010) Polymorphonuclear leukocytes mediate *Staphylococcus aureus* Panton-valentine leukocidin-induced lung inflammation and injury. Proc Natl Acad Sci U S A 107(12):5587–5592. https://doi.org/10.1073/pnas.0912403107
165. Cremieux AC, Dumitrescu O, Lina G, Vallee C, Côté J-F, Muffat-Joly M et al (2009) Panton-valentine leukocidin enhances the severity of community-associated methicillin-resistant *Staphylococcus aureus* rabbit osteomyelitis. PloS One 4(9):e7204. https://doi.org/10.1371/journal.pone.0007204
166. Graves SF, Kobayashi SD, Braughton KR, Whitney A, Sturdevant DE, Rasmussen DL et al (2012) Sublytic concentrations of *Staphylococcus aureus* Panton-valentine leukocidin alter human PMN gene expression and enhance bactericidal capacity. J Leukoc Biol 92(2):361–374. https://doi.org/10.1189/jlb.1111575

167. Kobayashi SD, Malachowa N, Whitney AR, Braughton KR, Gardner DJ, Long D et al (2011) Comparative analysis of USA300 virulence determinants in a rabbit model of skin and soft tissue infection. J Infect Dis 204(6):937–941. https://doi.org/10.1093/infdis/jir441
168. Chi CY, Lin CC, Liao IC, Yao YC, Shen FC, Liu CC et al (2014) Panton-valentine leukocidin facilitates the escape of *Staphylococcus aureus* from human keratinocyte endosomes and induces apoptosis. J Infect Dis 209(2):224–235. https://doi.org/10.1093/infdis/jit445
169. Boakes E, Kearns AM, Ganner M, Perry C, Hill RL, Ellington MJ (2011) Distinct bacteriophages encoding Panton-valentine leukocidin (PVL) among international methicillin-resistant *Staphylococcus aureus* clones harboring PVL. J Clin Microbiol 49(2):684–692. https://doi.org/10.1128/JCM.01917-10
170. Gillet Y, Issartel B, Vanhems P, Fournet J-C, Lina G, Bes M et al (2002) Association between *Staphylococcus aureus* strains carrying gene for Panton-valentine leukocidin and highly lethal necrotising pneumonia in young immunocompetent patients. Lancet 359(9308):753–759. https://doi.org/10.1016/S0140-6736(02)07877-7
171. DuMont AL, Yoong P, Liu X, Day CJ, Chumber NM, James DBA et al (2014) Identification of a crucial residue required for *Staphylococcus aureus* LukAB cytotoxicity and receptor recognition. Infect Immun 82(3):1268–1276. https://doi.org/10.1128/IAI.01444-13
172. Badarau A, Rouha H, Malafa S, Logan DT, Håkansson M, Stulik L et al (2015) Structure-function analysis of heterodimer formation, oligomerization, and receptor binding of the *Staphylococcus aureus* bi-component toxin LukGH. J Biol Chem 290(1):142–156. https://doi.org/10.1074/jbc.M114.598110
173. Ventura CL, Malachowa N, Hammer CH, Nardone GA, Robinson MA, Kobayashi SD et al (2010) Identification of a novel *Staphylococcus aureus* two-component leukotoxin using cell surface proteomics. PloS One 5(7):e11634. https://doi.org/10.1371/journal.pone.0011634
174. Malachowa N, Kobayashi SD, Braughton KR, Whitney AR, Parnell MJ, Gardner DJ et al (2012) *Staphylococcus aureus* leukotoxin GH promotes inflammation. J Infect Dis 206(8):1185–1193. https://doi.org/10.1093/infdis/jis495
175. DuMont AL, Yoong P, Day CJ, Alonzo F 3rd, McDonald WH, Jennings MP et al (2013) *Staphylococcus aureus* LukAB cytotoxin kills human neutrophils by targeting the CD11b subunit of the integrin mac-1. Proc Natl Acad Sci U S A 110(26):10794–10799. https://doi.org/10.1073/pnas.1305121110
176. Dozois A, Thomsen I, Jimenez-Truque N, Soper N, Pearson A, Mohamed-Rambaran P et al (2015) Prevalence and molecular characteristics of methicillin-resistant *Staphylococcus aureus* among skin and soft tissue infections in an emergency department in Guyana. Emerg Med J 32(10):800–803. https://doi.org/10.1136/emermed-2013-203373
177. Thomsen IP, Dumont AL, James DB, Yoong P, Saville BR, Soper N et al (2014) Children with invasive *Staphylococcus aureus* disease exhibit a potently neutralizing antibody response to the cytotoxin LukAB. Infect Immun 82(3):1234–1242. https://doi.org/10.1128/IAI.01558-13
178. Chadha AD, Thomsen IP, Jimenez-Truque N, Soper NR, Jones LS, Sokolaw AG et al (2016) Host response to *Staphylococcus aureus* cytotoxins in children with cystic fibrosis. J Cyst Fibros 15(5):597–604. https://doi.org/10.1016/j.jcf.2015.12.023
179. DuMont AL, Yoong P, Surewaard BG, Benson MA, Nijland R, van Strijp JAG et al (2013) *Staphylococcus aureus* elaborates leukocidin AB to mediate escape from within human neutrophils. Infect Immun 81(5):1830–1841. https://doi.org/10.1128/IAI.00095-13
180. Gravet A, Colin DA, Keller D, Girardot R, Monteil H, Prevost G (1998) Characterization of a novel structural member, LukE-LukD, of the bi-component staphylococcal leucotoxins family. FEBS Lett 436(2):202–208. https://doi.org/10.1016/s0014-5793(98)01130-2
181. Liu C, Chen ZJ, Sun Z, Gao S, Jiang G, Zhou Z et al (2015) Molecular characteristics and virulence factors in methicillin-susceptible, resistant, and heterogeneous vancomycin-intermediate *Staphylococcus aureus* from Central-Southern China. J Microbiol Immunol Infect 48(5):490–496. https://doi.org/10.1016/j.jmii.2014.03.003
182. Haveri M, Roslof A, Rantala L, Pyorala S (2007) Virulence genes of bovine *Staphylococcus aureus* from persistent and nonpersistent intramammary infections

with different clinical characteristics. J Appl Microbiol 103(4):993–1000. https://doi.org/10.1111/j.1365-2672.2007.03356.x
183. Alonzo F 3rd, Benson MA, Chen J, Novick RP, Shopsin B, Torres VJ (2012) *Staphylococcus aureus* leucocidin ED contributes to systemic infection by targeting neutrophils and promoting bacterial growth in vivo. Mol Microbiol 83(2):423–435. https://doi.org/10.1111/j.1365-2958.2011.07942.x
184. Reyes-Robles T, Alonzo F 3rd, Kozhaya L, Lacy DB, Unutmaz D, Torres VJ (2013) *Staphylococcus aureus* leukotoxin ED targets the chemokine receptors CXCR1 and CXCR2 to kill leukocytes and promote infection. Cell Host Microbe 14(4):453–439. https://doi.org/10.1016/j.chom.2013.09.005
185. Alonzo F 3rd, Kozhaya L, Rawlings SA, Reyes-Robles T, Dumont AL, Myszka DG et al (2013) CCR5 is a receptor for *Staphylococcus aureus* leukotoxin ED. Nature 493(7430):51–55. https://doi.org/10.1038/nature11724
186. Reyes-Robles T, Lubkin A, Alonzo F 3rd, Lacy DB, Torres VJ (2016) Exploiting dominant-negative toxins to combat *Staphylococcus aureus* pathogenesis. EMBO Rep 17(5):780. https://doi.org/10.15252/embr.201670010
187. Fromageau A, Cunha P, Gilbert FB, Rainard P (2011) Purified *Staphylococcus aureus* leukotoxin LukM/F' does not trigger inflammation in the bovine mammary gland. Microb Pathog 51(6):396–401. https://doi.org/10.1016/j.micpath.2011.09.005
188. Vrieling M, Koymans KJ, Heesterbeek DA, Aerts PC, Rutten VPMG, de Haas CJC et al (2015) Bovine *Staphylococcus aureu*s secretes the leukocidin LukMF' to kill migrating neutrophils through CCR1. MBio 6(3):e00335. https://doi.org/10.1128/mBio.00335-15
189. Koop G, Vrieling M, Storisteanu DM, Lok LSC, Monie T, van Wigcheren G et al (2017) Identification of LukPQ, a novel, equid-adapted leukocidin of *Staphylococcus aureus*. Sci Rep 7:40660. https://doi.org/10.1038/srep40660
190. Merriman JA, Klingelhutz AJ, Diekema DJ, Leung DY, Schlievert PM (2015) Novel *Staphylococcus aureus* secreted protein alters keratinocyte proliferation and elicits a proinflammatory response in vitro and in vivo. Biochemistry 54(31):4855–4862. https://doi.org/10.1021/acs.biochem.5b00523
191. Mehlin C, Headley CM, Klebanoff SJ (1999) An inflammatory polypeptide complex from *Staphylococcus epidermidis*: isolation and characterization. J Exp Med 189(6):907–918. https://doi.org/10.1084/jem.189.6.907
192. Wang R, Braughton KR, Kretschmer D, Bach T-HL, Queck SY, Li M et al (2007) Identification of novel cytolytic peptides as key virulence determinants for community-associated MRSA. Nat Med 13(12):1510–1514. https://doi.org/10.1038/nm1656
193. Peschel A, Otto M (2013) Phenol-soluble modulins and staphylococcal infection. Nat Rev Microbiol 11(10):667–673. https://doi.org/10.1038/nrmicro3110
194. Towle KM, Lohans CT, Miskolzie M, Acedo JZ, van Belkum MJ, Vederas JC (2016) Solution structures of phenol-soluble modulins α1, α3, and β2, virulence factors from *Staphylococcus aureus*. Biochemistry 55(34):4798–4806. https://doi.org/10.1021/acs.biochem.6b00615
195. Otto M (2014) Phenol-soluble modulins. Int J Med Microbiol 304(2):164–169. https://doi.org/10.1016/j.ijmm.2013.11.019
196. Rasigade JP, Trouillet-Assant S, Ferry T, An Diep B, Sapin A, Lhoste Y et al (2013) PSMs of hypervirulent *Staphylococcus aureus* act as intracellular toxins that kill infected osteoblasts. PloS One 8(5):e63176. https://doi.org/10.1371/journal.pone.0063176
197. Surewaard BG, de Haas CJ, Vervoort F, Rigby KM, DeLeo FR, Otto M et al (2013) Staphylococcal alpha-phenol soluble modulins contribute to neutrophil lysis after phagocytosis. Cell Microbiol 15(8):1427–1437. https://doi.org/10.1111/cmi.12130
198. Kretschmer D, Gleske AK, Rautenberg M, Wang R, Köberle M, Bohn E et al (2010) Human formyl peptide receptor 2 senses highly pathogenic *Staphylococcus aureus*. Cell Host Microbe 7(6):463–473. https://doi.org/10.1016/j.chom.2010.05.012

199. Periasamy S, Joo HS, Duong AC, Bach T-HL, Tan VY, Chatterjee S et al (2012) How *Staphylococcus aureus* biofilms develop their characteristic structure. Proc Natl Acad Sci U S A 109(4):1281–1286. https://doi.org/10.1073/pnas.1115006109
200. Yamaguchi T, Hayashi T, Takami H, Nakasone K, Ohnishi M, Nakayama K et al (2000) Phage conversion of exfoliative toxin a production in *Staphylococcus aureus*. Mol Microbiol 38(4):694–705. https://doi.org/10.1046/j.1365-2958.2000.02169.x
201. Yamaguchi T, Nishifuji K, Sasaki M, Fudaba Y, Aepfelbacher M, Takata T et al (2002) Identification of the *Staphylococcus aureus etd* pathogenicity Island which encodes a novel exfoliative toxin, ETD, and EDIN-B. Infect Immun 70(10):5835–5845. https://doi.org/10.1128/IAI.70.10.5835-5845.2002
202. Bukowski M, Wladyka B, Dubin G (2010) Exfoliative toxins of *Staphylococcus aureus*. Toxins (Basel) 2(5):1148–1165. https://doi.org/10.3390/toxins2051148
203. Vath GM, Earhart CA, Rago JV, Kim MM, Bohach GA, Schlievert PM et al (1997) The structure of the superantigen exfoliative toxin a suggests a novel regulation as a serine protease. Biochemistry 36(7):1559–1566. https://doi.org/10.1021/bi962614f
204. Amagai M, Yamaguchi T, Hanakawa Y, Nishifuji K, Sugai M, Stanley JR (2002) Staphylococcal exfoliative toxin B specifically cleaves desmoglein 1. J Invest Dermatol 118(5):845–850. https://doi.org/10.1046/j.1523-1747.2002.01751.x
205. Ladhani S (2001) Recent developments in staphylococcal scalded skin syndrome. Clin Microbiol Infect 7(6):301–307. https://doi.org/10.1046/j.1198-743x.2001.00258.x
206. Amagai M, Matsuyoshi N, Wang ZH, Andl C, Stanley JR (2000) Toxin in bullous impetigo and staphylococcal scalded-skin syndrome targets desmoglein 1. Nat Med 6(11):1275–1277. https://doi.org/10.1038/81385
207. Cribier B, Piemont Y, Grosshans E (1994) Staphylococcal scalded skin syndrome in adults. A clinical review illustrated with a new case. J Am Acad Dermatol 30(2 Pt 2):319–324. https://doi.org/10.1016/s0190-9622(94)70032-x
208. Melish ME, Glasgow LA (1970) The staphylococcal scalded-skin syndrome. N Engl J Med 282(20):1114–1119. https://doi.org/10.1056/NEJM197005142822002
209. Watanabe S, Ito T, Takeuchi F, Endo M, Okuno E, Hiramatsu K (2005) Structural comparison of ten serotypes of staphylocoagulases in *Staphylococcus aureus*. J Bacteriol 187(11):3698–3707. https://doi.org/10.1128/JB.187.11.3698-3707.2005
210. Peetermans M, Verhamme P, Vanassche T (2015) Coagulase activity by *Staphylococcus aureus*: a potential target for therapy? Semin Thromb Hemost 41(4):433–444. https://doi.org/10.1055/s-0035-1549849
211. Cheng AG, McAdow M, Kim HK, Bae T, Missiakas DM, Schneewind O (2010) Contribution of coagulases towards *Staphylococcus aureus* disease and protective immunity. PLoS Pathog 6(8):e1001036. https://doi.org/10.1371/journal.ppat.1001036
212. McAdow M, DeDent AC, Emolo C, Cheng AG, Kreiswirth BN, Missiakas DM et al (2012) Coagulases as determinants of protective immune responses against *Staphylococcus aureus*. Infect Immun 80(10):3389–3398. https://doi.org/10.1128/IAI.00562-12
213. Zapotoczna M, McCarthy H, Rudkin JK, O'Gara JP, O'Neill E (2015) An essential role for coagulase in *Staphylococcus aureus* biofilm development reveals new therapeutic possibilities for device-related infections. J Infect Dis 212(12):1883–1893. https://doi.org/10.1093/infdis/jiv319
214. Horii T, Yokoyama K, Barua S, Odagiri T, Futamura N, Hasegawa T et al (2000) The staphylokinase gene is located in the structural gene encoding N-acetylmuramyl-L-alanine amidase in methicillin-resistant *Staphylococcus aureus*. FEMS Microbiol Lett 185(2):221–224. https://doi.org/10.1111/j.1574-6968.2000.tb09065.x
215. Grella DK, Castellino FJ (1997) Activation of human plasminogen by staphylokinase. Direct evidence that preformed plasmin is necessary for activation to occur. Blood 89(5):1585–1589. https://doi.org/10.1182/blood.V89.5.1585

216. Tang J, Zhou R, Shi X, Kang M, Wang H, Chen H (2008) Two thermostable nucleases coexisted in *Staphylococcus aureus*: evidence from mutagenesis and in vitro expression. FEMS Microbiol Lett 284(2):176–183. https://doi.org/10.1111/j.1574-6968.2008.01194.x
217. Olson ME, Nygaard TK, Ackermann L, Watkins RL, Zurek OW, Pallister KB et al (2013) *Staphylococcus aureus* nuclease is an SaeRS-dependent virulence factor. Infect Immun 81(14):1316–1324. https://doi.org/10.1128/IAI.01242-12
218. Hu C, Xiong N, Zhang Y, Rayner S, Chen S (2012) Functional characterization of lipase in the pathogenesis of *Staphylococcus aureus*. Biochem Biophys Res Commun 419(4):617–620. https://doi.org/10.1016/j.bbrc.2012.02.057
219. Cadieux B, Vijayakumaran V, Bernards MA, McGavin MJ, Heinrichs DE (2014) Role of lipase from community-associated methicillin-resistant *Staphylococcus aureus* strain USA300 in hydrolyzing triglycerides into growth-inhibitory free fatty acids. J Bacteriol 196(23):4044–4056. https://doi.org/10.1128/JB.02044-14

Chapter 4
Staphylococcus aureus Pathogenesis Based on Genetic Background

Yusuke Sato'o

Abstract *Staphylococcus aureus* is a common pathogen. Although it is always present in or near humans and animals, it can sometimes unleash its pathogenic potential and lead to diseases, including death. Consequently, it is considered "one of the most familiar pathogens" among pathogenic bacteria. Moreover, this pathogen is highly flexible. That is, it appears at first glance to be a complex and elusive combination of factors. While some pathogenic bacteria possess limited or defined characteristics, such as causing disease exclusively in specific hosts or through specific virulent factors, *S. aureus* displays a wide spectrum of diseases across various hosts, ranging from mild to fatal. This ubiquitous phenomenon is one of the distinctive features of *S. aureus*. Regarding pathogenicity, microbial virulence typically manifests in two modes: one characterized by a dominant and well-defined virulence factor, and another lacking such clarity. The former often presents a clear pathogenesis to the host, while the latter tends to exhibit a more varied or ambiguous pathogenesis. *S. aureus* falls into the latter category, lacking a singular definitive virulence factor. Instead, it possesses various combinations of virulence factors, resulting in diverse pathological outcomes. This versatility stems from the acquisition of mobile genetic elements like plasmids and bacteriophages, which complement virulence factors in the core genome. Since its discovery in the nineteenth century, there has been continuous interest in how virulence factors, like toxins and adherence factors produced by *S. aureus*, influence disease development. The absence of a conclusive answer has posed challenges in developing vaccines against this bacterium. The absence of a commercial vaccine has spurred intense research efforts worldwide. In this review, the flexibility and adaptability of *S. aureus* as pathogenic agent will be described.

Keywords Toxin · Enzyme · Adhesin · Biofilm · Mobile genetic element · Gene regulation

Y. Sato'o (✉)
Laboratory of Infection Control and Immunology, School of Veterinary Medicine, Azabu University, Sagamihara, Kanagawa, Japan
e-mail: yus-satoh@azabu-u.ac.jp

A. Nakane, K. Asano (eds.), *Staphylococcus aureus*,
https://doi.org/10.1007/978-981-99-9428-1_4

Abbreviations

ACME	Arginine catabolic mobile element
AIP	Autoinducing peptide
ALI	Acute lung injury
Bap	Biofilm-associated protein
ClfA	Clumping factor A
Eap	Extracellular adherence protein
ECM	Extracellular matrix
Edin	Epidermal cell differentiation inhibitor
ET	Exfoliative toxin
Fnbp	Fibronectin binding proteins
Ica	Intracellular adhesin
IEC	Immune evasion cluster
MGE	Mobile genetic element
MLST	Multilocus sequence typing
MRSA	Methicillin-resistant *Staphylococcus aureus*
MSCRAMM	Microbial surface components recognizing adhesive matrix molecule
NETs	Neutrophil extracellular traps
Phage	Bacteriophage
PIA	Polysaccharide intercellular adhesin
PICI	Phage inducible chromosomal island
PMN	Polymorphonuclear neutrophils
PNAG	Poly-beta(1–6)-N-acetylglucosamine
PSMs	Phenol-soluble modulins
PVL	Panton-Valentine leucocidin
S. aureus	*Staphylococcus aureus*
SaPI	*Staphylococcus aureus* pathogenicity island
Sar	Staphylococcal accessory regulator
SCC*mec*	Staphylococcal cassette chromosome *mec*
Sdr	Serine aspartate repeat-containing protein
SEs	Staphylococcal enterotoxins
SFP	Staphylococcal food poisoning
SSSS	Staphylococcal scalding skin syndrome
SSTIs	Skin and soft-tissue infections
TA	Toxin/antitoxin
TCS	Two-component regulatory system
TSS	Toxic shock syndrome
TSST-1	Toxic shock syndrome toxin-1
WTA	Wall teichoic acids

4.1 Introduction

The Gram-stain image of staphylococci appears to be a cluster of purple spheres, exactly like a bunch of grapes. The common characteristics of this genus are catalase-positive and the ability to grow in the presence of high concentrations of salt condition. However, even within the same genus, the virulence of each species varies. As of December 2021, there are a total of 59 species of Genus *Staphylococcus* on National Center for Biotechnology Information. *Staphylococcus aureus* (*S. aureus*) is one of the most notorious and common pathogens causing a variety of diseases in humans and animals [1, 2]. Its diversity includes where it causes disease and what kind of diseases it causes. The organism causes disease throughout the human body, locally in the soft tissues of the skin, urinary and respiratory systems, and systemically in septicemia [3]. Toxins produced by the bacterium can affect the intestinal tract and result in staphylococcal food poisoning (SFP) [4]. There is also a great diversity of diseases: pyogenic diseases, necrotic diseases, toxic shock syndrome, and food poisoning. It also causes a variety of diseases in animals. For example, it is a major cause of mastitis in cows [5]. Mastitis in cow results in economic losses due to reduced milk production, and contamination of milk with this organism can be a food risk [6].

S. aureus is a bacterial species that lies between pathogenic and commensal bacteria. It is a member of the commensal microbial flora on the surface of the body, but it can cause a variety of diseases as a pathogen. In humans, about 30–50% of people have this organism as a commensal bacterium in their nose [7]. The types of *S. aureus* carrier can be divided into three groups: persistent colonization (up to 20%), intermittent colonization (up to 30%), and non-colonization (up to 50%) based on monitoring over time. However, there have been a number of cases in which this supposedly commensal bacterium is suspected to have entered the host body and caused the disease [8–12]. For this reason, in Japan, patients who are nasal carriers of *S. aureus* are sometimes decolonized by antibiotics [13, 14]. This is also the case in other countries and many protocols and methods have been developed [10, 15]. Not only patients, but also healthcare workers who are carrying the bacteria are a potential risk to (the easily infected) hosts but there is still no clear scientific evidence for this [16–18]. The same is true for animals. This common pathogen for humans is also causing infectious diseases in live stocks and companion animals in human living areas [2, 19]. It has also been reported to be transmitted not only from humans to animals but also from animals to humans, resulting in a zoonotic disease [20, 21]. One of the characteristics of the pathogenicity of this organism is that it can behave harmlessly at first as a member of the community, but then cause disease and kill the host, like a “Trojan horse”.

The diversity of disease and the Trojan horse nature of *S. aureus* may be due to its lack of a specific virulence factor. Some pathogens do have a defined virulence

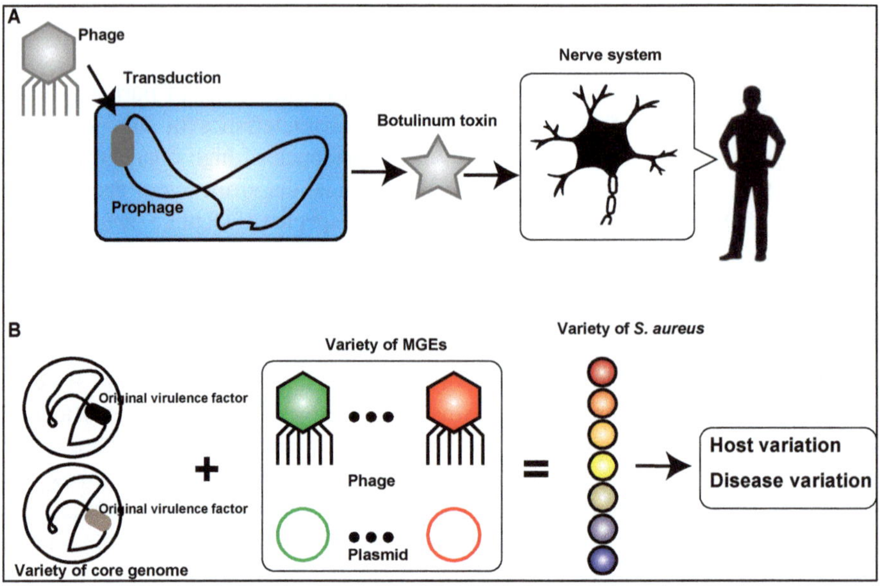

Fig. 4.1 Bacteria and virulence factors. There are two types of pathogens: those with a clear virulence factor and those without. (**a**) In botulism types C and D, bacteriophages carry toxin genes and transduce them to become toxin-producing bacteria. These toxins are taken up by humans and cause the disease. (**b**) *Staphylococcus aureus* and its diseases. The bacteria originally possess virulence factors including cytotoxins and adhesion factors. In addition, they acquire additional virulence factors from various MGEs exogenously. As a result, there are staphylococci with diverse virulence factor patterns, leading to a diversity of hosts and diseases

factor, i.e., *Clostridium botulinum* and *C. tetani* produce neurotoxins that cause diseases (Fig. 4.1) [22]. In other cases, *Bacillus anthracis*, protective antigen, edema factors, lethal factors and capsule on the plasmids define the four major virulence of the organism, and the loss of the plasmid weakens the pathogenicity [22, 23]. The virulence factors also serve as candidates for vaccines. Tetanus toxin from *C. tetani* and capsular antigen from *Haemophilus influenzae* type B have been used as vaccines [24, 25]. On the other hand, there are some bacteria that do not have specific virulence factors. Staphylococci are not known to have distinct virulence factors that correspond one-to-one with disease as in the aforementioned pathogens. This is also true for enterococci and fungi [26, 27]. This lack of a clear virulence factor may be one reason why a staphylococcal vaccine has not yet been developed. *S. aureus* also harbors a variety of virulence factors in different patterns. *S. aureus* carries a variety of virulence factors in different patterns and the combination of these factors determines the type of disease it is likely to cause. In this chapter, the wide variety of virulence factors of *S. aureus* will be described, divided by type and nature, and the repertoires of virulence factors will be discussed.

4.2 Why Do Pathogenic Bacteria Cause Disease?

It is challenging for microorganisms to cause disease in a healthy host [28, 29]. The host has a variety of barriers to the outside world that prevent microbial infection. However, if the barrier is breached or the microorganism possesses invasive potential in host cells or tissues, it can penetrate the host and initiate an infection. Looking at the process, bacteria must first adhere to the surface of the host body, invade the host's interior, and grow to establish an infection. After the infection is established, the substances derived from the microorganism cause abnormalities in the tissues and organs of hosts. Depending on the location and extent of the abnormality, symptoms appear, and an infectious disease is developed. Simultaneously, the host tries to eliminate the pathogen by its own immune system, but the pathogen cleverly uses the evasion mechanism to survive in the host for an extended period. These processes involve multiple factors rather than a single one, with gene regulators precisely controlling virulence factors playing a significant role.

The loci of virulence factors exhibit considerable variability. Taking *S. aureus* as an example, its genome can be divided into three types: core genome, core variable, and mobile genetic elements (MGEs) [30, 31]. Core genome accounts for about 75% of the genes in *S. aureus* and is the gene region that most strains (>95%) in the species possess. Core variable accounts for about 10% of *S. aureus*. This is a unique gene region found in specific groups within a species. In multilocus sequence typing (MLST) analysis, this gene region is specific to clonal complex(es) and can be found in very localized groups or shared by multiple groups. MGEs are factors that move between strains and comprise about 15% of the genes in *S. aureus*. Recently, it was also reported that the core genome moves via MGEs [32]. These can move across the above groups, and the combination of core variants and MGEs is one of the factors that give rise to the diversity of the organism [33]. Virulence factors are distributed across all three genomic elements in *S. aureus* [30, 31]. Virulence factors present in the core genome include β-hemolysins and protein A. The core variant contains *S. aureus* surface protein G (SasG)-encoding gene and enterotoxin gene cluster (*egc*). Finally, there are many virulence factors that reside in mobile genetic elements, such as Panton-Valentine Leukocidin (PVL) and biofilm-associated protein (Bap). These will be discussed in the following sections.

4.3 Classes of Pathogenic Factors

4.3.1 Toxins and Other Secretive Factors

Bacteria produce a variety of toxins [22, 23, 30, 34]. Among these, exotoxins are proteins produced and secreted by bacteria that affect cells and molecules in the host tissues and cause diseases. Their target tissues include all parts of the body, such as the intestinal tract and nerves. Their activities are quite diverse, such as activating or inhibiting biological signals and disrupting cells. Thus, toxins and secretagogues are closely related to pathogenesis, which is one of the major research themes that intrigue bacteriologists. It is well known that *S. aureus* produces many virulence factors, among which toxins and secreted factors are abundant [30, 35, 36]. One of the primary function of the toxins produced by *S. aureus* is cell membrane disruption (lysis), puncturing and damaging cell membranes, resulting in leakage of cell contents and ultimately cell death. Destroying immune cells such as neutrophils is thought to help them evade host immune system. Toxins such as staphylococcal enterotoxins (SEs) and toxic shock syndrome toxin-1 (TSST-1) can also act as superantigens, disturbing the immune cells and contributing to the development of toxic shock syndrome (TSS). Enterotoxins also act on intestinal tissues and are involved in SFP.

"Toxin" is defined as a substance with toxic properties. In other words, a bacterial toxin is a molecule that can affect the host cell or its constituent molecules directly and cause harmful effects. However, certain enzymes and other factors, which do not fit the definition of a so-called toxin, are considered to contribute to immune disruption and play an important role in the pathogenesis of the organism [36, 37]. These secreted virulence factors will be discussed and classified according to their actions.

4.3.1.1 Cytotoxin

The cell membrane is one of the main targets of bacterial toxins [38]. Cell-targeted toxins create pores in biological membranes and release macromolecules, metabolites, ions (which alter osmotic pressure), ATP, and other substances necessary for the maintenance of cell life, finally resulting in cell death. These toxins are classified according to the type of target cell being damaged. The protein or lipid toxins that target red blood cells are called hemolysins. α-hemolysin from *S. aureus* is secreted as a monomer, and subsequently recognizes the host receptor to form a stable pore by forming a seven-dimer complex [39]. On the other hand, β-hemolysin has a different function. This toxin does not form pores on the cell membrane but has phospholipase C activity, which hydrolyzes sphingomyelin present in the cell membrane to ceramide and phosphorylcholine, causing cell lysis [40]. These hemolysins can target not only red blood cells, but also other blood cells [41, 42]. Toxins that target leukocytes are called leukotoxins, such as PVL and leucocidin DE (LukDE) [43, 44]. These are classified into two-component toxins. In PVL, for example, LukS binds to its receptor (Complement 5a receptor 1; C5aR1) first, and then LukF approaches to CD45 and forms a dimer [45]. Subsequent oligomerization results in the formation of a pore penetrating the cell membrane. Since leukotoxins target a variety of leukocytes, including phagocytes, natural killer cells, and lymphocytes, they are thought to interfere with both innate and acquired immunity to the bacterium in the host. Initially, community-acquired methicillin-resistant *S. aureus* (MRSA) such as MW2 and USA300 possessed this toxin exogenously, and its association with high virulence was suggested [43, 46, 47]. However, the importance of toxins presenting in the core genome and core variants of MRSA, such as α-toxin and Phenol-soluble modulin (PSM), has been pointed out [48]. Therefore, the relationship between these cytotoxins and the virulence potential remains to be elucidated and is still a subject of great interest. The lytic activity of these cells is thought to be important not only for immune evasion by the bacteria from the host, but also for nutrient deprivation. In bacterial infections, iron plays an important role [49, 50]. However, free ferrous ions in the body are usually kept at low concentrations. Therefore, *S. aureus* and other pathogenic bacteria are expected to acquire iron, which is an essential factor, by lysing red blood cells with heme containing iron.

It is already known that each of the above-mentioned toxins has its own specific receptor. For example, a disintegrin and metalloprotease 10 (ADMA10) serve as receptors for α-hemolysin, and C5aR serves as a receptor for Panton-Valentine leukocidin (PVL) [41, 43]. On the other hand, there are some toxins that attack cells indiscriminately without having a specific receptor. This includes a family of small peptides, including PSMα1–4, PSMβ1–2, delta-toxin (PSMδ), and PSM-mec [51, 52]. This family of peptides has features of diverse functions such as biofilm formation and colony spreading as well as non-specific lysis of host cells including polymorphonuclear neutrophils (PMN) as toxins. Of the above, only PSM-mec resides on staphylococcal cassette chromosome *mec* (SCC*mec*), which is a mobile genetic element, while the other PSMs reside on the bacterial core genome. Unlike the above toxins, these families exhibit non-specific cytolytic activity. That is, the peptide molecule itself has both hydrophobic and hydrophilic parts, and has a surfactant effect, which disrupts the normal structure of the cell membrane. Although the cytotoxicity of PSM is non-specific due to these molecular characteristics, the activity (affinity for cell membranes) varies among PSMs, and the activity of PSMα is generally considered to be higher than that of PSMβ. In addition to *S. aureus*, other coagulase-negative staphylococci are also reported to produce PSMs and show virulence [53, 54].

4.3.1.2 Exotoxins Except Cytotoxin

SE is a 20–30 kD protein toxin that affects two types of cells (tissues): intestinal tract and immune cells [4, 55, 56]. When it acts on the intestinal tract, food poisoning occurs, inducing nausea and vomiting. Diarrhea is not essential, and fever is rare. This food poisoning is classified as foodborne-intoxication. In other words, after the contamination of *S. aureus* on foods, the bacterium produces SEs. By consuming this SEs-contaminated food, the SE enters the body. SEs taken into the body affect the intestinal tissues (cells). Until recently, this mechanism of action in the intestinal tract was not well understood, but the full picture is becoming clearer. First, SEs are taken up by goblet cells in the intestinal mucosal epithelium [57]. Subsequently, the toxin causes mast cell degranulation. Histamine and serotonin activate the vagus nerve, and the stimulation reaches the vomiting center, resulting in vomiting [58, 59]. SEs are also superantigen toxins that cause activation and proliferation of a tremendous number of T cells by successive external cross-linking of the class II major histocompatibility complex, MHC II, with T cell receptors [60, 61]. It is important to note that this process does not require specific antigen presentation by antigen-presenting cells. TSST-1, a toxin similar to SEs, was originally reported as staphylococcal pyrogenic exotoxin C/SEF in tampon-associated TSS in the 1980s. Subsequent studies showed that it did not have emetic activity, and the toxin was renamed TSST-1 [62]. This toxin also acts as a superantigen and is thought to play a role in the pathogenesis of TSS and immune disruption.

SEB has aspects of biological warfare; its use was studied in the 1960s in the U.S., and it is one of the substances that are feared to be available for use by terrorists [63]. People who inhale SEB through the respiratory tract show symptoms such as acute lung injury (ALI), breathing difficulties and shock within a few days after exposure, and in the worst case, death can happen. This ALI is due to the activation of immune cells by SEB [64]. SEH has also been suggested to be involved in the pathogenesis of mastitis by inducing apoptosis of bovine mammary cells [65].

In addition to the cytotoxins and enterotoxins (superantigen toxins) mentioned above, *S. aureus* also secretes factors that interfere with the host's immune system. For example, CHIPS (chemotaxis inhibitory protein of *S. aureus*) is one of the most studied factors, which is single 14 kDa protein and resists elimination by innate immunity [66]. This factor can bind to the C5aR and formylated peptide receptor (FPR) receptors and interfere with immune cell activation by suppressing the complement system.

4.3.1.3 Other Secretory Factors

PSM is a multifunctional peptide as described above; the first function identified for PSM is an inflammation inducer of *S. epidermidis*, not *S. aureus* [67] but these factors have been found in many staphylococci. PSMs can act on host cells to induce the production of various cytokines [68, 69], resulting in the recruitment of PMNs such as neutrophils to the site of infection [70]. *S. aureus* is a non-motile bacterium, and the spread of this organism was previously considered to be other forces, including physical extrusion associated with bacterial growth and riding the host's circulation. However, after the discovery that *Staphylococcus* undergoes colony spreading [71], this phenomenon has been studied and PSM has shown to play a role in this translocation phenomenon [72, 73]. It has been reported that the virulence of bacteria with loss of colony spreading is reduced [74], and this participation of PSMs in migration is also considered to contribute to virulence. In addition to cytolysis and cytokine induction, PSMs have also been reported to have specific effects on neutrophils, inducing the release of neutrophil extracellular traps and contributing to immune disruption [75]. PSMs have also been reported to be associated with biofilm formation, and certain PSMs are responsible for biofilm maturation and deprivation [51, 76]. PSM-mec acts as a toxic peptide by itself and its mRNA also acts as a regulator as described below [52]. Thus, PSMs are multifunctional virulent factors, and it has been suggested that PSM production may contribute to the severity of infection in the pathogenesis of bacteria [77, 78]. In other words, several reports have shown that PSM production is higher in community-acquired MRSA (CA-MRSA) than in healthcare-associated MRSA (HA-MRSA), and is considered to be one of the factors shaping the high virulence of CA-MRSA. Among the PSMs, PSMδ is known to act on mast cells and cause degranulation [79]. In atopic dermatitis, its bacterial flora is dominated by *S. aureus*, and PSMδ is thought to be one factor involved in the severity of this condition [80].

4.3.1.4 Enzymes

S. aureus also secretes several enzymes [36]. Many of these are catabolic enzymes that degrade the components of the host, including proteases, lipases, and nucleases. These enzymes are involved in expanding the site of infection, obtaining nutrition, etc. Some enzymes are clearly involved in pathogenicity. For example, exfoliative toxin (ET), also known as exfoliatin, is a serine protease produced by *S. aureus*. This enzyme specifically cleaves cadherin (desmoglein 1; Dsg1), a protein that is present in the host's skin and is required for skin tissue adhesion and maintenance of tissue structure [81–83]. In mild cases, blistering occurs, while in severe cases, Staphylococcal scalding skin syndrome (SSSS) occurs, especially in infants. The pathology is similar to that of pemphigus vulgaris, an autoimmune disease caused by autoantibodies to the same cadherin [84]. Coagulase and staphylokinase act on plasma proteins [36]. Coagulase production is one of the discriminatory features of staphylococci. Coagulase-positive staphylococci are represented by *S. aureus* and *S. argenteus*, while coagulase-negative staphylococci are represented by *S. epidermidis* and *S. saprophyticus*. Although the name of coagulase is "enzyme," the protein itself has no enzymatically catalytic activity or active center and work as co-factor. When coagulase binds to prothrombin and forms a complex with staphylothrombin, the active center is exposed by changing the structure of the closed prothrombin [85]. The complex converts the fibrinogen in the plasma to fibrin. On the other hand, staphylokinase is also a co-factor [86]. Staphylokinase-plasminogen complex degrades fibrin, which serves as a scaffold for bacterial growth. The decoration of the surface of *Staphylococcus* by fibrin is considered to assist the bacterium in resisting elimination by host phagocytes and to promote surface colonization. In addition, this enzyme also works as an anti-opsonic factor and cleaves IgG and complement component C3b [87].

4.3.2 Adhesin

Attachment is the first step required for the multiplication of bacteria that invade from the outside. Bacteria gaining a foothold inside the susceptible host then multiply and cause infection by inducing changes in the host such as tissue damage and immune response. For this reason, the research on bacterial attachment and the adhesins is one of the major topics in bacteriology. Adhesins are structures on the bacterial surface that promote attachment to other cells, the extracellular matrix (ECM), organic materials such as plasma proteins, or inorganic surfaces such as plastics on medical device surfaces. Pilus, polysaccharides, and proteins function as adhesins [88–90]. In *S. aureus*, proteins and polysaccharides play important roles among these. Many Adhesins are encoded in the genome and details are described below.

Most of the proteinaceous adhesins of *S. aureus* are classified as microbial surface components recognizing adhesive matrix molecules (MSCRAMMs) [91]. These are a group of cell wall-anchored proteins that are secreted via the Sec system, the extracellular transport system of the bacteria, and are present on the cell wall. This family has sec-dependent signal sequence at N-terminal and LPXTG motif at C-terminal within peptide. Sortase A, a transpeptidase, recognizes this sequence and anchors MSCRAMM to the cell wall by forming an amide bond between the C-terminal side of threonine and the N-terminal side of peptidoglycan. This MSCRAMM can bind to extracellular matrix such as collagen, fibrinogen, and fibronectin. Some MSCRAMMs recognize multiple host-derived substances. Fibronectin binding proteins A and B (FnbpA and FnbpB) not only bind to fibronectin, but also recognize and bridge the $\alpha5\beta1$ integrin on the host cells [92]. Serine aspartate repeat-containing protein D (SdrD) and clumping factor A (ClfA) are also MSCRAMM proteins that bind to desmoglein I and fibrinogen I, respectively [93, 94]. MSCRAMMs contribute not only to adhesion but also to cell (tissue) invasion. After recognition of integrins, FnbpA and FnbpB induce a signal transduction pathway from integrins into the host cell [92]. Bacteria are then internalized into the cell by remodeling of the actin cytoskeleton. In fact, it has been observed that strains lacking Fnbp are unable to enter the cell, indicating that this factor is a major factor in cell invasion [95–97].

S. aureus has other adhesins classified as secretable expanded repertoire adhesive molecules (SERAMs) [98]. Extracellular adherence protein (Eap) is a representative protein, which has no LPXTG motif and is secreted out of the bacterium. Then, it recognizes matrix proteins such as collagen, fibrinogen and extracellular DNA (eDNA) [99, 100]. This protein cross-links with the bacterial cells, allowing them to attach. This protein not only acts as an adhesion factor, but also interferes with immunity and inhibits the formation of neutrophil extracellular traps (NETs). As will be described in detail in the next section, attachment by adhesion molecules is not limited to the host body, but also involves attachment to non-living surfaces such as plastics, causing biofilm formation on medical devices and biofilm infections. MSCRAMMs, polysaccharides [wall teichoic acids (WTA)], and eDNA are involved in the adhesion of materials to surfaces and are important for the early attachment of the planktonic cells in biofilm formation. Polysaccharide intercellular adhesin (PIA) and extracellular nucleic acids are involved in the maintenance of the biofilm structure after its formation.

4.3.3 Biofilm

Biofilm is a community composed of microorganisms and microbial-derived materials [101–103]. Formed by the growth of one or more species of bacteria forming a community, biofilms are one of the bacterial defense systems. Biofilms act as a barrier, blocking the effects of antimicrobial chemotherapy, phagocytosis, pattern recognition, and other immune systems from the outside, favoring the survival of the bacteria [104, 105]. In addition, biofilms are not stable, but attach and detach intermittently. Biofilms such as *S. epidermidis* and *S. aureus*, which are problematic in healthcare, form on medical devices such as indwelling catheters, implants, and artificial joints in the body and can cause subsequent infections. They can also form on tissue surfaces, such as heart valves in the case of infective endocarditis [106]. Biofilm formation proceeds in three stages: attachment, maturation, and detachment, and the detached bacteria can create new areas of infection. The detailed molecular mechanism is as follows (Fig. 4.2).

In the initial attachment to biological or non-biological surfaces, the substances present on the surface layer of the organism play an important role. In vivo, MSCRAMMs play a major role, binding to host matrix components such as collagen and fibronectin; FnbpA, FnbpB, Clf, and Sdr family proteins are known to be

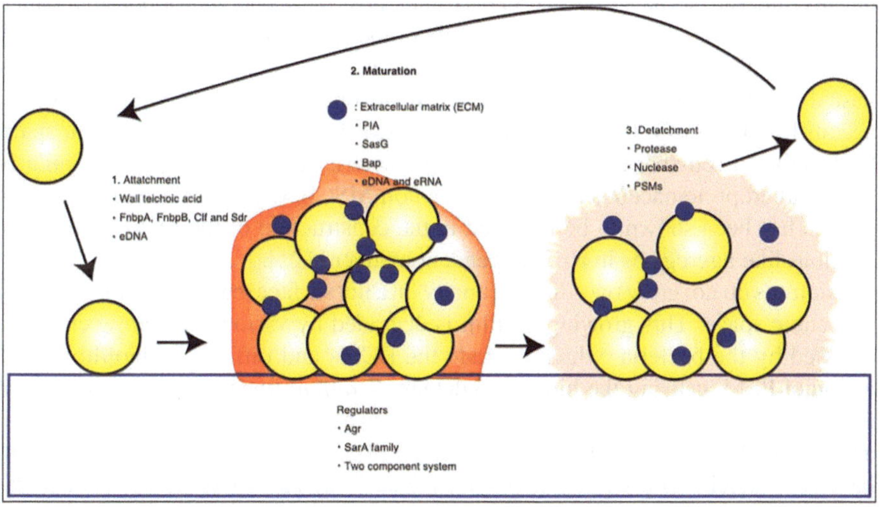

Fig. 4.2 Biofilm life cycle and associated factors in *Staphylococcus aureus*. This figure shows the process of biofilm formation on the surface of a medical device such as a catheter. Biofilm formation in *S. aureus*, as in other bacteria, occurs in three stages: attachment, maturation, and detachment. Typical factors involved in each stage are shown in the figure. Biofilms are composed of extracellular matrices such as eDNA and PIA as well as viable and dead bacteria. The regulatory factors that control these factors are also involved in biofilm formation and virulence of bacteria

involved in attachment to host matrix, as mentioned above. On the other hand, WTA plays an important role on inorganic surfaces [107]. WTA is a component found in the cell walls of Gram-positive bacteria, and is a polymer composed of glycerol phosphate or ribitol phosphate linked by phosphodiester bonds. Since WTA is negatively charged, it maintains a negative charge on the cell wall [108]. The plastic surface has a positive charge, so the electrical force promotes adhesion. FnbpA, FnbpB, and SdrC have also been suggested to promote adhesion [109, 110], although they are non-living organisms. eDNA is extracellular DNA released from the bacteria by autolysis. It has been suggested that this eDNA may also be involved in the initial adhesion [111].

This is followed by the formation of tiny micro-colonies and biofilm maturation by formation of ECM that fills the spaces between the cells (clumps) and bacterial growth. The components of the extracellular matrix are polysaccharides such as PIA, nucleic acids such as eDNA and extracellular RNA (eRNA), and SasG and its similar proteins. PIA is poly-beta(1–6)-N-acetylglucosamine (PNAG), which is a polymer of sugar [112]. This substance is positively charged and is electrically attracted to the WTA mentioned above, thereby bridging the gap between bacterial cells. A series of enzymes, the IcaADBC operon (Intracellular adhesin ADCB operon) and its regulator, IcaR, are involved in this synthesis. In addition to PIA, extracellular nucleic acids also function as ECM, including eDNA and eRNA [113, 114]. eDNA is released during bacterial lysis. eDNA and the cell wall maintain a negative charge, so bacteria hold eDNA via a positively charged lipoprotein called SaeP [115]. It has also been suggested that other proteins also can bind to eDNA [116, 117]. Recently, the importance of eRNA as well as eDNA has also been suggested, while its actual functional role remain unclear [118]. A famous proteinaceous ECM is SasG. It is one of the MSCRAMMs and fills the intercellular space by forming a rod-shape structure between bacterial cells [119, 120]. A similarly well-studied proteinaceous ECM is Bap, a non-MSCRAMMs surface protein. This protein has been shown to be involved in the construction of the ECM by forming amyloid-like aggregates in a pH-dependent manner [121].

Biofilms do not grow two-dimensionally on a flat surface, but three-dimensionally in a cubic space. This growing biofilm is degraded by various degrading enzymes secreted by the bacteria, which degrade the ECMs and diffuse the biofilm by the action of PSM, causing the biofilm to detach. *S. aureus* encodes a large number of proteases in its genome. Some of these enzymes act on specific/unspecific proteinaceous ECMs to disperse the biofilm [122, 123]. Similarly, *S. aureus* is known to have two nucleic acid-degrading enzymes, Nuc, which are involved in the degradation of extracellular diffusion [124, 125].

Biofilm formation is regulated by many factors [112]. The Agr system, a bacterial quorum sensing system, plays an important role in controlling biofilm maturation and detachment. In addition, it has been reported that IcaR, various two-component regulatory systems, and SarA family proteins are involved in this process.

4.3.4 *Genetic Map of* S. aureus *and Location of Virulence Factors*

The virulence factors of *S. aureus* are scattered throughout the genome, rather than clustered somewhere in the genome. The genome of *S. aureus* was first reported in 2001 [126]. Since then, many genomes have been registered in the database, and the number of genomes exceeds 13,000 in the NCBI Genome Assembly and Annotation report (as of the end of 2021). Through this comparative genome analysis, it was found that the chromosomes of this bacterium exhibit a mosaic structure. In other words, exogenous factors are incorporated into the core genome like a patchwork [127]. The main exogenous genetic factors on the chromosome are prophages, *Staphylococcus aureus* pathogenicity islands (SaPIs), and SCC*mec*, and these regions contain virulence factors. In addition, many virulence factors exist not only on the chromosomes but also on the plasmids. These MGEs are involved in the acquisition of virulence because they can be transferred between strains. Some other regions, core-variable, also contain virulence genes that are associated with specific groups of bacteria. Figure 4.3 illustrates an example of the movement of this mobile genetic factor and the acquisition of virulence factors using SEs, which exist in various forms within *S. aureus*. These are described in more detail below.

Bacteriophages (phages) are microbes (viruses) whose hosts are microorganisms. Phages are the most ubiquitous and most diverse organisms on the planet (there is some controversy about treating viruses as living organisms, but they are considered living organisms here). They infect and replicate in bacteria, and are ubiquitous wherever these hosts exist. They are found in soil, seawater, fresh water (rivers, lakes, etc.), and in our daily lives (drinking water, sewage, etc.). Estimates suggest that there are over 10^{31-32} phages on Earth, surpassing the combined count of all living organisms, from bacteria to humans [129]. After injecting their genome into the bacterial cytoplasm, phages replicate within it [130, 131]. In the lytic cycle, the virus synthesizes a copy of itself in the bacterial cell, and then the bacterial cell is destroyed by lytic enzymes and killed. The progeny of these released phages infects new hosts and produce further generations of the virus. In the lysogenic cycle, on the other hand, the host cell can survive for a time. In this cycle, the viral genome is incorporated into the host chromosomal DNA and becomes quiescent, replicating together with the host genome in a (relatively) harmless manner. This dormant viral genome is called prophage, and the bacteria that harbor the prophage are called lysogenic bacteria. This prophage is activated by the host SOS response to produce viral particles, which eventually cause lysis of the host. The newly formed daughter phage then infects the neighboring susceptible bacterium, causing them to either lyse or become lysogenic again. Some phages that have only a lytic cycle are called virulent phages, while some phages that have both lytic and lysogenic cycles are called lysogenic phages. Some lysogenic phages have toxin gene(s) on their genomes [130–132]. When these phages are integrated into the bacterial genome as prophages, new toxic genes are occasionally obtained in the host genome giving the bacteria new virulence. This is an important problem in medicine,

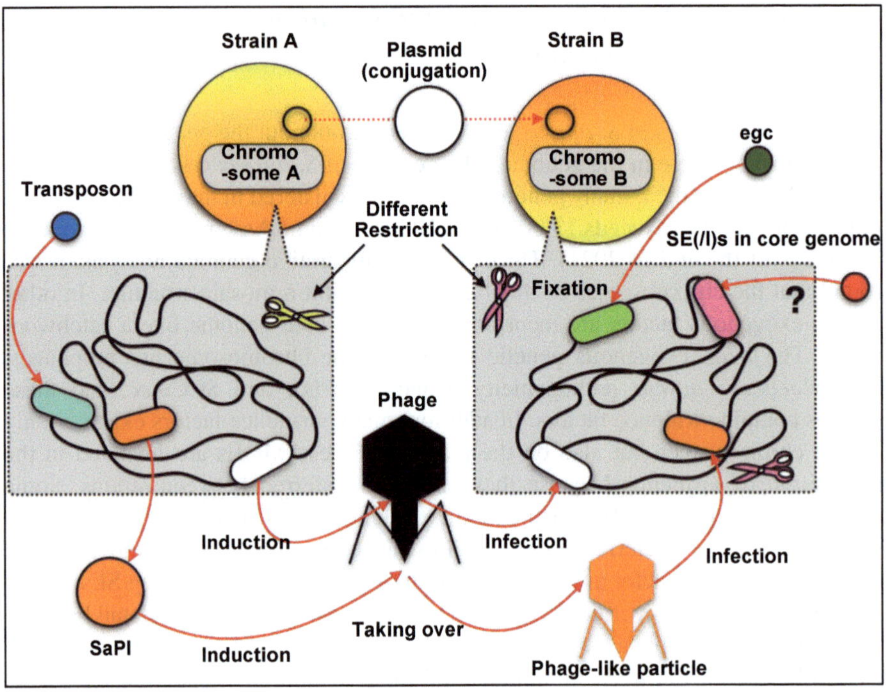

Fig. 4.3 Acquisition mechanism of virulence factors in *S. aureus* with examples of SEs family members. SEs/SEls are found in various regions of *S. aureus*. In the chromosome, some SEs/SEls are isolated on the core genome, while others are located on mobile genetic factors such as phage and SaPI. They are also present on plasmids, which are mobile genetic factors located outside the chromosome. These mobile genetic factors can move between strains, leading to the acquisition of new toxins by the bacteria. (This figure is an English translation with some modifications from a previous article by the author. Ref. [157])

veterinary medicine, and public health because it can lead to the emergence of highly toxic strains. When a phage is incorporated into the genome, it not only imparts new properties to the host, but also may cause the host to lose its inherent properties. In other words, phage is integrated inside of some gene, resulting in the failure to produce normal proteins. In the case of *S. aureus,* the gene region of β-hemolysin is also called φSa3. This is the region where phages such as φSa3mw2 become integrated [48]. In this case, the gene for β-hemolysin is disrupted, and negative conversion occurs, in which the hemolytic toxin is no longer synthesized. On the other hand, since φSa3mw2 also contains the enterotoxin genes SEA, SEK, and SEQ, this lysogenization also causes positive conversion. Thus, lysogenization of a certain type of phage causes double conversion, positive and negative. This phenomenon had been already known before the first whole genome report of *S. aureus*, concerning β-hemolysin and staphylokinase [133]. There is another way to acquire genes other than the lysogenic cycle [134]. A phage can accidentally take up host DNA at random and propagate the DNA encoding a new trait to another host cell. There are phages that infect multiple *Staphylococcus* species [135]. Therefore, phages play a major role in the evolution of bacteria because the exchange of genes described above takes place among a large number of bacteria. Pathogenic factors on phages include SEA, SEE, SEP, PVL, and exfoliative toxin type B (ETB). The immune evasion cluster (IEC) within the phage's late operon includes SEA, staphylokinase, chemotaxis inhibitory protein, and the staphylococcal complement inhibitor [136, 137].

SaPIs are a class of phage inducible chromosomal islands (PICIs) [138, 139]. PICIs are mobile genetic elements found in Gram-positive bacteria and resemble phages. Like lysogenic phages, SaPI integrate their own DNA into the host chromosome after entering the host bacterial cell. The insertion region is a specific region like phage. However, unlike lysogenic phages, SaPI cannot migrate to other strains by itself once it is integrated into the host genome. A prophage present in the genome or a newly infected phage must act as a helper phage. The insertion of SaPI DNA into the head generated by the helper phage results in the formation of a phage-like particle, which infects the surrounding bacteria. SaPIs can take over phage by interfering with the head structure of the phage, reducing the size of the head to a size where the phage DNA cannot enter but the SaPI DNA can, and changing a small terminase subunit (TerS) from the phage-derived TerS to its own TerS, which is necessary for DNA porting. Like phages, SaPIs are often present in one or several copies in the genome of *S. aureus*, and contribute to the pathogenicity and evolution of the bacterium [31, 140]. Toxins such as SEB and SEC, biofilm-associated factors called Bap are present on SaPIs [136, 137].

Plasmids are DNA molecules other than bacterial chromosomes that are replicated in the bacterial cytoplasm and passed on to daughter cells during binary fission [141]. This plasmid DNA replicates autonomously and independently of the nucleoid DNA. The plasmid DNA is generally double-stranded circular DNA and is generally relatively short compared to the chromosomes. Chromosomes carry all the genetic information necessary for the bacteria to survive in their habitat, while plasmids encode additional genes that give the bacteria an advantage in certain conditions, environments, and places other than their original niche. For example, plasmids contain genes that are useful for bacterial survival strategies, such as drug resistance genes and heavy metal resistance genes [142, 143]. Plasmids have a replicon, a sequence that serves as a starting point for their independent replication in the cell [141]. This sequence contains the Rep gene, a protein that initiates replication specific to that plasmid, repetitive units called iterons, and AT-rich sequences in the vicinity. Replication of the plasmid occurs independently of the chromosome, but is strictly controlled with respect to its number (copy number) in the cytoplasm [144]. Depending on the type of replicon and the length of the plasmid, the copy number varies. Generally, there is an inverse correlation with larger plasmids producing fewer copies and smaller plasmids producing more copies. There is also a uniform control on the distribution during cell division. In plasmids with low or single copy numbers, partition factors (ParABS systems) and other mechanisms function to ensure equal distribution to both dividing daughter cells [145]. In some plasmids, the toxin/antitoxin system (TA system) serves to maintain the plasmid

[146]. This system consists of a stable toxin that is harmful to the cell and an unstable antitoxin that neutralizes it. Both are generally short genes and are found in contiguous sequences. In daughter cells that retain the plasmid, the antitoxin continues to neutralize the toxin and the bacterial cell survives. However, daughter cells that fail to maintain the plasmid will die or become dormant due to the action of toxin, since the unstable antitoxin is no longer supplied after its disappearance. This system can be divided into different types based on the combination of toxin and antitoxin, which can be RNA or protein. Bacteria can harbor multiple plasmids but plasmids sometimes show incompatibility [147]. If multiple plasmids with the same replication mechanism are present in a cell, only one will typically be preserved, and the others will be eliminated. Some plasmids are conjugative [148]. On this type of plasmid, there are a series of factors, such as sexual pili, that are involved in the transfer of the plasmid between two cells, causing the conjugation of the plasmid. Such conjugation enables the addition of various genetic information to bacteria, and together with phage, plays an important role in the evolution of bacteria. Although the general theory of plasmids has been described above, this phenomenon is also commonly observed in staphylococci. Genes on plasmids in *S. aureus* includes many drug resistance genes, heavy metal resistance genes, and enzymes. It is also known that ETA, SED and SER are present [136, 137].

SCC*mec* is a well-known mobile genetic element with *mecA* encoding PBP2'. SCC*mec* has been classified into 14 types (I-XIV) based on the combination of *mec* complex and *ccr* complex types [149, 150]. While this element has primarily been investigated in relation to drug resistance and epidemiology, virulence factors are also found within SCC*mec* for example PSM-mec and Plasmin-sensitive surface protein [52, 91].

The mobile genetic elements described above are those that can still be transferred between strains. There are other former mobile factors that are thought to have been immobilized. νSaα, β and γ are representative genomic islands [46, 151]. νSaα is a region consisting of exotoxin genes (*set* or *ssl*) and lipoprotein-like genes (*lpl*) [152, 153]. Ssl is a secreted protein similar to SEs and TSST-1. The function of these proteins remains largely unknown, but it has been suggested that they interfere with immune cells and disrupt the innate immune system. Lpl has been suggested to be involved in host immunity and tissue invasiveness. νSaβ is a region that contains regions related to (former) phages, and is commonly presented in *S. aureus* [154]. However, there are variations in the genes contained in this region, and multiple types exist. This region contains a cluster of SEs and SE-like toxins called the enterotoxin gene cluster, as well as cytotoxins and proteases such as LukDE. The νSaγ region, formerly known as *etd*PI, contains ETD, α-hemolysin, epidermal cell differentiation inhibitor B (Edin-B), and PSMβ [83, 155]. The role of Edin-B in infection is still unclear, but it disrupts host keratinocyte signaling and inhibits differentiation. Arginine catabolic mobile element (ACME) is located in the neighborhood of SCC*mec* and is a hallmark of USA300, a representative strain of community-acquired MRSA [47, 156]. Within this region, it encodes the enzymes of arginine deiminase pathway, which metabolizes L-arginine to carbon dioxide, ATP, and ammonia.

In addition to the above current/previous mobile genetic factors, *S. aureus* has a number of virulence factors in its core genome. These are virulence factors common to most *S. aureus* and scattered throughout the genome. Such virulence factors include protein A, coagulase, and β-hemolysin [127].

However, it is not all diversity, there is also convergence [128]. Mobile genetic factors are capable of transferring between bacteria, but the restriction modification system serves as a barrier. Five types of restriction enzymes (I-V) are known so far, and in staphylococci, types I and IV play a major role [128, 157–159]. These type I and IV restriction enzymes are part of the core genome of bacteria, and the patterns of restriction enzyme possession and DNA sequence recognition domains within these enzymes vary from one bacterium to another. It is known that there is diversity in restriction enzymes and their possession patterns. However, this "diversity of mobile genetic factors and diversity of restriction systems" has produced a convergence. In other words, the selection of mobile genetic factors by restriction enzymes takes place in the bacterium, and each core bacterium tends to have its own unique mobile genetic factors [127, 160–163]. Therefore, using classification methods based on core genome, such as Multi locus sequence typing, it became clear that each Clonal complex tends to have its own unique mobile genetic repertoire. This fact is an unexpected convergence in the diversity revealed in the genome analysis.

4.3.5 Regulation of Virulence Factors

S. aureus has a variety type of gene regulators. Representative examples are the accessary gene regulator (Agr) and the SarA family (Fig. 4.4). Agr is one of the two-component regulatory systems (TCS), initially found to be involved in the secretion of exotoxins [164]. This TCS allows bacteria to monitor their own surroundings and change their own gene expression [165, 166]. In general, TCS consists of two proteins. Membrane-bound histidine kinase located in the cell membrane is activated by external stimuli, resulting in autophosphorylation. The phosphorylation relay is passed on to the response regulator inside the cell, which in turn regulates gene regulation. Agr is composed of AgrBDCA in RNAII region and RNAIII, and is involved in quorum sensing to detect the density of bacteria [167]. AgrD is a precursor protein that is secreted out of the cell as a mature protein, AIP (Autoinducing peptide), by the catalytic activity (protease activity) of AgrB. In *S. aureus*, AIPs are cyclic thiolactone peptides, and four types are known. This difference is due to variation in the AgrBDC region (AgrI-IV) and different types of Agr cause antagonistic effects on each other. For example, AIP-I is known to promote Agr-I, but antagonize Agr-II, -III, and -IV. When AIP concentration increases as a result of increased bacterial density, AgrC on the surface of the bacterial cell membrane is activated and auto-phosphorylated. This AgrC phosphorylates and activates AgrA. Phosphorylated AgrA induces the transcription of RNAIII, which itself encodes a protein as δ-hemolysin [168, 169], and acts as a functional RNA and a regulator of translation in the cell. In general, during the logarithmic growth phase,

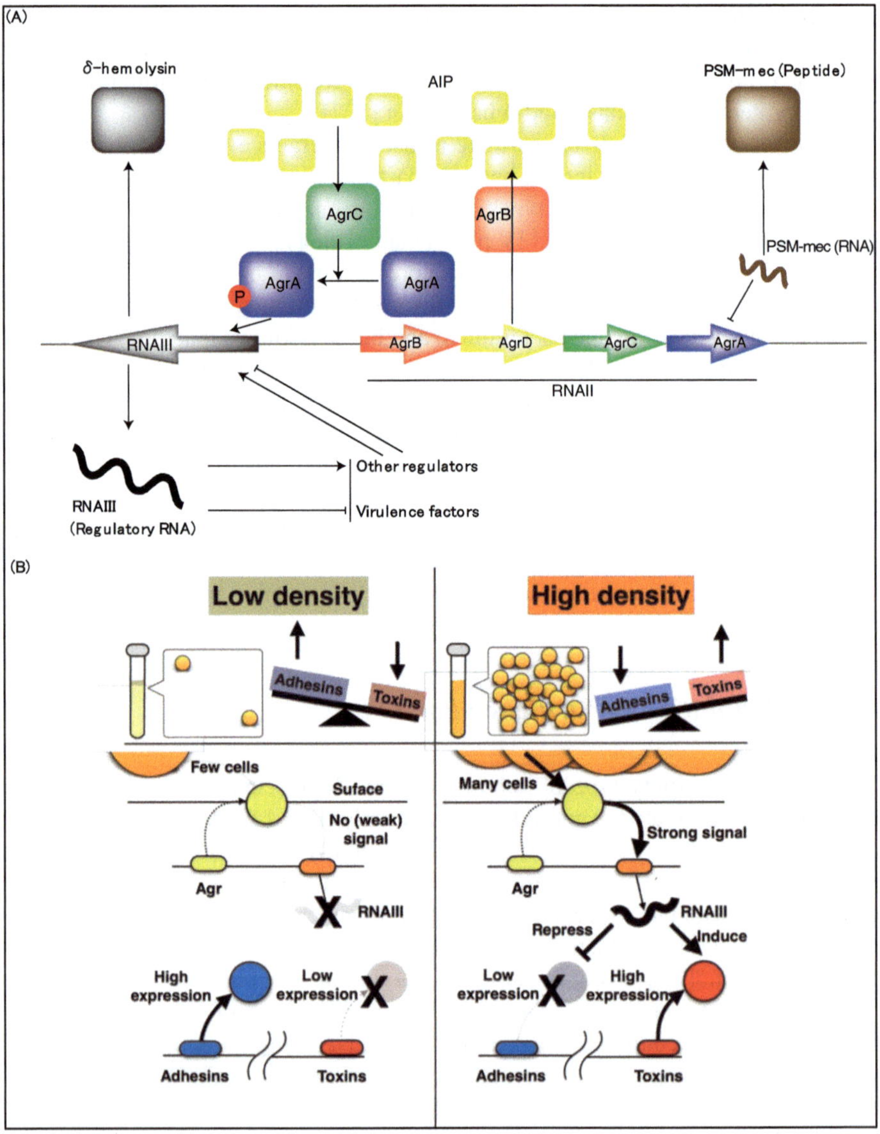

Fig. 4.4 Gene regulation by Agr in *S. aureus*. (**a**) AIP-mediated quorum sensing. AgrD is matured into AIP by AgrB. This AIP activates AgrC, which in turn phosphorylates AgrA. This triggers the transcription of RNAIII, which in turn regulates the translation of other genes. Multiple factors such as other regulatory factors and virulence factors are affected as its targets. (**b**) Genes are switched on and off depending on bacterial density. When the bacterial density is low, RNAIII does not work, but when the bacterial density increases, RNAIII becomes active, and the expression of various genes is switched on and off. (This figure is an English translation with some modifications from a previous article by the author. Ref. [128])

when the bacterial density is low and growth is active, the Agr switch is turned off and the expression of adhesion factors is high. In the stationary phase, when the bacterial density is high and growth has plateaued, the Agr switch is turned on and toxin production increases. In addition, at least 16 other two-component regulatory factors have been reported in *S. aureus* [170]. The SaeRS system, which stands for *S. aureus* exoprotein expression, was discovered as a regulator of secretory proteins. SaeRS has been found to promote the expression of various toxin genes and coagulase [171–174]. SrrAB, which stands for staphylococcal respiratory regulator, is a regulator that responds to oxygen concentration. It is involved in the repression of toxin and surface protein expression, especially under hypoxic conditions [175, 176]. Strains lacking SrrAB and nitric oxide synthase (NOS) show reduced viability in vivo, suggesting that it contributes to the survival of the bacterium [177]. ArlRS stands for autolysis regulated locus, a gene regulator that is involved not only in autolysis but also production of toxin and Protein A [178, 179].

In *S. aureus*, the SarA (staphylococcal accessory regulator A) family is a large regulator clade as well as TCSs [180]. This family consists of SarA, SarS, MgrA, Rot, and others. Among these, SarA is the most representative and has been studied well. SarA is a protein that binds to the specific sites (26 bp and 7 bp sequence) present in the transcription promoters of various mRNA and small RNA [181]. The protein is a potent negative regulator of proteases [182]. Structural changes in DNA have been proposed as functions of the molecular mechanisms controlled by SarA and the other SarA family proteins [183]. However, even for the same virulence factor, its control is sometimes the opposite, and this is because SarA acts not only on virulence factors but also on other regulatory factors such as Agr [183]. Moreover, since there are many SarA binding sites in the genome, the effect of SarA is reflected in the phenotype as a butterfly effect. SarS is a member of the SarA family, which is repressed by SarA and Agr [184, 185]; it promotes the expression of Spa in the neighborhood of SarS [18], but represses the expression of SEH and Hla [174, 185]. Rot stands for Repressor of toxin. This factor represses the production of toxins such as α-hemplysin and SEB and enhances Spa production [185, 186]. The mode of repression may be direct binding to the promoter of the toxin gene or may be mediated by other regulatory factors. However, our previous studies have reported that Rot plays an important role in food poisoning outbreaks by binding to the promoter of SEH and facilitating its expression by allowing access to the RNA promoter [174]. Therefore, it is not a repressor of all toxins.

The Agr system described above is a two-component regulatory system, and RNAIII is a dual function RNA [169]. In other words, the RNA regulates translation as a functional RNA and encodes δ-hemolysin as an mRNA. A similar example is PSM-mec it exhibits dual properties as both a regulator and a toxin. As mentioned above, PSM-mec exhibits a variety of toxic effects [149]. Approximately 150 bp of this small regulator RNA inhibits the quorum sensing by repressing AgrA translation, resulting in decreased virulence. This is one explanation for the higher virulence of CA-MRSA compared to HA-MRSA [187].

4.3.6 Vaccines

Vaccines are the mainstay in the prevention of infectious diseases. Especially in the absence of effective therapeutic agents, vaccines play an important role in preventing serious cases in patients and controlling the spread of the disease in the population. Currently, most vaccines in use target viruses. For example, as of July 2021, there are 35 types of monovalent and multivalent human vaccines produced in Japan, covering a total of 21 pathogens. Of these, 14 are viruses and 3 are toxoids [188]. On the other hand, the number of successful vaccines against bacteria is limited, and a vaccine against *S. aureus* remains a great challenge. However, successful vaccines against bacteria are limited, and developing a vaccine against *S. aureus* remains challenging due to the prevalence of drug-resistant and highly pathogenic strains. Successful bacterial vaccines often target clear virulence factors. For toxigenic diseases such as diphtheria and tetanus, the acquisition of antibodies by the toxoid has been very effective in preventing the diseases [189, 190]. The capsules of *Streptococcus pneumoniae* and *Haemophilus influenzae* play a major role in the development of infection, and vaccines designed to stimulate immunity against them are effective in preventing such diseases [191, 192]. For those pathogenic bacteria, there is a clear mechanism between the virulence factor and the establishment of disease. However, in the case of *S. aureus*, it is difficult to show a clear relationship between bacterial factors and diseases. There are some cases where the disease and the causative toxin are clear, such as TSS and superantigen, ET and staphylococcal scalded skin syndrome (SSSS), SEs and food poisoning. However, in *S. aureus* populations, SEs and TSST-1 are common toxins, and many bacteria carry one or more. Currently, there are more than 20 types of toxins in these families, showing different antigenicity [55]. These facts make it difficult to target these virulence factors for toxoid vaccines. All *S. aureus* strains are believed to cause symptoms to various degrees. It is also believed that many factors are involved in this mechanism. Attachment factors on the surface of the bacteria recognize ECM components such as collagen and fibrinogen. In addition, protein A, another cell wall protein, binds to the Fc region of antibodies to evade immunity. In addition, the iron uptake mechanism present on the cell surface allows the bacteria to survive in vivo. Looking at secreted proteins, coagulase and staphylokinase facilitate the formation of localized infection sites, immune evasion, and subsequent spreading. The same is true for various toxins, which contribute to the destruction of host cells by cytotoxins and immune disruption by superantigens. Thus, the pathogenesis of this organism is complicated by the fact that multiple factors are involved in the pathogenesis and a single factor has multiple effects. In addition, although mice are widely used for research on virulence factors and vaccines, differences between humans and mice have been reported, and animal research cannot be simply extrapolated to humans [193–195]. So far, whole bacterial cell, toxoid, polysaccharide, adhesion and other factors have been investigated, but without success [196, 197]. Vaccine development by researchers is expected to continue in the future, but the results of pathological research will be indispensable for this.

4.4 Conclusion

Among bacterial pathogens, *S. aureus* has attracted the attention of many researchers. This is not unrelated to the fact that though the bacterium was discovered more than 100 years ago, the actual nature of its infections is still unclear. When the full extent of the disease is understood with the enthusiasm of these researchers, we should be able to obtain the results that society has been waiting for, such as vaccines and therapeutic agents (including treatment methods such as guidelines).

References

1. Lowy FD (1998) *Staphylococcus aureus* infections. N Engl J Med 339:520–532. https://doi.org/10.1056/NEJM199808203390806
2. Leonard FC, Markey BK (2008) Meticillin-resistant *Staphylococcus aureus* in animals: a review. Vet J 75:27–36. Epub 2007 Jan 9. https://doi.org/10.1016/j.tvjl.2006.11.008
3. Tong SY, Davis JS, Eichenberger E, Holland TL, Fowler VG Jr (2015) *Staphylococcus aureus* infections: epidemiology, pathophysiology, clinical manifestations, and management. Clin Microbiol Rev 28:603–661. PMID: 26016486; PMCID: PMC4451395. https://doi.org/10.1128/CMR.00134-14
4. Hennekinne JA, De Buyser ML, Dragacci S (2012) *Staphylococcus aureus* and its food poisoning toxins: characterization and outbreak investigation. FEMS Microbiol Rev 36:815–836. Epub 2011 Nov 8. https://doi.org/10.1111/j.1574-6976.2011.00311.x
5. Barkema HW, Schukken YH, Zadoks RN (2006) Invited review: the role of cow, pathogen, and treatment regimen in the therapeutic success of bovine *Staphylococcus aureus* mastitis. J Dairy Sci 89:1877–1895. https://doi.org/10.3168/jds.S0022-0302(06)72256-1
6. Heikkilä AM, Liski E, Pyörälä S, Taponen S (2018) Pathogen-specific production losses in bovine mastitis. J Dairy Sci 101:9493–9504. Epub 2018 Aug 16. https://doi.org/10.3168/jds.2018-14824
7. Mulcahy ME, McLoughlin RM (2016) Host-bacterial crosstalk determines *Staphylococcus aureus* nasal colonization. Trends Microbiol 24:872–886. https://doi.org/10.1016/j.tim.2016.06.012
8. Levy PY, Ollivier M, Drancourt M, Raoult D, Argenson JN (2013) Relation between nasal carriage of *Staphylococcus aureus* and surgical site infection in orthopedic surgery: the role of nasal contamination. A systematic literature review and meta-analysis. Orthop Traumatol Surg Res 99:645–651. https://doi.org/10.1016/j.otsr.2013.03.030
9. Luzar MA, Coles GA, Faller B, Slingeneyer A, Dah GD, Briat C et al (1990) *Staphylococcus aureus* nasal carriage and infection in patients on continuous ambulatory peritoneal dialysis. N Engl J Med 32:505–509. https://doi.org/10.1056/NEJM199002223220804
10. Perl TM, Cullen JJ, Wenzel RP, Zimmerman MB, Pfaller MA, Sheppard D et al (2002) Intranasal mupirocin to prevent postoperative *Staphylococcus aureus* infections. N Engl J Med 346:1871–1877. https://doi.org/10.1056/NEJMoa003069
11. Wertheim HF, Melles DC, Vos MC, van Leeuwen W, van Belkum A, Verbrugh HA et al (2005) The role of nasal carriage in *Staphylococcus aureus* infections. Lancet Infect Dis 5:751–762. https://doi.org/10.1016/S1473-3099(05)70295-4
12. Zimakoff J, Bangsgaard Pedersen F, Bergen L, Baagø J, Daldorph B et al (1996) *Staphylococcus au*reus carriage and infections among patients in four haemo- and peritoneal-dialysis centres in Denmark. The Danish study Group of Peritonitis in dialysis (DASPID). J Hosp Infect 33:289–300. https://doi.org/10.1016/s0195-6701(96)90015-8

13. Ogura M, Ohkusu K, Ezaki T, Yano H, Sato M, Nakamura A et al (2013) Comparative analysis of MRSA strains isolated from cases of mupirocin ointment treatment in which eradication was successful and in which eradication failed. J Infect Chemother 19:196–201. https://doi.org/10.1007/s10156-012-0445-0
14. MRSA kansensyo no gideline kaiteiban (2019) Nihon Kagakuryouhou Gakkai (Japanese)
15. Septimus EJ (2019) Nasal decolonization: what antimicrobials are most effective prior to surgery? Am J Infect Control 47S:A53–A57. https://doi.org/10.1016/j.ajic.2019.02.028
16. Brady RR, McDermott C, Graham C, Harrison EM, Eunson G, Fraise AP et al (2009) A prevalence screen of MRSA nasal colonisation amongst UK doctors in a non-clinical environment. Eur J Clin Microbiol Infect Dis 28:991–995. https://doi.org/10.1007/s10096-009-0718-4
17. Emaneini M, Jabalameli F, Rahdar H, Leeuwen WBV, Beigverdi R (2017) Nasal carriage rate of methicillin resistant *Staphylococcus aureus* among Iranian healthcare workers: a systematic review and meta-analysis. Rev Soc Bras Med Trop 50:590–597. https://doi.org/10.1590/0037-8682-0534-2016
18. Hawkins G, Stewart S, Blatchford O, Reilly J (2011) Should healthcare workers be screened routinely for meticillin-resistant *Staphylococcus aureus*? A review of the evidence. J Hosp Infect 77:285–289. https://doi.org/10.1016/j.jhin.2010.09.038
19. Morris DO, Rook KA, Shofer FS, Rankin SC (2006) Screening of *Staphylococcus aureus*, *Staphylococcus intermedius*, and *Staphylococcus schleiferi* isolates obtained from small companion animals for antimicrobial resistance: a retrospective review of 749 isolates (2003-04). Vet Dermatol 17:332–337. https://doi.org/10.1111/j.1365-3164.2006.00536.x
20. Gómez P, Ruiz-Ripa L, Fernández-Fernández R, Gharsa H, Slama KB, Höfle U et al (2021) Genomic analysis of *Staphylococcus aureus* of the lineage CC130, including *mecC*-carrying MRSA and MSSA isolates recovered of animal, human, and environmental origins. Front Microbiol 12:655994. https://doi.org/10.3389/fmicb.2021.655994
21. Smith TC, Wardyn SE (2015) Human infections with *Staphylococcus aureus* CC398. Curr Environ Health Rep 2:41–51. https://doi.org/10.1007/s40572-014-0034-8
22. Adams V, Li J, Wisniewski JA, Uzal FA, Moore RJ, McClane BA et al (2014) Virulence plasmids of spore-forming bacteria. Microbiol Spectr 2:2.6.04. https://doi.org/10.1128/microbiolspec.PLAS-0024-2014
23. Pilo P, Frey J (2018) Pathogenicity, population genetics and dissemination of *Bacillus anthracis*. Infect Genet Evol 64:115–125. https://doi.org/10.1016/j.meegid.2018.06.024
24. Arístegui J, Dal-Ré R, Díez-Delgado J, Marés J, Casanovas JM, García-Corbeira P et al (2003) Comparison of the reactogenicity and immunogenicity of a combined diphtheria, tetanus, acellular pertussis, hepatitis B, inactivated polio (DTPa-HBV-IPV) vaccine, mixed with the *Haemophilus influenzae* type b (Hib) conjugate vaccine and administered as a single injection, with the DTPa-IPV/Hib and hepatitis B vaccines administered in two simultaneous injections to infants at 2, 4 and 6 months of age. Vaccine 21:3593–3600. https://doi.org/10.1016/s0264-410x(03)00420-1
25. Bar-On ES, Goldberg E, Hellmann S, Leibovici L (2012) Combined DTP-HBV-HIB vaccine versus separately administered DTP-HBV and HIB vaccines for primary prevention of diphtheria, tetanus, pertussis, hepatitis B and *Haemophilus influenzae* B (HIB). Cochrane Database Syst Rev (4):CD005530. https://doi.org/10.1002/14651858.CD005530.pub3
26. Fisher K, Phillips C (2009) The ecology, epidemiology and virulence of *Enterococcus*. Microbiology 155:1749–1757. https://doi.org/10.1099/mic.0.026385-0
27. Hogan LH, Klein BS, Levitz SM (1996) Virulence factors of medically important fungi. Clin Microbiol Rev 9:469–488. https://doi.org/10.1128/CMR.9.4.469
28. Balzan S, de Almeida QC, de Cleva R, Zilberstein B, Cecconello I (2007) Bacterial translocation: overview of mechanisms and clinical impact. J Gastroenterol Hepatol 22:464–471. https://doi.org/10.1111/j.1440-1746.2007.04933.x
29. Ki V, Rotstein C (2008) Bacterial skin and soft tissue infections in adults: a review of their epidemiology, pathogenesis, diagnosis, treatment and site of care. Can J Infect Dis Med Microbiol 19:173–184. https://doi.org/10.1155/2008/846453

30. Grumann D, Nübel U, Bröker BM (2014) *Staphylococcus aureus* toxins—their functions and genetics. Infect Genet Evol 21:583–592. https://doi.org/10.1016/j.meegid.2013.03.013
31. Lindsay JA, Holden MT (2004) *Staphylococcus aureus*: superbug, super genome? Trends Microbiol 12:378–385. https://doi.org/10.1016/j.tim.2004.06.004
32. Everitt RG, Didelot X, Batty EM, Miller RR, Knox K, Young BC (2014) Mobile elements drive recombination hotspots in the core genome of *staphylococcus aureus*. Nat Commun 5:3956. https://doi.org/10.1038/ncomms4956
33. Jamrozy D, Coll F, Mather AE, Harris SR, Harrison EM, MacGowan A et al (2017) Evolution of mobile genetic element composition in an epidemic methicillin-resistant *Staphylococcus aureus*: temporal changes correlated with frequent loss and gain events. BMC Genomics 18:684. https://doi.org/10.1186/s12864-017-4065-z
34. Rietschel ET, Kirikae T, Schade U, Mamat U, Schmidt G, Loppnow H et al (1994) Bacterial endotoxin: molecular relationships of structure to activity and function. FASEB J 8:217–225. https://doi.org/10.1096/fasebj.8.2.8119492
35. Dinges MM, Orwin PM, Schlievert PM (2000) Exotoxins of *Staphylococcus aureus*. Clin Microbiol Rev 13:16–34. https://doi.org/10.1128/CMR.13.1.16
36. Tam K, Torres VJ (2019) *Staphylococcus aureus* secreted toxins and extracellular enzymes. Microbiol Spectr 7:7.2.16. https://doi.org/10.1128/microbiolspec.GPP3-0039-2018
37. Foster TJ (2005) Immune evasion by staphylococci. Nat Rev Microbiol 3:948–958. https://doi.org/10.1038/nrmicro1289
38. Gilbert RJC (2002) Pore-forming toxins. Cell Mol Life Sci 59:832–844. https://doi.org/10.1007/s00018-002-8471-1
39. Bhakdi S, Tranum-Jensen J (1991) Alpha-toxin of *Staphylococcus aureus*. Microbiol Rev 55:733–751. https://doi.org/10.1128/mr.55.4.733-751.1991
40. Doery HM, Magnusson BJ, Cheyne IM, Gulasekharam J (1963) A phospholipase in staphylococcal toxin which hydrolyses sphingomyelin. Nature 198:1091–1092. https://doi.org/10.1038/1981091a0
41. Berube BJ, Wardenburg JB (2013) *Staphylococcus aureus* α-toxin: nearly a century of intrigue. Toxins 5:1140–1166. https://doi.org/10.3390/toxins5061140
42. Vandenesch F, Lina G, Henry T (2012) *Staphylococcus aureus* hemolysins, bi-component leukocidins, and cytolytic peptides: a redundant arsenal of membrane-damaging virulence factors? Front Cell Infect Microbiol 2:12. https://doi.org/10.3389/fcimb.2012.00012
43. Labandeira-Rey M, Couzon F, Boisset S, Brown EL, Bes M, Benito Y et al (2007) *Staphylococcus aureus* Panton-Valentine leukocidin causes necrotizing pneumonia. Science 315:1130–1133. https://doi.org/10.1126/science.1137165
44. Morinaga N, Kaihou Y, Noda M (2003) Purification, cloning and characterization of variant LukE-LukD with strong leukocidal activity of staphylococcal bi-component leukotoxin family. Microbiol Immunol 47:81–90. https://doi.org/10.1111/j.1348-0421.2003.tb02789.x
45. Tromp AT, Van Gent M, Abrial P, Martin A, Jansen JP, De Haas CJ et al (2018) Human CD45 is an F-component-specific receptor for the staphylococcal toxin Panton–Valentine leukocidin. Nat Microbiol 3:708–717. https://doi.org/10.1038/s41564-018-0247-y
46. Baba T, Takeuchi F, Kuroda M, Yuzawa H, Aoki KI, Oguchi A et al (2002) Genome and virulence determinants of high virulence community-acquired MRSA. Lancet 359:1819–1827. https://doi.org/10.1016/S0140-6736(02)08713-5
47. Diep BA, Gill SR, Chang RF, Phan TH, Chen JH, Davidson MG et al (2006) Complete genome sequence of USA300, an epidemic clone of community-acquired meticillin-resistant *Staphylococcus aureus*. Lancet 367:731–739. https://doi.org/10.1016/S0140-6736(06)68231-7
48. Kobayashi SD, Melachowa N, Whitney AK, Braughton KR, Gardner DJ, Long D et al (2011) Comparative analysis of USA300 virulence determinants in a rabbit model of skin and soft tissue infection. J Infect Dis 204:937–941. https://doi.org/10.1093/infdis/jir441
49. Wooldridge KG, Williams PH (1993) Iron uptake mechanisms of pathogenic bacteria. FEMS Microbiol Rev 12:325–348. https://doi.org/10.1111/j.1574-6976.1993.tb00026.x

50. Weinberg ED (2009) Iron availability and infection. Biochim Biophys Acta 1790:600–605. https://doi.org/10.1016/j.bbagen.2008.07.002
51. Peschel A, Otto M (2013) Phenol-soluble modulins and staphylococcal infection. Nat Rev Microbiol 11:667–673. https://doi.org/10.1038/nrmicro3110
52. Qin L, McCausland JW, Cheung GY, Otto M (2016) PSM-mec-a virulence determinant that connects transcriptional regulation, virulence, and antibiotic resistance in staphylococci. Front Microbiol 7:1293. https://doi.org/10.3389/fmicb.2016.01293
53. Qin L, Da F, Fisher EL, Tan DC, Nguyen TH, Fu CL et al (2017) Toxin mediates sepsis caused by methicillin-resistant *Staphylococcus epidermidis*. PLoS Pathog 13:e1006153. https://doi.org/10.1371/journal.ppat.1006153
54. Otto M, O'Mahoney DS, Guina T, Klebanoff SJ (2004) Activity of *Staphylococcus epidermidis* phenol-soluble modulin peptides expressed in *Staphylococcus carnosus*. J Infect Dis 190:748–755. https://doi.org/10.1086/422157
55. Hu DL, Nakane A (2014) Mechanisms of staphylococcal enterotoxin-induced emesis. Eur J Pharmacol 722:95–107. https://doi.org/10.1016/j.ejphar.2013.08.050
56. Spaulding AR, Salgado-Pabón W, Kohler PL, Horswill AR, Leung DY, Schlievert PM (2013) Staphylococcal and streptococcal superantigen exotoxins. Clin Microbiol Rev 26:422–447. https://doi.org/10.1128/CMR.00104-12
57. Hirose S, Ono HK, Omoe K, Hu D-L, Asano K, Yamamoto Y et al (2016) Goblet cells are involved in translocation of staphylococcal enterotoxin a in the intestinal tissue of house musk shrew (*Suncus murinus*). J Appl Microbiol 120:781–789. https://doi.org/10.1111/jam.13029
58. Ono HK, Hirose S, Narita K, Sugiyama M, Asano K, Hu D-L et al (2019) Histamine release from intestinal mast cells induced by staphylococcal enterotoxin a (SEA) evokes vomiting reflex in common marmoset. PLoS Pathog 15:e1007803. https://doi.org/10.1371/journal.ppat.1007803
59. Ono HK, Nishizawa M, Yamamoto Y, Hu D-L, Nakane A, Shinagawa K et al (2012) Submucosal mast cells in the gastrointestinal tract are a target of staphylococcal enterotoxin type a. FEMS Immunol Med Microbiol 64:392–402. https://doi.org/10.1111/j.1574-695X.2011.00924.x
60. Kappler J, Kotzin B, Herron L, Gelfand EW, Bigler RD, Boylston A et al (1989) Vβ-specific stimulation of human T cells by staphylococcal toxins. Science 244:811–813. https://doi.org/10.1126/science.2524876
61. White J, Herman A, Pullen AM, Kubo R, Kappler JW, Marrack P (1989) The Vβ-specific superantigen staphylococcal enterotoxin B: stimulation of mature T cells and clonal deletion in neonatal mice. Cell 56:27–35. https://doi.org/10.1016/0092-8674(89)90980-x
62. Bergdoll MS, Schlievert PM (1984) Toxic-shock syndrome toxin. Lancet 324:691
63. Ahanotu E, Alvelo-Ceron D, Ravita T, Gaunt E (2006) Staphylococcal enterotoxin B as a biological weapon: recognition, management, and surveillance of Staphylococcal enterotoxin. Appl Biosaf 11:120–126. https://doi.org/10.1177/153567600601100303
64. Rieder SA, Nagarkatti P, Nagarkatti M (2011) CD1d-independent activation of invariant natural killer T cells by Staphylococcal enterotoxin B through major histocompatibility complex class II/T cell receptor interaction results in acute lung injury. Infect Immun 79:3141–3148. https://doi.org/10.1128/IAI.00177-11
65. Liu Y, Chen W, Ali T, Alkasir R, Yin G, Liu G et al (2014) Staphylococcal enterotoxin H induced apoptosis of bovine mammary epithelial cells in vitro. Toxins (Basel) 6:3552–3567. https://doi.org/10.3390/toxins6123552
66. de Haas CJ, Veldkamp KE, Peschel A, Weerkamp F, van Wamel WJB, Heezius ECJM et al (2004) Chemotaxis inhibitory protein of *Staphylococcus aureus*, a bacterial antiinflammatory agent. J Exp Med 199:687–695. https://doi.org/10.1084/jem.20031636
67. Mehlin C, Headley CM, Klebanoff SJ (1999) An inflammatory polypeptide complex from *Staphylococcus epidermidis*: isolation and characterization. J Exp Med 189:907–918. https://doi.org/10.1084/jem.189.6.907
68. Armbruster NS, Richardson JR, Schreiner J, Klenk J, Günter M, Autenrieth SE (2016) *Staphylococcus aureus* PSM peptides induce tolerogenic dendritic cells upon treatment with

ligands of extracellular and intracellular TLRs. Int J Med Microbiol 306:666–674. https://doi.org/10.1016/j.ijmm.2016.09.002
69. Syed AK, Reed TJ, Clark KL, Boles BR, Kahlenberg JM (2015) *Staphylococcus aureus* phenol-soluble modulins stimulate the release of proinflammatory cytokines from keratinocytes and are required for induction of skin inflammation [published correction appears in infect Immun. 2015;83:4450]. Infect Immun 83:3428–3437. https://doi.org/10.1128/IAI.00401-15
70. Kobayashi SD, Voyich JM, Burlak C, DeLeo FR (2005) Neutrophils in the innate immune response. Arch Immunol Ther Exp (Warsz) 53:505–517
71. Kaito C, Sekimizu K (2007) Colony spreading in *Staphylococcus aureus*. J Bacteriol 189:2553–2557. https://doi.org/10.1128/JB.01635-06
72. Kizaki H, Omae Y, Tabuchi F, Saito Y, Sekimizu K, Kaito C (2016) Cell-surface phenol soluble modulins regulate *Staphylococcus aureus* colony spreading. PLoS One 11:e0164523. https://doi.org/10.1371/journal.pone.0164523
73. Tsompanidou E, Denham EL, Becher D, de Jong BG, van Oosten M et al (2013) Distinct roles of phenol-soluble modulins in spreading of *Staphylococcus aureus* on wet surfaces. Appl Environ Microbiol 7:886–895. https://doi.org/10.1128/AEM.03157-12
74. Kaito C, Omae Y, Matsumoto Y, Nagata M, Yamaguchi H, Aoto T et al (2008) A novel gene, *fudoh*, in the SCC*mec* region suppresses the colony spreading ability and virulence of *Staphylococcus aureus*. PLoS One 3:e3921. https://doi.org/10.1371/journal.pone.0003921
75. Björnsdottir H, Rudin DA, Klose FP, Elmwall J, Welin A, Stylianou M et al (2017) Phenol-soluble modulin α peptide toxins from aggressive *Staphylococcus aureus* induce rapid formation of neutrophil extracellular traps through a reactive oxygen species-independent pathway. Front Immunol 8:257. https://doi.org/10.3389/fimmu.2017.00257
76. Zaman M, Andreasen M (2020) Cross-talk between individual phenol-soluble modulins in *Staphylococcus aureus* biofilm enables rapid and efficient amyloid formation. Elife 9:e59776. https://doi.org/10.7554/eLife.59776
77. Rasigade JP, Trouillet-Assant S, Ferry T, Diep BA, Sapin A, Lhoste Y et al (2013) PSMs of hypervirulent *Staphylococcus aureus* act as intracellular toxins that kill infected osteoblasts. PLoS One 8:e63176. https://doi.org/10.1371/journal.pone.0063176
78. Wang R, Braughton KR, Kretschmer D, Bach T-HL, Queck SY, Li M et al (2007) Identification of novel cytolytic peptides as key virulence determinants for community-associated MRSA. Nat Med 13:1510–1514. https://doi.org/10.1038/nm1656
79. Nakamura Y, Oscherwitz J, Cease KB, Chan SM, Muñoz-Planillo R, Hasegawa M et al (2013) *Staphylococcus* δ-toxin induces allergic skin disease by activating mast cells. Nature 503:397–401. https://doi.org/10.1038/nature12655
80. Weidinger S, Novak N (2016) Atopic dermatitis. Lancet 387:1109–1122. https://doi.org/10.1016/S0140-6736(15)00149-X
81. Amagai M, Matsuyoshi N, Wang ZH, Andl C, Stanley JR (2000) Toxin in bullous impetigo and staphylococcal scalded-skin syndrome targets desmoglein 1. Nat Med 6:1275–1277. https://doi.org/10.1038/81385
82. Amagai M, Yamaguchi T, Hanakawa Y, Nishifuji K, Sugai M, Stanley JR (2002) Staphylococcal exfoliative toxin B specifically cleaves desmoglein 1. J Invest Dermatol 118:845–850. https://doi.org/10.1046/j.1523-1747.2002.01751.x
83. Yamaguchi T, Nishifuji K, Sasaki M, Fudaba Y, Aepfelbacher M, Takata T et al (2002) Identification of the *Staphylococcus aureus etd* pathogenicity island which encodes a novel exfoliative toxin, ETD, and EDIN-B. Infect Immun 70:5835–5845. https://doi.org/10.1128/IAI.70.10.5835-5845.2002
84. Amagai M, Karpati S, Prussick R, Klaus-Kovtun V, Stanley JR (1992) Autoantibodies against the amino-terminal cadherin-like binding domain of pemphigus vulgaris antigen are pathogenic. J Clin Invest 90:919–926. https://doi.org/10.1172/JCI115968

85. Peetermans M, Verhamme P, Vanassche T (2015) Coagulase activity by *Staphylococcus aureus*: a potential target for therapy? Semin Thromb Hemost 41:433–444. https://doi.org/10.1055/s-0035-1549849
86. Bokarewa MI, Jin T, Tarkowski A (2006) *Staphylococcus aureus*: Staphylokinase. Int J Biochem Cell Biol 38:504–509. https://doi.org/10.1016/j.biocel.2005.07.005
87. Rooijakkers SH, van Wamel WJ, Ruyken M, van Kessel KP, van Strijp JA (2005) Anti-opsonic properties of staphylokinase. Microbes Infect 7:476–484. https://doi.org/10.1016/j.micinf.2004.12.014
88. Fronzes R, Remaut H, Waksman G (2008) Architectures and biogenesis of non-flagellar protein appendages in gram-negative bacteria. EMBO J 27:2271–2280. https://doi.org/10.1038/emboj.2008.155
89. Bazaka K, Crawford RJ, Nazarenko EL, Ivanova EP (2011) Bacterial extracellular polysaccharides. Adv Exp Med Biol 715:213–226. https://doi.org/10.1007/978-94-007-0940-9_13
90. Klemm P, Vejborg RM, Hancock V (2010) Prevention of bacterial adhesion. Appl Microbiol Biotechnol 88:451–459. https://doi.org/10.1007/s00253-010-2805-y
91. Foster TJ, Geoghegan JA, Ganesh VK, Höök M (2014) Adhesion, invasion and evasion: the many functions of the surface proteins of *Staphylococcus aureus*. Nat Rev Microbiol 12:49–62. https://doi.org/10.1038/nrmicro3161
92. Sinha B, François PP, Nüsse O, Foti M, Hartford OM, Vaudaw P et al (1999) Fibronectin-binding protein acts as *Staphylococcus aureus* invasin via fibronectin bridging to integrin alpha5beta1. Cell Microbiol 1:101–117. https://doi.org/10.1046/j.1462-5822.1999.00011.x
93. Askarian F, Ajayi C, Hanssen AM, van Sorge NM, Petterson I, Diep DB et al (2016) The interaction between *Staphylococcus aureus* SdrD and desmoglein 1 is important for adhesion to host cells. Sci Rep 6:22134. https://doi.org/10.1038/srep22134
94. McDevitt D, Francois P, Vaudaux P, Foster TJ (1994) Molecular characterization of the clumping factor (fibrinogen receptor) of *Staphylococcus aureus*. Mol Microbiol 11:237–248. https://doi.org/10.1111/j.1365-2958.1994.tb00304.x
95. Jett BD, Gilmore MS (2002) Internalization of *Staphylococcus aureus* by human corneal epithelial cells: role of bacterial fibronectin-binding protein and host cell factors. Infect Immun 70:4697–4700. https://doi.org/10.1128/IAI.70.8.4697-4700.2002
96. Shinji H, Yosizawa Y, Tajima A, Iwase T, Sugimoto S, Seki K et al (2011) Role of fibronectin-binding proteins a and B in in vitro cellular infections and in vivo septic infections by *Staphylococcus aureus*. Infect Immun 79:2215–2223. https://doi.org/10.1128/IAI.00133-11
97. Brouillette E, Grondin G, Shkreta L, Lacasse P, Talbot BG (2003) In vivo and in vitro demonstration that *Staphylococcus aureus* is an intracellular pathogen in the presence or absence of fibronectin-binding proteins. Microb Pathog 35:159–168. https://doi.org/10.1016/s0882-4010(03)00112-8
98. Chavakis T, Wiechmann K, Preissner KT, Herrmann M (2005) *Staphylococcus aureus* interactions with the endothelium: the role of bacterial "secretable expanded repertoire adhesive molecules" (SERAM) in disturbing host defense systems. Thromb Haemost 94:278–285. https://doi.org/10.1160/TH05-05-0306
99. Hansen U, Hussain M, Villone D et al (2006) The anchorless adhesin Eap (extracellular adherence protein) from *Staphylococcus aureus* selectively recognizes extracellular matrix aggregates but binds promiscuously to monomeric matrix macromolecules. Matrix Biol 25(4):252–260. https://doi.org/10.1016/j.matbio.2006.01.005
100. Eisenbeis J, Saffarzadeh M, Peisker H, Herrmann M, Robenek H, Peters G et al (2018) The *Staphylococcus aureus* extracellular adherence protein Eap is a DNA binding protein capable of blocking neutrophil extracellular trap formation. Front Cell Infect Microbiol 8:235. https://doi.org/10.3389/fcimb.2018.00235
101. Bjarnsholt T, Alhede M, Alhede M, Eickhardt-Sørensen SR, Moser C, Kühl M et al (2013) The in vivo biofilm. Trends Microbiol 21:466–474. https://doi.org/10.1016/j.tim.2013.06.002
102. Mah TF, O'Toole GA (2001) Mechanisms of biofilm resistance to antimicrobial agents. Trends Microbiol 9:34–39. https://doi.org/10.1016/s0966-842x(00)01913-2

103. Vestby LK, Grønseth T, Simm R, Nesse LL (2020) Bacterial biofilm and its role in the pathogenesis of disease. Antibiotics (Basel) 9:59. https://doi.org/10.3390/antibiotics9020059
104. del Pozo JL, Patel R (2007) The challenge of treating biofilm-associated bacterial infections. Clin Pharmacol Ther 82:204–209. https://doi.org/10.1038/sj.clpt.6100247
105. Thurlow LR, Hanke ML, Fritz T, Angle A, Aldrich A, Williams SH et al (2011) *Staphylococcus aureus* biofilms prevent macrophage phagocytosis and attenuate inflammation in vivo. J Immunol 186:6585–6596. https://doi.org/10.4049/jimmunol.1002794
106. Elgharably H, Hussain ST, Shrestha NK, Blackstone EH, Pettersson GB (2016) Current hypotheses in cardiac surgery: biofilm in infective endocarditis. Semin Thorac Cardiovasc Surg 28:56–59. https://doi.org/10.1053/j.semtcvs.2015.12.005
107. Gross M, Cramton SE, Götz F, Peschel A (2001) Key role of teichoic acid net charge in *Staphylococcus aureus* colonization of artificial surfaces. Infect Immun 69:3423–3426. https://doi.org/10.1128/IAI.69.5.3423-3426.2001
108. Xia G, Kohler T, Peschel A (2010) The wall teichoic acid and lipoteichoic acid polymers of *Staphylococcus aureus*. Int J Med Microbiol 300:148–154. https://doi.org/10.1016/j.ijmm.2009.10.001
109. O'Neill E, Pozzi C, Houston P, Humphreys H, Robinson DA, Loughman A et al (2008) A novel *Staphylococcus aureus* biofilm phenotype mediated by the fibronectin-binding proteins, FnBPA and FnBPB. J Bacteriol 190:3835–3850. https://doi.org/10.1128/JB.00167-08
110. Barbu EM, Mackenzie C, Foster TJ, Höök M (2014) SdrC induces staphylococcal biofilm formation through a homophilic interaction. Mol Microbiol 94:172–185. https://doi.org/10.1111/mmi.12750
111. Mann EE, Rice KC, Boles BR, Endres JL, Ranijt D, Chandramohan L et al (2009) Modulation of eDNA release and degradation affects *Staphylococcus aureus* biofilm maturation. PLoS One 4:e5822. https://doi.org/10.1371/journal.pone.0005822
112. O'Gara JP (2007) *Ica* and beyond: biofilm mechanisms and regulation in *Staphylococcus epidermidis* and *Staphylococcus aureus*. FEMS Microbiol Lett 270:179–188. https://doi.org/10.1111/j.1574-6968.2007.00688.x
113. Izano EA, Amarante MA, Kher WB, Kaplan JB (2008) Differential roles of poly-N-acetylglucosamine surface polysaccharide and extracellular DNA in *Staphylococcus aureus* and *Staphylococcus epidermidis* biofilms. Appl Environ Microbiol 74:470–476. https://doi.org/10.1128/AEM.02073-07
114. Dakheel KH, Abdul Rahim R, Neela VK, Al-Obaidi JR, Hun TG, Yusoff K (2016) Methicillin-resistant *Staphylococcus aureus* biofilms and their influence on bacterial adhesion and cohesion. Biomed Res Int 2016:4708425–4708414. https://doi.org/10.1155/2016/4708425
115. Kavanaugh JS, Flack CE, Lister J, Ricker EB, Ibberson CB, Jenul C et al (2019) Identification of extracellular DNA-binding proteins in the biofilm matrix. MBio 10:e01137–e01119. https://doi.org/10.1128/mBio.01137-19
116. Huseby MJ, Kruse AC, Digre J, Ricker EB, Ibberson CB, Jenul C et al (2010) Beta toxin catalyzes formation of nucleoprotein matrix in staphylococcal biofilms. Proc Natl Acad Sci U S A 107:14407–14412. https://doi.org/10.1073/pnas.0911032107
117. Mackey-Lawrence NM, Potter DE, Cerca N, Jefferson KK (2009) *Staphylococcus aureus* immunodominant surface antigen B is a cell-surface associated nucleic acid binding protein. BMC Microbiol 9:61. https://doi.org/10.1186/1471-2180-9-61
118. Karygianni L, Ren Z, Koo H, Thurnheer T (2020) Biofilm matrixome: extracellular components in structured microbial communities. Trends Microbiol 28:668–681. https://doi.org/10.1016/j.tim.2020.03.016
119. Formosa-Dague C, Speziale P, Foster TJ, Geoghegan JA, Dufrêne YF (2016) Zinc-dependent mechanical properties of *Staphylococcus aureus* biofilm-forming surface protein SasG. Proc Natl Acad Sci U S A 113:410–415. https://doi.org/10.1073/pnas.1519265113
120. Gruszka DT, Wojdyla JA, Bingham RJ, Turkenburg JP, Manfiel IW, Steward A et al (2012) Staphylococcal biofilm-forming protein has a contiguous rod-like structure. Proc Natl Acad Sci U S A 109:E1011–E1018. https://doi.org/10.1073/pnas.1119456109

121. Taglialegna A, Navarro S, Ventura S, Garnett JA, Matthews S, Penades JR et al (2016) Staphylococcal bap proteins build amyloid scaffold biofilm matrices in response to environmental signals. PLoS Pathog 12:e1005711. https://doi.org/10.1371/journal.ppat.1005711
122. Boles BR, Horswill AR (2008) *Agr*-mediated dispersal of *Staphylococcus aureus* biofilms. PLoS Pathog 4:e1000052. https://doi.org/10.1371/journal.ppat.1000052
123. Park JH, Lee JH, Cho MH, Herzberg M, Lee J (2012) Acceleration of protease effect on *Staphylococcus aureus* biofilm dispersal. FEMS Microbiol Lett 335:31–38. https://doi.org/10.1111/j.1574-6968.2012.02635.x
124. Kiedrowski MR, Kavanaugh JS, Malone CL, Mootz JM, Voyich JM, Smeltzer MS et al (2011) Nuclease modulates biofilm formation in community-associated methicillin-resistant *Staphylococcus aureus*. PLoS One 6:e26714. https://doi.org/10.1371/journal.pone.0026714
125. Kiedrowski MR, Crosby HA, Hernandez FJ, Malone CL, McNamara JO 2nd, Horswill AR (2014) *Staphylococcus aureus* Nuc2 is a functional, surface-attached extracellular nuclease. PLoS One 9:e95574. https://doi.org/10.1371/journal.pone.0095574
126. Kuroda M, Ohta T, Uchiyama I, Baba T, Yuzawa H, Kobayashi I et al (2001) Whole genome sequencing of meticillin-resistant *Staphylococcus aureus*. Lancet 21(357):1225–1240. https://doi.org/10.1016/s0140-6736(00)04403-2
127. Lindsay JA (2014) *Staphylococcus aureus* genomics and the impact of horizontal gene transfer. Int J Med Microbiol 304:103–109. https://doi.org/10.1016/j.ijmm.2013.11.010
128. Sato'o Y, Sugai M (2018) Budou kyukin. Igaku to Yakugaku 75:753–768. (Japanese)
129. UC San Diego Health. Novel phage therapy saves patient with multidrug-resistant bacterial infection. https://today.ucsd.edu/story/novel_phage_therapy_saves_patient_with_multidrug_resistant_bacterial_infect
130. Faruque SM, Asadulghani, Alim AR, Albert MJ, Islam KM, Mekalanos JJ (1998) Induction of the lysogenic phage encoding cholera toxin in naturally occurring strains of toxigenic *Vibrio cholerae* O1 and O139. Infect Immun 66:3752–3757. https://doi.org/10.1128/IAI.66.8.3752-3757.1998
131. Gamage SD, Patton AK, Hanson JF, Weiss AA (2004) Diversity and host range of Shiga toxin-encoding phage. Infect Immun 72:7131–7139. https://doi.org/10.1128/IAI.72.12.7131-7139.2004
132. Sumby P, Waldor MK (2003) Transcription of the toxin genes present within the staphylococcal phage phiSa3ms is intimately linked with the phage's life cycle. J Bacteriol 185:6841–6851. https://doi.org/10.1128/JB.185.23.6841-6851.2003
133. Mason RE, Allen WE (1975) Characteristics of *Staphylococcus aureus* associated with lysogenic conversion to loss of beta-hemolysin production. Can J Microbiol 21:1113–1116. https://doi.org/10.1139/m75-161
134. Penadés JR, Chen J, Quiles-Puchalt N, Carpena N, Novick RP (2015) Bacteriophage-mediated spread of bacterial virulence genes. Curr Opin Microbiol 23:171–178. https://doi.org/10.1016/j.mib.2014.11.019
135. Göller PC, Elsener T, Lorgé D, Radulovic N, Bernardi V, Naumann A et al (2021) Multi-species host range of staphylococcal phages isolated from wastewater. Nat Commun 12:6965. https://doi.org/10.1038/s41467-021-27037-6
136. Alibayov B, Baba-Moussa L, Sina H, Zdeňková K, Demnerová K (2014) *Staphylococcus aureus* mobile genetic elements. Mol Biol Rep 41:5005–5018. https://doi.org/10.1007/s11033-014-3367-3
137. Malachowa N, DeLeo FR (2010) Mobile genetic elements of *Staphylococcus aureus*. Cell Mol Life Sci 67:3057–3071. https://doi.org/10.1007/s00018-010-0389-4
138. Novick RP, Subedi A (2007) The SaPIs: mobile pathogenicity islands of *Staphylococcus*. Chem Immunol Allergy 93:42–57. https://doi.org/10.1159/000100857
139. Penadés JR, Christie GE (2015) The phage-inducible chromosomal islands: a family of highly evolved molecular parasites. Annu Rev Virol 2:181–201. https://doi.org/10.1146/annurev-virology-031413-085446

140. Sato'o Y, Omoe K, Ono HK, Nakane A, Hu DL (2013) A novel comprehensive analysis method for *Staphylococcus aureus* pathogenicity islands. Microbiol Immunol 57:91–99. https://doi.org/10.1111/1348-0421.12007
141. del Solar G, Giraldo R, Ruiz-Echevarría MJ, Espinosa M, Díaz-Orejas R (1998) Replication and control of circular bacterial plasmids. Microbiol Mol Biol Rev 62:434–464. https://doi.org/10.1128/MMBR.62.2.434-464.1998
142. Robicsek A, Jacoby GA, Hooper DC (2006) The worldwide emergence of plasmid-mediated quinolone resistance. Lancet Infect Dis 6:629–640. https://doi.org/10.1016/S1473-3099(06)70599-0
143. Silver S, Misra TK (1988) Plasmid-mediated heavy metal resistances. Annu Rev Microbiol 42:717–743. https://doi.org/10.1146/annurev.mi.42.100188.003441
144. Scott JR (1984) Regulation of plasmid replication. Microbiol Rev 48:1–23. https://doi.org/10.1128/mr.48.1.1-23.1984
145. Pilla G, Tang CM (2018) Going around in circles: virulence plasmids in enteric pathogens. Nat Rev Microbiol 16:484–495. https://doi.org/10.1038/s41579-018-0031-2
146. Van Melderen L (2010) Toxin-antitoxin systems: why so many, what for? Curr Opin Microbiol 13:781–785. https://doi.org/10.1016/j.mib.2010.10.006
147. Novick RP (1987) Plasmid incompatibility. Microbiol Rev 51:381–395. https://doi.org/10.1128/mr.51.4.381-395.1987
148. Willetts N, Wilkins B (1984) Processing of plasmid DNA during bacterial conjugation. Microbiol Rev 48:24–41. https://doi.org/10.1128/mr.48.1.24-41.1984
149. Hiramatsu K, Ito T, Tsubakishita S, Sasaki T, Takeuchi F, Morimoto Y et al (2013) Genomic basis for methicillin resistance in *Staphylococcus aureus*. Infect Chemother 45:117–136. https://doi.org/10.3947/ic.2013.45.2.117
150. Zuo H, Uehara Y, Lu Y, Sasaki T, Hiramatsu K (2021) Genetic and phenotypic diversity of methicillin-resistant *Staphylococcus aureus* among Japanese inpatients in the early 1980s. Sci Rep 11:5447. https://doi.org/10.1038/s41598-021-84481-6
151. Turner NA, Sharma-Kuinkel BK, Maskarinec SA, Eichenberger EM, Shah PP, Carugati M et al (2019) Methicillin-resistant *Staphylococcus aureus*: an overview of basic and clinical research. Nat Rev Microbiol 17:203–218. https://doi.org/10.1038/s41579-018-0147-4
152. Bestebroer J, van Kessel KP, Azouagh H, Walenkamp AM, Boer IG, Romijn RA et al (2009) Staphylococcal SSL5 inhibits leukocyte activation by chemokines and anaphylatoxins. Blood 113:328–337. https://doi.org/10.1182/blood-2008-04-153882
153. Nguyen MT, Kraft B, Yu W, Demircioglu DD, Hertlein T, Burian M et al (2015) The νSaα specific lipoprotein like cluster (*lpl*) of *S. aureus* USA300 contributes to immune stimulation and invasion in human cells. PLoS Pathog 11:e1004984. Erratum in: PLoS Pathog. 2015;11(9):e1005189. https://doi.org/10.1371/journal.ppat.1004984
154. Moon BY, Park JY, Hwang SY, Robinson DA, Thomas JC, Fitzgerald JR et al (2015) Phage-mediated horizontal transfer of a *Staphylococcus aureus* virulence-associated genomic Island. Sci Rep 5:9784. https://doi.org/10.1038/srep09784
155. Gill SR, Fouts DE, Archer CL, Mongodin EF, DeBoy RT, Ravel J et al (2005) Insights on evolution of virulence and resistance from the complete genome analysis of an early methicillin-resistant *Staphylococcus aureus* strain and a biofilm-producing methicillin-resistant *Staphylococcus epidermidis* strain. J Bacteriol 187:2426–2438. https://doi.org/10.1128/jb.187.7.2426-2438.2005
156. Otto M (2013) How colonization factors are linked to outbreaks of methicillin-resistant *Staphylococcus aureus*: the roles of SasX and ACME. Biomol Concepts 4:533–537. https://doi.org/10.1515/bmc-2013-0025
157. Corvaglia AR, François P, Hernandez D, Perron K, Linder P, Schrenzel J (2010) A type III-like restriction endonuclease functions as a major barrier to horizontal gene transfer in clinical *Staphylococcus* aureus strains. Proc Natl Acad Sci U S A 107:11954–11958. https://doi.org/10.1073/pnas.1000489107

158. Cooper LP, Roberts GA, White JH, Luyten YA, Bower EKM, Morgan RD et al (2017) DNA target recognition domains in the type I restriction and modification systems of *Staphylococcus aureus*. Nucleic Acids Res 45:3395–3406. https://doi.org/10.1093/nar/gkx067
159. Xu SY, Corvaglia AR, Chan SH, Zheng Y, Linder P (2011) A type IV modification-dependent restriction enzyme SauUSI from *Staphylococcus aureus* subsp. *aureus* USA300. Nucleic Acids Res 39:5597–5610. https://doi.org/10.1093/nar/gkr098
160. McCarthy AJ, Lindsay JA (2010) Genetic variation in *Staphylococcus aureus* surface and immune evasion genes is lineage associated: implications for vaccine design and host-pathogen interactions. BMC Microbiol 10:173. https://doi.org/10.1186/1471-2180-10-173
161. McCarthy AJ, Lindsay JA (2012) The distribution of plasmids that carry virulence and resistance genes in *Staphylococcus aureus* is lineage associated. BMC Microbiol 12:104. https://doi.org/10.1186/1471-2180-12-104
162. McCarthy AJ, Witney AA, Lindsay JA (2012) *Staphylococcus aureus* temperate bacteriophage: carriage and horizontal gene transfer is lineage associated. Front Cell Infect Microbiol 2:6. https://doi.org/10.3389/fcimb.2012.00006
163. McCarthy AJ, Loeffler A, Witney AA, Gould KA, Lloyd DH, Lindsay JA (2014) Extensive horizontal gene transfer during *Staphylococcus aureus* co-colonization in vivo. Genome Biol Evol 6:2697–2708. https://doi.org/10.1093/gbe/evu214
164. Recsei P, Kreiswirth B, O'Reilly M, Schlievert P, Gruss A, Novick RP (1986) Regulation of exoprotein gene expression in *Staphylococcus aureus* by agar. Mol Gen Genet 202:58–61. https://doi.org/10.1007/BF00330517
165. Beier D, Gross R (2006) Regulation of bacterial virulence by two-component systems. Curr Opin Microbiol 9:143–152. https://doi.org/10.1016/j.mib.2006.01.005
166. Mitrophanov AY, Groisman EA (2008) Signal integration in bacterial two-component regulatory systems. Genes Dev 22:2601–2611. https://doi.org/10.1101/gad.1700308
167. Novick RP, Geisinger E (2008) Quorum sensing in staphylococci. Annu Rev Genet 42:541–564. https://doi.org/10.1146/annurev.genet.42.110807.091640
168. Verdon J, Girardin N, Lacombe C, Berjeaud JM, Héchard Y (2009) Delta-hemolysin, an update on a membrane-interacting peptide. Peptides 30:817–823. https://doi.org/10.1016/j.peptides.2008.12.017
169. Raina M, King A, Bianco C, Vanderpool CK (2018) Dual-function RNAs. Microbiol Spectr 6. https://doi.org/10.1128/microbiolspec.RWR-0032-2018
170. Villanueva M, García B, Valle J, Rapún B, de Los R, Mozos I, Solano C et al (2018) Sensory deprivation in *Staphylococcus aureus*. Nat Commun 9:523. https://doi.org/10.1038/s41467-018-02949-y
171. Baroja ML, Herfst CA, Kasper KJ, Xu SX, Gillett DA, Li J et al (2016) The SaeRS two-component system is a direct and dominant transcriptional activator of toxic shock syndrome toxin 1 in *Staphylococcus aureus*. J Bacteriol 198:2732–2742. https://doi.org/10.1128/JB.00425-16
172. Guo H, Hall JW, Yang J, Ji Y (2017) The SaeRS two-component system controls survival of *Staphylococcus aureus* in human blood through regulation of coagulase. Front Cell Infect Microbiol 7:204. https://doi.org/10.3389/fcimb.2017.00204
173. Kato F, Kadomoto N, Iwamoto Y, Bunai K, Komatsuzawa H, Sugai M (2011) Regulatory mechanism for exfoliative toxin production in *Staphylococcus aureus*. Infect Immun 79:1660–1670. https://doi.org/10.1128/IAI.00872-10
174. Sato'o Y, Hisatsune J, Nagasako Y, Ono HK, Omoe K, Sugai M (2015) Positive regulation of staphylococcal enterotoxin H by rot (repressor of toxin) protein and its importance in clonal complex 81 Ssubtype 1 lineage-related food poisoning. Appl Environ Microbiol 81:7782–7790. https://doi.org/10.1128/AEM.01936-15
175. Yarwood JM, McCormick JK, Schlievert PM (2001) Identification of a novel two-component regulatory system that acts in global regulation of virulence factors of *Staphylococcus aureus*. J Bacteriol 183:1113–1123. https://doi.org/10.1128/JB.183.4.1113-1123.2001

176. Ulrich M, Bastian M, Cramton SE, Ziegler K, Pragman AA, Bragonzi A et al (2007) The staphylococcal respiratory response regulator SrrAB induces Ica gene transcription and polysaccharide intercellular adhesin expression, protecting *Staphylococcus aureus* from neutrophil killing under anaerobic growth conditions. Mol Microbiol 65:1276–1287. Erratum in: Mol Microbiol 2007;66:278. https://doi.org/10.1111/j.1365-2958.2007.05863.x
177. James KL, Mogen AB, Brandwein JN, Orsini SS, Ridder MJ, Markiewicz MA et al (2019) Interplay of nitric oxide synthase (NOS) and SrrAB in modulation of *Staphylococcus aureus* metabolism and virulence. Infect Immun 87:e00570–e00518. https://doi.org/10.1128/IAI.00570-18
178. Fournier B, Hooper DC (2000) A new two-component regulatory system involved in adhesion, autolysis, and extracellular proteolytic activity of *Staphylococcus aureus*. J Bacteriol 182:3955–3964. https://doi.org/10.1128/JB.182.14.3955-3964.2000
179. Fournier B, Klier A, Rapoport G (2001) The two-component system ArlS-ArlR is a regulator of virulence gene expression in *Staphylococcus aureus*. Mol Microbiol 41:247–261. https://doi.org/10.1046/j.1365-2958.2001.02515.x
180. Jenul C, Horswill AR (2019) Regulation of *Staphylococcus aureus* virulence. Microbiol Spectr 7. https://doi.org/10.1128/microbiolspec.GPP3-0031-2018
181. Oriol C, Cengher L, Manna AC, Mauro T, Pinel-Marie ML, Felden B et al (2021) Expanding the *Staphylococcus aureus* SarA regulon to small RNAs. mSystems 6:e0071321. https://doi.org/10.1128/mSystems.00713-21
182. Chan PF, Foster SJ (1998) Role of SarA in virulence determinant production and environmental signal transduction in *Staphylococcus aureus*. J Bacteriol 180:6232–6241. https://doi.org/10.1128/JB.180.23.6232-6241.1998
183. Cheung AL, Nishina KA, Trotonda MP, Tamber S (2008) The SarA protein family of *Staphylococcus aureus*. Int J Biochem Cell Biol 40:355–361. https://doi.org/10.1016/j.biocel.2007.10.032
184. Cheung AL, Schmidt K, Bateman B, Manna AC (2001) SarS, a SarA homolog repressible by *agr*, is an activator of protein a synthesis in *Staphylococcus aureus*. Infect Immun 69:2448–2455. https://doi.org/10.1128/iai.69.4.2448-2455.2001
185. Bronner S, Monteil H, Prévost G (2004) Regulation of virulence determinants in *Staphylococcus aureus*: complexity and applications. FEMS Microbiol Rev 28:183–200. https://doi.org/10.1016/j.femsre.2003.09.003
186. Tseng CW, Stewart GC (2005) Rot repression of enterotoxin B expression in *Staphylococcus aureus*. J Bacteriol 187:5301–5309. https://doi.org/10.1128/JB.187.15.5301-5309.2005
187. Kaito C, Saito Y, Ikuo M, Omae Y, Mao H, Nagano G et al (2013) Mobile genetic element SCC*mec*-encoded *PSM-mec* RNA suppresses translation of *agrA* and attenuates MRSA virulence. PLoS Pathog 9:e1003269. https://doi.org/10.1371/journal.ppat.1003269
188. Nihon Wakutin Sangyo. http://www.wakutin.or.jp/faq/. (Japanese)
189. Blencowe H, Lawn J, Vandelaer J, Roper M, Cousens S (2010) Tetanus toxoid immunization to reduce mortality from neonatal tetanus. Int J Epidemiol 39:i102–i109. https://doi.org/10.1093/ije/dyq027
190. Truelove SA, Keegan LT, Moss WJ, Chaisson LH, Macher E, Azman AS et al (2020) Clinical and epidemiological aspects of diphtheria: a systematic review and pooled analysis. Clin Infect Dis 71:89–97. https://doi.org/10.1093/cid/ciz808
191. Musher DM, Chapman AJ, Goree A, Jonsson S, Briles D, Baughn RE (1986) Natural and vaccine-related immunity to *Streptococcus pneumoniae*. J Infect Dis 154:245–256. https://doi.org/10.1093/infdis/154.2.245
192. Adams WG, Deaver KA, Cochi SL, Plikaytis BD, Zell ER, Broome CV et al (1993) Decline of childhood *Haemophilus influenzae* type b (Hib) disease in the Hib vaccine era. JAMA 269:221–226. https://doi.org/10.1001/jama.1993.03500020055031
193. Löffler B, Hussain M, Grundmeier M, Brück M, Holzinger D, Varga G et al (2010) *Staphylococcus aureus* panton-valentine leukocidin is a very potent cytotoxic factor for human neutrophils. PLoS Pathog 6:e1000715. https://doi.org/10.1371/journal.ppat.1000715

194. Rooijakkers SH, Ruyken M, Roos A, Daha MR, Presanis JS, Sim RB et al (2005) Immune evasion by a staphylococcal complement inhibitor that acts on C3 convertases. Nat Immunol 6:920–927. https://doi.org/10.1038/ni1235
195. Langley R, Wines B, Willoughby N, Basu I, Proft T, Fraser JD (2005) The staphylococcal superantigen-like protein 7 binds IgA and complement C5 and inhibits IgA-fc alpha RI binding and serum killing of bacteria. J Immunol 174:2926–2933. https://doi.org/10.4049/jimmunol.174.5.2926
196. Giersing BK, Dastgheyb SS, Modjarrad K, Moorthy V (2016) Status of vaccine research and development of vaccines for *Staphylococcus aureus*. Vaccine 34:2962–2966. https://doi.org/10.1016/j.vaccine.2016.03.110
197. Mirzaei B, Babaei R, Zeighami H, Dadar M, Soltani A (2021) *Staphylococcus aureus* putative vaccines based on the virulence factors: a mini-review. Front Microbiol 12:704247. https://doi.org/10.3389/fmicb.2021.704247

Chapter 5
Immune Response to *Staphylococcus aureus*

Akio Nakane and Phawinee Subsomwong

Abstract Although *Staphylococcus aureus* is not a high virulent pathogen in healthy persons, this bacterium often persists in various tissues and mucosa and attacks immunocompromised hosts. *S. aureus* produces various virulence factors containing adhesion molecules and exotoxins. *S. aureus* tries to colonize using microbial surface components recognizing adhesive matrix molecules (MSCRAMMs) such as fibrinogen binding proteins and clumping factors located on the surface of this bacterium. The proliferating *S. aureus* cells produce exotoxins containing staphylococcal enterotoxins (SEs), toxic shock syndrome toxin-1 (TSST-1), Panton-Valentine leucocidin, hemolysins, and phenol-soluble modulins. These virulence factors and bacterial constituents containing cell wall components can stimulate host immune system. Typically, SEs and TSST-1 act as "superantigen" and induce massive proinflammatory cytokines. Thus, staphylococcal virulence factors set up active stimulation of host immune system. On the other hand, some factors interrupt performance of immune responses: destruction of leukocytes and trap of IgG by protein A. Although *S. aureus* seems not to be an aggressive pathogen, vigorous and continuous offense and defense must be executed between *S. aureus* and host.

Keywords Innate immunity · Antibodies · T-helper cell subsets · Allergy

Abbreviations

γδT	Gamma delta T

A. Nakane (✉)
Hirosaki University Graduate School of Medicine,
Hirosaki, Aomori, Japan
e-mail: a27k03n0@hirosaki-u.ac.jp

P. Subsomwong
Department of Microbiology and Immunology, Hirosaki University Graduate School of Medicine, Hirosaki, Aomori, Japan

A. Nakane, K. Asano (eds.), *Staphylococcus aureus*,
https://doi.org/10.1007/978-981-99-9428-1_5

AD	Atopic dermatitis
AD-HIES	Autosomal dominant hyper IgE syndrome
agr	Accessory gene regulator
AMP	Antimicrobial peptide
APC	Antigen-presenting cell
Arg-1	Arginase-1
BCG	Bacillus Calmette Guérin
BMDM	Bone marrow-derived macrophage
Breg	Regulatory B
C5a	Complement cascade component 5a
CA-MRSA	Community-associated methicillin-resistant *Staphylococcus aureus*
CHIPS	Chemotaxis inhibitory protein of *Staphylococcus aureus*
Clf	Clumping factor
CXCL	CXC-chemokine ligand
DC	Dendritic cell
Fib	Fibrinogen-binding protein
FLIPr	Formyl peptide-like 1 inhibitor
FLIPrL	Formyl peptide-like 1 inhibitor-like
FPR2	Formyl peptide receptor 2
G-MDSC	Granulocyte-myeloid-derived suppressor cell
hBD	Human β-defensin
IFN	Interferon
IFNAR	Interferon-α/β receptor
Ig	Immunoglobulin
IL	Interleukin
IL-1R	Interleukin-1 receptor
ILC2	Innate lymphoid cell type 2
iNKT	Invariant natural killer T
lncRNA	Long noncoding RNA
LTA	Lipoteichoic acid
LTB_4	Leukotiriene B_4
mAb	Monoclonal antibody
MAIT	Mucosa-associated invariant T
MDSC	Myeloid-derived suppressor cell
MHC	Major histocompatibility complex
MiRNA	micro RNA
M-MDSC	Macrophage-myeloid-derived suppressor cell
MMP9	Matrix metalloproteinase-9
MRSA	Methicillin-resistant *Staphylococcus aureus*
MSCRAMM	Microbial surface components recognizing adhesive matrix molecule
NET	Neutrophil extracellular trap
NKT	Natural killer T
NLR	NOD-like receptor
NLRP	NOD-, LLR- and pyrin domain-containing
NOD	Nucleotide-binding oligomerization domain-containing protein

PAMP	Pathogen-associated molecular pattern
PGN	Peptidoglycan
PGRP-S	Peptidoglycan recognition protein short
PRR	Pattern recognition receptor
PSM	Phenol-soluble modulin
PVL	Panton-Valentine leucocidin
ROS	Reactive oxygen species
SaCP	*Staphylococcus aureus* containing phagosome
SE	Staphylococcal enterotoxin
SpA	*Staphylococcus aureus* protein A
SSL	Staphylococcal superantigen-like protein
TCR	T-cell receptor
TGF-β	Transforming growth factor-β
Th	T-helper
TLR	Toll-like receptor
TNFR1	TNF receptor 1
TNF-α	Tumor necrosis factor-α
Treg	Regulatory T
TSST-1	Toxic shock syndrome toxin-1
WTA	Wall teichoic acid

5.1 Introduction

Staphylococcus aureus is an opportunistic and sometimes life-threatening human pathogen in healthcare-, community- and livestock-associated infections. *S. aureus* colonizes the mucosal surfaces, especially the anterior nares are the primary reservoirs of the bacterium. Asymptomatic colonization occurs in approximately 20% of the normal population, and 60% are transiently colonized, while 20% appear to be rarely or never colonized [1]. *S. aureus* develops numerous infections, including skin and soft-tissue infections (carbuncle, furuncle, felon, stye, impetigo), surgical and trauma wounds, urinary tract infections, gastrointestinal tract infections, pneumonia, osteomyelitis, endocarditis, thrombophlebitis, mastitis, meningitis, infections by inserted medical devices, toxic shock syndrome, and septicemia [2, 3]. The high-risk populations are low-birth-weight infants, children, the elderly and patients with indwelling catheters, endotracheal intubation, implantation of foreign bodies, trauma, surgical procedures, hemodialysis, diabetes or immunosuppression by cancer therapy [4].

Although *S. aureus* is not a high virulent pathogen in healthy persons, this bacterium often persists in various tissues and mucosa and attacks immunocompromised hosts. *S. aureus* highly acquires resistance to antibiotics such as methicillin-resistant *S. aureus* (MRSA) and evades host defenses. *S. aureus* produces various virulence factors such as capsule, coagulase, lipase, hyaluronidase, protein A, fibronectin-binding proteins (Fibs), and secreted toxins such as staphylococcal enterotoxins

(SEs), toxic shock syndrome toxin-1 (TSST-1), Panton-Valentine leucocidin (PVL), hemolysins, and phenol-soluble modulins (PSMs) [5–8]. Host immune responses to the bacterial components and various virulence factors is critical to defend and eliminate *S. aureus* from the host. However, host immune system is often attenuated by evasion systems of *S. aureus*. Acquisition of virulence factors to combat host immune system may have resulted in parasitism in humans and animals.

5.2 Innate Immune Response to *S. aureus*

5.2.1 Physiological Barriers

S. aureus colonizes the mucosal surfaces such as the anterior nares. Most of *S. aureus* infections from other hosts and contaminated materials occur contagious infection. Human bodies are protected by skin and mucosa as physiological barriers from attack by external microorganisms [9]. Additionally, on the surface of the skin, commensal microorganisms occupy niches and interfere pathogen growth by creating an environment including a low temperature and pH and bacterial metabolites such as fatty acids [10]. Antimicrobial peptides (AMPs) are also crucial factors for resistance to *S. aureus*. Keratinocytes can produce AMPs including human β-defensin (hBD) 2, hBD3, cathelicidin (LL-37) and RNase 7, which have bacteriostatic or bactericidal activity against *S. aureus* during infection [10–12]. *S. aureus* attempts to adhere and colonize using adhesion molecules, so-called microbial surface components recognizing adhesive matrix molecules (MSCRAMMs) containing Fibs, clumping factors (Clfs) A and B, and collagen binding proteins located on the surface of this bacterium. MSCRAMMs recognize host extracellular matrix components such as fibrinogen and fibronectin, and are anchored to the bacterial cell wall peptidoglycan (PGN) to adhere and invade into host cells [13]. Occasionally, they can produce biofilm, which consists of an exopolysaccharidic matrix that is involved in adherence and colonization to the epithelium. Alternatively, *S. aureus* directly injuries epithelial cells by α-toxin, a prototypic pore-forming toxin. Many clinical isolates of *S. aureus* produce this toxin and human epithelial cells express disintegrin and metalloprotease 10 (ADAM-10), a high-affinity protein receptor for α-toxin, which makes epithelial cells intrinsically sensitive to the cytotoxic effects of α-toxin [14].

5.2.2 Recognition of S. aureus *by Innate Immune Sensors*

Innate immune responses are triggered by bacteria, substantially by their components called pathogen-associated molecular patterns (PAMPs) and bacterial secretory products via pattern recognition receptors (PRRs) on host cells (Fig. 5.1) [36]. Preferentially, staphylococcal PGN and lipoteichoic acid (LTA) are recognized by

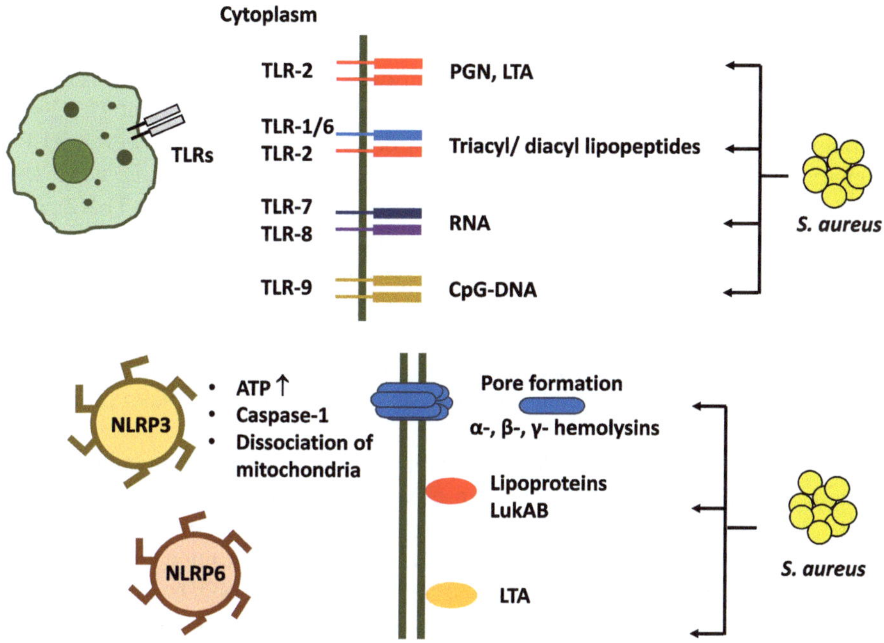

Fig. 5.1 Recognition of *S. aureus* by innate immune sensors. TLR2 recognizes staphylococcal PGN and LTA [15]. TLR2 heterodimerizes with TLR1 or TLR6 to recognize bacterial triacyl or diacyl lipopeptides, respectively [16]. TLR6 can cooperate with TLR2 in the recognition of *S. aureus* [17, 18]. TLR2 also senses LTA with TLR1 [19]. PVL directly binds TLR2 and activates TLR2 signaling genes [20]. PSM is also recognized by TLR2 with TLR6 [21]. In contrast, SSL3 inhibits ligand binding by blocking TLR2 or prevents TLR2-TLR6 association [22–24]. TLR7 and TLR8 sense *S. aureus* RNA [25, 26], and TLR9 binds *S. aureus* CpG-DNA and induces type I IFN signaling [27, 28]. NOD2, a cytoplasmic PRR, recognizes muramyl-dipeptide derived from staphylococcal PGN. Staphylococcal PGN activates NLRP3 inflammasome [29]. Mechanisms of NLRP3 activation by *S. aureus* might include triggering of the purinergic receptor P2X7 by increased levels of ATP that is released from stressed or damaged cells by infection [30], pore formation by the α-, β- and γ-hemolysins of *S. aureus* [31], α- and β-hemolysins induce caspase-1 activation via NLRP3 in the presence of *S. aureus* lipoproteins [32]. NLRP6 inflammasome senses *S. aureus* LTA [33]. LukAB activates NLRP3 inflammasome [34]. α-Toxin triggers dissociation of mitochondria from internalized *S. aureus* in macrophages by activation of NLRP3 inflammasome [35]

Toll-like receptor (TLR) 2 and induce secretion of cytokines and chemoattractants including tumor necrosis factor-α (TNF-α), interleukin (IL)-1β, IL-6, IL-10, IL-12, CXC-chemokine ligand (CXCL) 8, CXCL1, complement cascade component 5a (C5a) and leukotriene B_4 (LTB_4) [15]. TLR2 heterodimerizes with TLR1 or TLR6 to recognize bacterial triacyl or diacyl lipopeptides, respectively [16]. TLR6 can cooperate with TLR2 in the recognition of *S. aureus* [17, 18]. TLR2 also senses LTA with TLR1 [19]. PVL, a critical cytotoxic factor, directly binds TLR2 and activates TLR2 signaling genes [20]. PSM is also recognized by TLR2 with TLR6 [21]. On the other hand, superantigen-like protein (SSL) 3 impairs heterodimer TLR2-TLR6 association [22–24]. Roles of TLR2 in *S. aureus* infection are contradictory, depending on condition of infection. TLR2-deficient mice are highly susceptible to intravenous inoculation with *S. aureus* [37]. Bacterial load is higher in local foci by subcutaneous injection or intranasal inoculation with *S. aureus* in TLR2-deficient mice compared with wild-type mice [38, 39]. These results suggest that TLR2 signaling is important for host defense against *S. aureus* infection. However, TLR2 signaling is reportedly dispensable for the protection against *S. aureus* when less virulent strains or high doses of *S. aureus* were used [40]. In human, TLR2 polymorphism is associated with increase of susceptibility to *S. aureus* infection [41]. TLR7 and TLR8 sense *S. aureus* RNA [25, 26], and TLR9 binds *S. aureus* CpG-DNA and induces type I interferon (IFN) signaling, resulting a strong inflammatory response [27, 28]. Nucleic acid-sensing TLRs are also involved in protection against *S. aureus* infection. TLR7, TLR9 and TLR13 (sensing bacterial rRNA) triple deficient mice are highly susceptible to *S. aureus* infection, although susceptibility of single deficiency is normal [42].

The cytoplasmic PRRs, nucleotide-binding oligomerization domain-containing protein 1 (NOD1) and NOD2, also sense bacterial PGN to induce inflammation, AMP production, and functions of phagocytes. NOD2 recognizes muramyl-dipeptide derived from staphylococcal PGN. NOD2-deficient mice are highly susceptible to *S. aureus* skin and systemic infections [43, 44]. NODs are included in NOD-like receptor (NLR) family. Some of the NLRs assemble macro-molecular protein complexes called inflammasome in the cytosol. NOD-, LRR- and pyrin domain-containing (NLRP) 1, NLRP3, NOD-, LRR- and CARD-containing 4 (NLRC4), and absent in melanoma 2 (AIM2) (not NLR family) form inflammasome, which consists of sensor, adaptor protein ASC (also known as PYCARD) and caspase-1 [45]. Activation of inflammasome results in secretion of processed forms of IL-1β and IL-18, and cell death by pyroptosis [45, 46]. *S. aureus* PGN activates NLRP3 inflammasome and IL-1β secretion [47]. Mechanisms of NLRP3 activation by *S. aureus* might include triggering of the purinergic receptor P2X7 by increased levels of ATP that is released from stressed or damaged cells by infection [29], pore formation by the α-, β- and γ-hemolysins of *S. aureus* [30], and the digestion of PGN by lysozyme [47]. α- and β-Hemolysins induce caspase-1 activation via NLRP3 in the presence of *S. aureus* lipoproteins [31]. Host defense against cutaneous *S. aureus* infection is impaired in ASC-deficient mice as well as IL-1β-deficient mice [32]. NLRP6 inflammasome containing ASC leads processing of caspase-11 to induce caspase-1 activation and senses *S. aureus* LTA [33]. Systemic *S. aureus* infection is deteriorated in NLRP6-deficient mice and caspase-11-deficient mice [33].

Innate immunity through innate immune receptors is negatively regulated by some *S. aureus* products. Staphylococcal superantigen-like protein (SSL) 3 inhibits ligand binding by blocking TLR2 or prevents TLR hetero-dimerization and downstream signaling, resulting in inhibition of cytokine production [23, 24]. *S. aureus* leukocidin A/B (LukAB) activates NLRP3 inflammasome and induces IL-1β but also induces cell death of monocytes in human [34]. Although NLRP3 inflammasome acidifies phagosomes through regulating NADPH oxidase NOX2 [48], *S. aureus* α-toxin triggers dissociation of mitochondria from internalized *S. aureus* in macrophages by activation of NLRP3 inflammasome [49]. On the other hand, modified PGN by *S. aureus* enzymes evades inflammasome activation: degradation of cell wall PGN is required for activation of NLRP3 inflammasome but PG O-acetyltransferase A makes PGN resistant to lysozyme, resulting in suppression of inflammasome activation [47].

5.2.3 Roles of Cytokines in Host Defense Against S. aureus *Infection*

5.2.3.1 IL-17

IL-17 including IL-17A and IL-17F plays a critical role in innate and adaptive host defense against *S. aureus* infections [35, 50]. In the early phase of *S. aureus* infection, IL-17A and IL-17F production is severely impaired in γδ T cell-deficient mice and cutaneous infection worsens in these mice but not in αβ T cell-deficient mice compared with wild-type mice [50]. It has been demonstrated that IL-17A and IL-17F production induced by PGN, LTA and wall teichoic acid (WTA) is dependent on MyD88, IL-1, TLR2, and IL-23 [15, 50, 51]. IL-17A and IL-17F have been shown to promote neutrophil recruitment through induction of CXCL1, CXCL2, CXCL5, CXCL8, granulocyte colony-stimulating factor (G-CSF), and granulocyte-macrophage CSF (GM-CSF) [52, 53]. In addition, both IL-17s induce production of antimicrobial peptides such as α-defensins, β-defensins and LL-37 [10].

5.2.3.2 IL-1β

IL-1β is a crucial cytokine in host defense against *S. aureus* infection [28]. TLR2 signaling induces pro-IL-1β production and active IL-1β is processed by inflammasomes such as NLRP3 inflammasome [47]. *S. aureus* hemolysins and cell wall components can activate NLRP3 inflammasome [29–31, 47]. MyD88 is a signaling adaptor protein for TLRs and IL-1 receptor (IL-1R) family including IL-1R and IL-18R. Mice deficient with IL-1R and MyD88, but not TLR2, have similar defects in bacterial clearance and neutrophil recruitment [28], suggesting that IL-1R-MyD88 signaling pathway is essential for host defense against *S. aureus* infection and other innate immune sensors including NOD2 cover TLR2 signaling.

5.2.3.3 TNF-α and IFN-γ

TNF-α and IFN-γ play both beneficial and detrimental roles in hosts exposed to bacterial infections. These cytokines reportedly play a protective role in host defense against facultative intracellular-growing bacteria such as *Listeria monocytogenes* and *Mycobacterium tuberculosis* [54, 55]. In contrast, TNF-α and IFN-γ mediate Gram-negative septic shock and endotoxin shock [56]. Systemic *S. aureus* infection worsens by administration of anti-TNF-α monoclonal antibody (mAb) [57] and also in TNF receptor 1 (TNFR1)-deficient mice [58]. These results suggest that TNF-α is critical innate host defense against *S. aureus* infection. TNF-α production is possibly induced by *S. aureus* cell wall components including PGN, LTA and lipoprotein via innate immune receptors. S. aureus protein A (SpA), surface protein and binds to immunoglobulin (Ig) G, induces TNF-α by shedding TNFR1 [58]. On the other hand, our previous study showed that injection of anti-TNF-α mAb accelerated death of mice at both nonlethal and lethal infection of *S. aureus* in the control mice but that anti-IFN-γ mAb injection did not significantly affect *S. aureus* infection at the nonlethal dose and rather inhibited host defense against the infected mice early in infection at the lethal dose [57]. Similarly, IFN-γ receptor-deficient mice developed severe sepsis and high mortality early in systemic *S. aureus* infection, but their mortality became significantly lower at the late stage of infection compared with wild-type mice [59]. Thus, IFN-γ seems to display the opposite roles depending on infectious doses and infectious periods in innate immune response to *S. aureus* infection.

5.2.3.4 Type I IFN

Type I IFN including IFN-α and IFN-β has more potent antiviral activity to limit viral replication and spread compared with IFN-γ (type II IFN) and is indispensable for host defense to viral infections. Roles of type I IFN is variable in bacterial infections, depending on the microorganisms. For example, type I IFN plays a protective role in infection with *Streptococcus pneumoniae* and *Salmonella enterica* subsp. *enterica* serovar Typhimurium, but a detrimental role in *L. monocytogenes* and *M. tuberculosis* [27]. *S. aureus* DNA and RNA induce type I IFN through TLR9 and TLR7/8 signaling, respectively [25, 60]. Type I IFN interacts with its receptor, IFN α/β receptor (IFNAR). Type I IFN signaling in response to *S. aureus* is detrimental to the host; IFNAR-deficient mice are significantly less likely to succumb to pulmonary infection than their wild-type mice [60]. Cyclic-di-AMP released from *S. aureus* biofilm induces type I IFN by macrophages via STING-dependent pathway and inhibits the elimination of bacteria by promoting intracellular survival of *S. aureus* in macrophages [61]. In contrast, subcutaneous injection of IFN-β protects mice from *S. aureus* infection via induction of CCL5 [62]. Alternatively, IFN-α protects lung cells from α-toxin-induced cell death [63].

5.2.3.5 IL-10

IL-10 is a representative of immunoregulatory cytokines produced by various immune cells including monocytes, macrophages, T cells and B cells [64]. IL-10 have shown anti-inflammatory activity and have played a role in protecting the host from inflammatory diseases including endotoxin shock [65] and septic shock [66]. We have showed that IL-10 plays a protective role in nonlethal systemic *S. aureus*

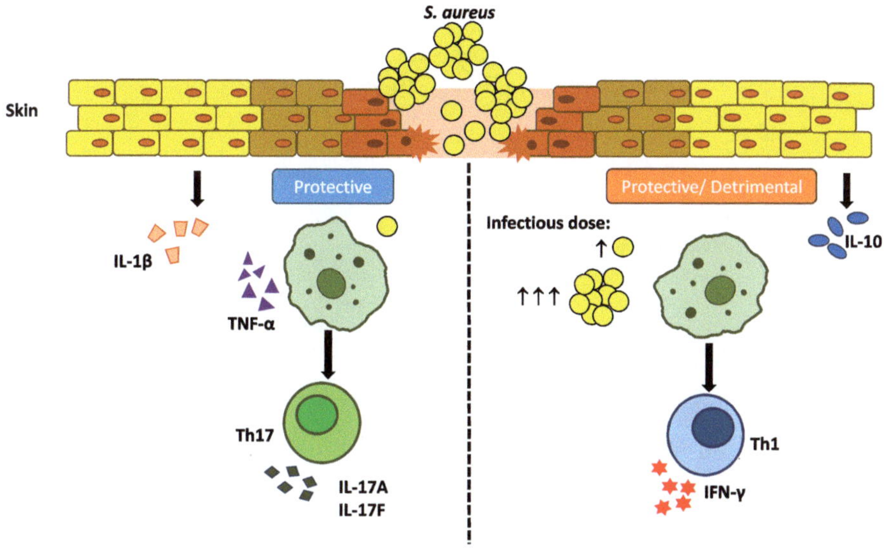

Fig. 5.2 Conflicting roles of cytokines in innate host defense against *S. aureus* infection. IL-17A, IL-17F [10, 50–54], IL-β [28] and TNF-α [57, 58] are reportedly protective, and type I IFN is detrimental in host resistance to *S. aureus* infection. However, IFN-γ [59] and IL-10 [67, 68] are protective and detrimental depending on infectious doses and infection periods

infection because administration of anti-IL-10 mAb inhibited the elimination of *S. aureus* from organs [67]. The results have been supported by Lizak et al. [68] in IL-10-deficient mice and they showed that IL-10 is produced by B1a cells depending on TLR2 and regulates excessive proinflammatory cytokines. On the contrary, they demonstrated that IL-10 plays a detrimental role in subcutaneous *S. aureus* infection by facilitating bacterial persistence by the similar mechanism [68]. The protective and detrimental roles of cytokines in *S. aureus* infection are summarized in Fig. 5.2.

5.2.4 *Host Defense by Neutrophils*

Host defense against *S. aureus* by neutrophils are summarized Figs. 5.3 and 5.4.

5.2.4.1 Recruitment

The primary defense against *S. aureus* infection is the innate immunity provided by neutrophils. In response to damage or the presence of invading *S. aureus*, a variety of host cells i.e. monocytes, macrophages, dendritic cells (DCs), mast cells, fibroblasts, keratinocytes, endothelial and epithelial cells produce potent inflammatory

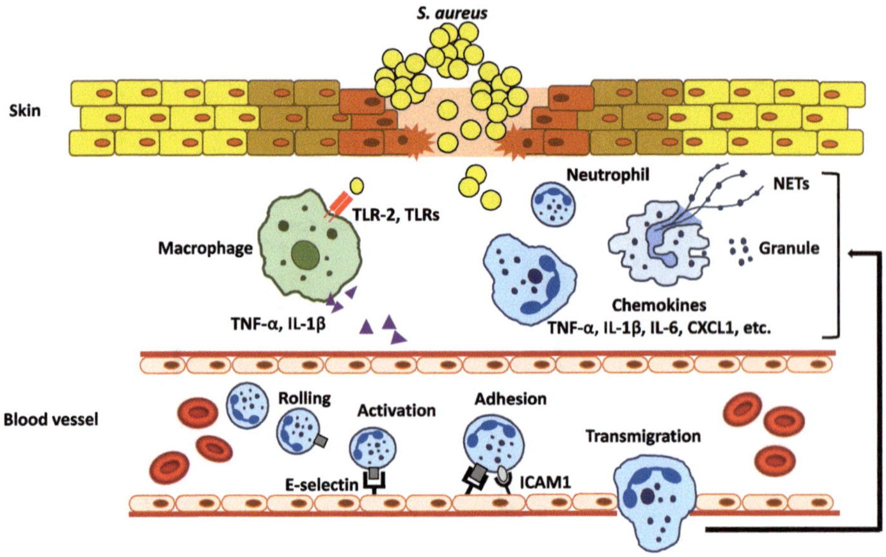

Fig. 5.3 Elimination of *S. aureus* cells by neutrophils. Invaded *S. aureus* cells through injured skin and mucous membrane are phagocytized by resident macrophages. Simultaneously, macrophages sense these bacteria by TLRs such as TLR2 and secrete proinflammatory cytokines including TNF-α and IL-1β. Proinflammatory cytokines, which penetrated tissues, up-regulate adhesion molecules including E-selectin, P-selectin, ICAM1 on endothelium and LFA1 on neutrophil of adjacent capillaries, resulting in rolling, adhesion and diapedesis of neutrophils. Leaked neutrophils are recruited to infectious foci mediated by chemoattractants including TNF-α, IL-1β, IL-6, CXCL1, LTB4, C5a and N-formyl peptides. Neutrophils are proceeded from phagocytosis, degranulation, and finally the release of NETs

mediators and neutrophil chemoattractants. *S. aureus* surface components such as LTA and capsular polysaccharide have been shown to induce IL-8, also known as CXCL8, production by monocytes, epithelial cells, and endothelial cells [72–74]. $CD4^+$ T cells stimulated with *S. aureus* capsular polysaccharide produce cytokines that recruit neutrophils to site of infection [75, 76]. Staphylococcal PGN has also been shown to activate the complement cascade and induce production of C5a, which is another potent neutrophil chemotactic molecule [77]. These cell surface components also induce bacterial-derived products such as N-formyl peptides and PSMs that have the ability to recruit neutrophils directly [77–79]. PSMs are a family of short formyl-peptides that are secreted via an ABC transporter [80] and activate formyl-peptide receptor 2 (FPR2) on human and mouse neutrophils and stimulate cytokine release [81]. Staphylococcal PGN and LTA induce secretion of cytokines and chemoattractants including TNF-α, IL-1β, IL-6, CXCL1, and LTB_4 that bind and engage specific surface receptors on neutrophils [15]. These signals direct neutrophil chemotactic movement out of intravascular circulation to sites of damage or infection within tissues, resulting in a rapid influx and accumulation of neutrophils. Briefly, these cytokines up-regulate adhesion molecules such as P-selectin, E-selectin and intercellular adhesion molecule 1 (ICAM1) on endothelium and

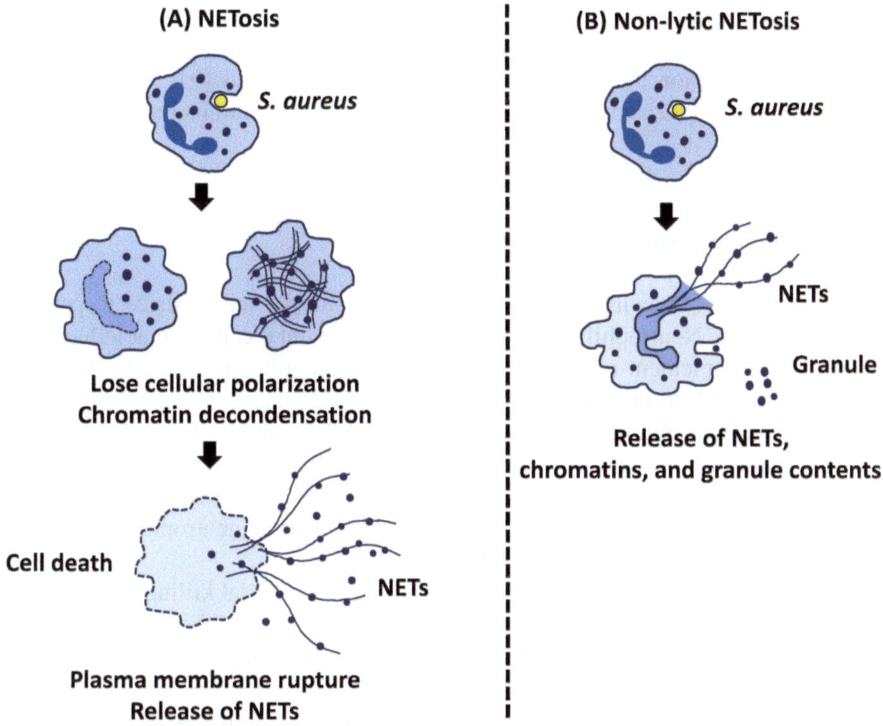

Fig. 5.4 Two ways of NET formation. (**a**) NETosis: After engulfing of *S. aureus*, the neutrophil loses cellular polarization, and causes chromatin decondensation and plasma membrane rupture. Extracellular fibers from dead neutrophils by NETosis can arrest extracellular *S. aureus* [69]. (**b**) Non-lytic NETosis: The rapid release of NETs within minutes of exposure to *S. aureus* and secretion of chromatin and granule contents through degranulation occurs in the absence of cell death [70, 71]

L-selectin and lymphocyte function-associated antigen 1 (LFA1) on neutrophils to promote neutrophil rolling, adhesion and diapedesis so that neutrophils can enter the infectious tissue [15].

Against to the responses, some virulence factors hamper neutrophil recruitment. PSMs lyse neutrophils, although they lead the recruitment and activation of these cells [79]. PVL is one of the biocomponent pore-forming toxins and forms octometric pores of the cell membranes of neutrophils and macrophages and is specific to humans and rabbits [82]. PVL is epidemiologically linked to invasive pneumonia caused by USA300 strains that are major in community-associated (CA)-MRSA [83–85]. Alternatively, this toxin is a strong inducer of IL-1β [86]. SSLs are a family of 14 secreted proteins with structural homology to staphylococcal superantigens but lack the functional T-cell receptor domain [87]. SSL5 and SSL11 bind P-selectin glycoprotein-1 (PSG-1), the ligand of P-selectin, and interfere neutrophil adhesion and rolling [88, 89]. SSL11 may also mediate motility arrest of neutrophils [90]. Neutrophil extravasation and chemotaxis are also interfered by chemotaxis inhibitory protein of *S. aureus* (CHIPS), formyl peptide receptor-like 1 inhibitor (FLIPr) and its homologue FLIPr-like (FLIPrL). CHIPS binds to human FPR1 and human C5a receptor, the complement receptor of neutrophils [91, 92]. FLIPr and FLIPrL bind to FPR2 and inhibit its signaling and FLIPrL inhibits the FPR1 signaling [93, 94].

5.2.4.2 Phagocytosis and Killing of *S. aureus*

Once neutrophils encounter *S. aureus*, they use multiple mechanisms to perform bacterial killing. Neutrophils express Fc and complement receptors that induce antibody- and complement-mediated phagocytosis to engulf opsonized bacteria and phagosomes [95]. In opposition to opsonin-mediated phagocytosis, *S. aureus* produces factors that inhibit phagocytosis. SpA, which binds the Fc portion of IgG, prevents antibody-mediated phagocytosis and complement activation [96–98]. Fib and ClfA bind fibrinogen and impair phagocytosis [96, 97]. SSL13 acts as a chemoattractant and induces degradation and oxidation burst in neutrophils [99]. SSL7 binds to C5 and therefore prevents C5a production [100].

Cytoplasmic granules of neutrophils are critical in bacterial killing. Four species of granules have been known in neutrophils depending on differentiation stages, included substances, and exocytosis. Neutrophil secretory organelles include azurophilic (primary), specific (secondary), and gelatinase (tertiary) granules and secretory vesicles [101, 102]. Azurophilic (primary) granules contain AMPs (α-defensins), proteases (proteinase-3, elastase, cathepsin G), and other cytotoxic molecules (azurocidin, lysozyme, myeloperoxidase). Specific (secondary) granules contain integral membrane components of NADPH oxidase (flavocytochrome *b558*), AMP (LL-37), lysozyme, collagenase and lactoferrin. Gelatinase (tertiary) granules contain gelatinase, lysozyme, leucocidin and matrix metalloproteinase-9 (MMP-9). Secretory granules contain complement receptors CD14 and CD16 and FPRs. Secretory granules are rapidly mobilized in response to weak stimulation by up-regulation of adhesion molecules and chemotactic receptors, including Mac-1 and CXCR2, which link degranulation with neutrophil recruitment. Secondary and tertiary granules are mobilized in response to stronger stimuli and contain FPR1, MMP-9, and LL-37. The recent study showed that endoplasmic reticulum stress sensor inositol-requiring enzyme 1 (IRE1) triggered by *S. aureus* infection induces mitochondria-derived reactive oxygen species (ROS) and are delivered to bacteria-containing phagosomes via mitochondria-derived vesicles [103]. The major bacterial killing occurs within mature phagosomes. However, secreted extracellular molecules such as group IIA phospholipase A2 have been shown to act synergistically with neutrophil NADPH oxidase on promotion of digestion of *S. aureus* phospholipids. Neutrophils kill *S. aureus* by collaboration with oxygen-dependent and oxygen-independent mechanisms [104].

5.2.4.3 Neutrophil Extracellular Traps (NETs)

Antibacterial functions of neutrophils are proceeded from phagocytosis, degranulation, and finally the release of neutrophil extracellular traps (NETs). At first, Brinkmann et al. reported that granule proteins and chromatin released from neutrophils form extracellular fibers that bind Gram-positive bacteria such as *S. aureus* and Gram-negative bacteria such as *S.* Typhimurium and *Shigella flexneri* [69]. NET release occurs primarily through a cell death process termed NETosis. To initiate this process, neutrophils arrest their actin dynamics and depolarize. Next, the nuclear envelope disassembles, and nuclear chromatin decondenses into the cytoplasm of intact cells, mixing with cytoplasmic and granule components. The plasma membrane then permeabilizes, and NETs expand into the extracellular space 3–8 h after neutrophil activation [105]. NETs consist nuclear components such as DNA and histones and are decorated by proteins from primary granules such as myeloperoxidase and neutrophil elastase, secondary granules such as lactoferrin and tertiary granules such as MMP9 and peptidoglycan recognition protein short (PGRP-S) [106]. Although the majority of DNA in NETs originates from the nucleus, these structures also contain mitochondrial DNA. NETs trap, neutralize and kill bacteria, fungi, viruses and parasites and are thought to prevent bacterial and fungal dissemination. NET-localized molecules have different functions, including microbial recognition by such as PGRP-S, antimicrobial activity by such as defensins and LL-37, bactericidal permeability-increasing protein, and tissue remodelling by elastase and MMP9 [106]. However, neutrophils can sense microbial size, and trap larger pathogens such as *Candida albicans* hyphae and aggregated bacteria but not small pathogens such as small yeast and single bacteria [107]. In addition to the cell lytic NETosis-related NET mechanism, an alternative mechanism termed non-lytic NETosis have been reported [70, 71]. The rapid release of NETs within minutes of exposure to *S. aureus* and secretion of chromatin and granule contents through degranulation occurs in the absence of cell death. These components assemble extracellularly and leave behind active anucleated cytoplasts that continue to ingest *S. aureus*. Interestingly, *S. aureus* takes advantage of NETs to evade immune response. *S. aureus* converts NETs to deoxyadenosine that triggers the caspase-3-mediated apoptosis of macrophages by bacteria-derived nuclease and adenosine synthase [108, 109].

5.2.5 *Roles of Macrophages*

Macrophages are major professional phagocytes to eliminate a variety of microorganisms. Meanwhile, some bacteria such as *M. tuberculosis*, *L. monocytogenes*, *Legionella pneumophila*, *S. enterica* and *Francisella tularensis* use macrophages as habitats. It is well known that *S. aureus* internalizes and can survive within these cells for extended periods in non-professional cells including epithelial cells and endothelial cells [110]. Moreover, it has been reported that *S. aureus* can survive and replicate by escape from phagosomal attack [111]. Phagocytized *S. aureus* resides in phagosome of macrophages. In some studies, *S. aureus* does not interfere initial maturation of *S. aureus* containing phagosome (SaCP). Fusion of SaCP and lysosome results in acidification, release of hydrolytic enzymes and AMPs, and produces ROS, reactive nitrogen species (RNS) and chlorine species [112]. *S. aureus* is able to tolerate acid environment and this characteristic is involved in survival within phagolysosomes [113]. Studies using CA-MRSA strain USA300 showed that a low phagosomal pH is required for survival and replication inside macrophages because acid condition up-regulates the accessory gene regulator (agr) system and GraRS two component system [114, 115]. Alternatively, *S. aureus* is reportedly to be able to deacidify the bacterial microenvironment by the generation of ammonia, arginine deiminase, urease, as well as nitrite and nitrate reductase [113]. *S. aureus* also resists or degrades ROS and RNS produced in phagolysosome of oxidative burst macrophages [111]. Virulence factors such as α-toxin and aureolysin contribute to persistence of *S. aureus* in macrophages [114, 116]. After bacterial replication, macrophages die by apoptosis, necroptosis or necrosis and release bacteria, leading to a continuous cycle by other macrophages [112, 117]. It is possible that *S. aureus*-infected macrophages induce systemic dissemination. More than 90% of *S. aureus* in the blood is incorporated into Kupffer cells, immobilized macrophages that reside in the vasculature of liver and replicate intracellularly, and escape across the mesothelium into the peritoneal cavity [117, 118]. Infected GATA-binding factor 6^+ peritoneal cavity macrophages can disseminate *S. aureus* to various peritoneal and retroperitoneal organs including kidneys [119]. Thus, macrophages and in part neutrophils plays as "Trojan Horse" for dissemination of *S. aureus* [120].

5.2.6 Roles of Unconventional T Cells

Unconventional T cells including Gamma delta (γδ) T cells, natural killer T (NKT) cells and mucosa-associated invariant T (MAIT) cells are involved in innate immunity to *S. aureus* infection.

γδ T cells express a T-cell receptor (TCR) composed of γ and δ chains that are distinct from the conventional αβ T cells. Antigen presentation with major histocompatibility complex (MHC) molecules is not required for activation of γδ T cells. These cells are not a major population in peripheral blood or lymphoid tissues but abundant in mucosal barriers. In cutaneous *S. aureus* infection, γδ T cell-deficient mice but not αβ-deficient mice develop larger skin lesions and higher bacterial loads compared with their wild-type mice. This impairment of host defense is associated with loss of IL-17 production by γδ T cells [50]. Various reports have documented that γδ T cells are a critical source of IL-17 in innate immune response. The significance of γδ T cells in host defense against *S. aureus* infection is also described in Sect. 5.3.

NKT cells are a subset of T cells restricted by CD1d, MHC class I-like molecule. These cells recognize lipid and glycolipid antigens presented on CD1d in a direct manner and produce various cytokines after activation. NKT cells are divided into two types: type I NKT cells, which are also called as invariant NKT (iNKT) cells, and type II NKT cells. iNKT cells express an invariant Vα chain of TCR and produce proinflammatory cytokines such as IL-17A and IFN-γ by stimulation with some glycolipids. Staphylococcal superantigens such as SEB can directly activate iNKT cells to produce proinflammatory cytokines [121] in a CD1d-independent manner. Another subpopulation is type II NKT cells, which use a diverse TCR and recognize sulfide (a self-glycosphingolipid derived from myelin). These cells decrease inflammatory responses. Indeed, activation of type II NKT cells by treatment with sulfide prevents the death from *S. aureus* sepsis by down-regulating proinflammatory cytokines such as TNF-α and IL-6 in blood without bacterial loads in blood and organs [122].

MAIT cells localize various tissues and are reportedly important mucosal immunity by producing IL-17, IL-22, and IFN-γ [123]. MAIT cells possess TCR and are activated by binding of TCR to MHC class I like molecule MR1 [123]. SEB and TSST-1 have been shown to be able to stimulate MAIT cells by cross-linking the TCR of these cells and MHC class II molecules [124, 125].

5.2.7 Role of Dendritic Cells

DCs are highly specialized phagocytes and they function as professional antigen-presenting cells (APCs) that present digested antigen pieces to CD4 T cells and CD8 T cells accompanied with MHC class II or class I molecules. DCs originate in the bone marrows and distribute through bloodstream into the tissues throughout bodies including skin, mucosal tissues and lymphoid organs [126]. Immature DCs continuously take up antigen in their microenvironment and initiate response to induction of the adaptive immune system by recognition of pathogen-associated and danger-associated signal. The DCs stop additional antigen incorporation and migrate from local tissues to secondary lymphoid organs, where they differentiate to mature DCs. Mature DCs express high levels of MHC class I and II, costimulatory molecules such as CD80 and CD86, and can act as APCs for T cells to differentiate naïve T cells to T-helper 1 (Th1), Th2, Th17, regulatory T (Treg) and killer T cells.

Some staphylococcal products hamper the action of DCs. Although DCs recognize lipoproteins, lipopeptides, WTA and LTA as well as PGN through TLR2, SSL3 and SSL4 block TLR2 by interfering both TLR2 and TLR1 or TLR6 heterodimerization and lipopeptide binding [47].

5.2.8 *Modulation of Host Defense Against* S. aureus *Infection by Regulatory Cells*

Myeloid-derived suppressor cells (MDSCs) are a heterogeneous population of myeloid progenitor cells that share attributes with monocytes (M-MDSCs) and granulocytes (G-MDSCs) [127]. The role of MDSCs in cancer has been well documented, and their functions has been expanded to bacterial, viral, fungal and parasitic infections [128]. Heim et al. have established chronic *S. aureus* biofilm in infected prosthetic joint and investigated a critical role for MDSCs in promoting *S. aureus* biofilm persistence [129]. They compared transcriptional profiles by transcriptosome sequencing (RNA-Seq) between biofilm associated MDSCs and polymorphonuclear cells versus monocytes, indicating that biofilm-associated MDSCs can be classified as G-MDSCs [129]. MDSCs recruited to the site of biofilm infection express IL-10, which is one contributing mechanism of immune suppression [130]. IL-10 released by recruited MDSCs is responsible for the inhibition of the antimicrobial effect of monocytes/macrophages [131]. Furthermore, the recruitment of MDSC to sites of *S. aureus* biofilm is mediated by IL-12 [132]. The collected results demonstrate that IL-10 production by MDSCs contributes to the persistence of *S. aureus* orthopedic biofilm infections. In addition, biofilmassociated MDSCs and macrophages up-regulate arginase-1 (Arg-1) expression which may further potentiate the anti-inflammatory responses during *S. aureus* biofilm infections [132]. The effects of Arg-1 are important in cutaneous infection but not biofilm infection [133]. In cutaneous infection, TLR2 activation in skin resident cells triggers IL-6 production and induces MDSCs. These suppressive cells are then recruited to the skin suppressing T cell-mediated recall responses such as dermatitis [134].

Treg cells play a role in the mechanisms of peripheral tolerance that control the development of allergies and autoimmune diseases. Alternatively, these cells have been known to inhibit host defense against various disorders including infectious disease and cancers. In mice and humans, Treg cells are characterized by high expression of the IL-2 receptor α -chain (CD25) and expression of the transcription factor Foxp3 in mice and FOXP3 in humans, which are required for their development, function and stability [135]. Treg cells can be broadly classified into two groups on the basis of their developmental origin. Thymic Treg cells, also known as 'natural Treg cells', are generated in the thymus as a separate lineage at the stage of $CD4^+$ single-positive thymocytes and are thought to show TCRs with high affinity for self-peptides [136]. Although the details of their suppressive mechanisms are still not completely understood, and their regulatory actions are most probably dependent on the microenvironment and the target population to be controlled, these cells reportedly perform their function by both cell-contact mechanisms involving specific cell-surface receptors and the secretion of inhibitory cytokines such as IL-10, TGF-β and IL-35 [137]. The second Treg group, peripherally derived Treg cells, also known as 'inducible Treg cells', develop from conventional $CD4^+$ T cells in the periphery after antigen encounter and in the presence of specific factors such as TGF-β and IL-2 [138]. Although there is no definitive protein marker that distinguishes between these two Treg cell populations, there are important differences in their epigenetic signatures, and particularly in the *foxp3* locus [138]. In *S. aureus*, PSMs induce IL-10 secretion by DCs, resulting in impairing the capacity of DCs to induce activation and proliferation of $CD4^+$ T cells, but increased frequency of $Foxp3^+$ Treg cells. These Treg cells secrete high amounts of IL-10, and their suppression capacity is dependent on IL-10 and TGF-β [8]. SEC1 can induce another Treg, $CD8^+CD25^+FOXP3^+$ Treg in human peripheral blood mononuclear cells [139]. Implication of Treg cells in systemic *S. aureus* infection was compared with MDSCs and it was suggested that contribution of $CD4^+CD25^+$ Treg cells is minor degree compared with MDSCs.

A population of suppressor B cells, which are called as regulatory B (Breg) cells, has been known since 1970s. Breg cells are immunosuppressive cells that support immunological tolerance through the production of IL-10, IL-35 and transforming growth factor (TGF)-β [140]. Importantly, Breg cells are involved in maintenance of Treg cells and suppress the differentiation of Th1 cells and TH17 cells [141]. Breg cells are known to originate from immature B cells, mature B cells and plasmablasts. In *S. aureus*, it was demonstrated that SpA can induce B cell proliferation and IL-10-secreting Breg cells via plasmacytoid DC pathway [140].

5.3 Trained Immunity (Innate Memory Immunity)

The immune system is divided into innate and adaptive immune responses. The innate immunity confronts rapidly and non-specifically (roughly antigen-specific via innate immune receptors) to pathogens, whereas adaptive immune immunity responds in a slower but precisely specific manner, with generation of long-lived immunological memory. The innate and adaptive immune processes are not independent but closely linked between them. Activation of the adaptive immune response and induction of classical immune memory in lymphocytes are dependent on the innate immune system, in particular on APCs such as DCs. However, recent evidence indicates that innate immune responses exhibit adaptive characteristics [142, 143]. It has been demonstrated that there is cross protection among infections with different pathogens [144]. For example, herpesvirus latency increases resistance to the bacterial pathogens *L. monocytogenes* and *Yersinia pestis* and the protection is achieved through enhanced production of IFN-γ and systemic activation of macrophages [145]. Alternatively, vaccination evokes protection against different pathogens. Tuberculosis vaccine Bacillus Calmette Guérin (BCG) induces T cell-independent protection against *C. albicans* and influenza virus in mice [146, 147]. These phenomena are reproducible in human data. BCG vaccination in human volunteers protects against an experimental infection with yellow fever vaccine virus [148]. Large epidemiological studies showed that BCG and measles virus vaccination protects heterologous infections [149, 150]. These results suggest that innate immunity is able to be modulated by previous stimulation with microbes or microbial products, and this property has been termed "trained immunity" or "innate memory immunity".

Adaptive immune responses occur by precisely recognizing different pathogens using gene recombination processes in the immunoglobulin gene family, and subsequently establish immunological memory. Epigenetic programming is achieved during these processes. Epigenetic mechanisms are essential for regulation of gene expression. DNA methyltransferases, chromatin remodeling, and histone modifying enzymes rearrange chromatin structure at gene regulatory elements and regulate accessibility of DNA for the transcriptional machinery. MicroRNAs (miRNAs) silence gene expression at the level of transcription or translation and long noncoding RNA (lncRNA) can enhance or inhibit chromatin interactions. In fact, DNA methylation plays a great role in differentiation of naïve lymphocytes including $CD4^+$ T cells, $CD8^+$ T cells and B cells [143]. Exposure of monocyte/macrophages to *C. albicans* or β-glucan modulates their subsequent response to stimulation with unrelated pathogens or PAMPs, and the changed functional landscape of the trained monocytes is accompanied by epigenetic reprogramming [151, 152].

The representative study on trained immunity of *S. aureus* is that β-glucan induces protection against *S. aureus* infection [153, 154]. A recent study using *S. aureus* skin infection [155] demonstrated that macrophages are able to acquire memory characteristic during *S. aureus* skin infection: 1) innate immune memory may be site specific to the affected tissue, 2) bone marrow-derived macrophages (BMDMs) primed with *S. aureus* infection in vivo and with heat-killed *S. aureus* in vitro revealed *S. aureus*-specific killing because the killing activity of the primed BMDMs was similar to that of the non-primed BMDMs against other bacteria including *S. epidermidis*, *Enterococcus faecalis* and *Escherichia coli*. 3) Site specific protection against *S. aureus* infection in the recipient mice by adoptively transferring of BMDMs that were primed with *S. aureus* infection in vivo and with live *S. aureus* in vitro. This innate memory in macrophages can also polarize other neighboring cells that drive antibacterial, Th17, and M1 responses, as well as bacterial clearance. This result suggests that macrophages not only retain memory of previous exposures but also affect the responses of other cells.

Basically, macrophage-dependent secondary protection against infection is nonspecific even in the innate memory phase. In contrast. NK cell-mediated immune memory may provide increased specificity. Three types of NK-cell mediated immune memory have been reported: hapten-specific NK memory cells, mouse cytomegalovirus-specific NK memory cells and cytokine-induced memory NK cells, although cytokine-induced memory NK cells are not antigen-specific [143, 156]. IL-17 is a critical cytokine for host defense against *S. aureus* infection in both innate and adaptive phases and γδ T cells are a critical source of IL-17 in innate immune response [50]. Clonally expanded γδ T cells are reportedly important in IL-17-mediated immune response against *S. aureus* infection [157–160]. Moreover, innate lymphoid cells group 2 (ILC2) that share the common lymphoid progenitor with NK cells, but they do not possess antigen receptors. They also show the potential to memorize their activation status and generate enhanced responses by secondary stimulation. Inhaled allergens stimulate ILC2s to produce IL-5 and IL-13 in an

IL-33-dependent manner during allergic response in the lung [161]. Sensitization of mice with IL-33 is sufficient to generate memory ILC2s responsive to allergen secondary stimulation, indicating the non-antigen specific character of ILC2 memory [162].

5.4 Adaptive Immune Responses to *S. aureus*

5.4.1 Humoral Immunity

It has been verified that antibodies contribute to clinical protection against *S. aureus* infection. It is known to be induced antibodies against toxins such as α-toxin, β and γ hemolysins, PVL and SEs, virulence factors such as aureolysin and iron-regulated surface determinant system (IsdA), cell surface proteins including ClfA/B and Fib, and non-protein antigens such as capsular polysaccharides, LTA and PGN [163]. High titers of specific antibodies are associated with a reduced risk of *S. aureus* infection and attenuate severe clinical course [164, 165]. Binding of IgG antibody to Fcγ receptor on neutrophils and macrophages acquires opsonizing the bacteria, and binding of IgG and IgM to bacteria can induce the activation of the complement cascade. Moreover, IgG and IgA antibodies are able to neutralize *S. aureus* toxins and other virulence factors [164]. SpA, which is expressed by all clinical isolates, modulates antibody responses as well as the antibody function. SpA binds to the Fcγ domain of IgG and blocks phagocytosis of bacteria, whereas it binds to the Fab domains and crosslink IgM and promotes B cell superantigenic activity [108]. It has been pointed out that *S. aureus* infection does not induce long-lived protective immunity and SpA is involved in this reason [166]. SpA alters the fate of plasmablasts and plasma cells by enhancing the short-lived extrafollicular response and reducing the pool of bone marrow-resident long-lived plasma cells. Reduction of the long-lived plasma cells is associated with a rapid decline in antigen-specific by class switch. Thus, SpA reduces the proliferative capacity of plasma cells that enter the bone marrow, diminishing the number of cells in the long-lived pool. In humans, antibodies generated against many *S. aureus* antigens are prevalent and long-lasting [33].

It should be pointed out that considerable discordance exists between the humoral response and protective immunity against *S. aureus*. For example, persistent carriers with *S. aureus* have significantly higher IgG levels for staphylococcal exotoxins than those of non-carriers [167]. However, epidemiological studies have documented that persistent carriers are increased the risk for recurrent MRSA infections [2, 168]. Moreover, individuals with defects in humoral immunity are not necessarily at increased risk for *S. aureus* infections [169]. These studies indicate that antibodies against virulence factors of *S. aureus* are insufficient for host defense [170]. In mouse models, B-cell deficient mice are not susceptible to systemic *S. aureus* infection compared with wild-type mice, while T cell-deficient mice are highly susceptible to the infection [171].

5.4.2 T-Cell Mediated Immunity

Helper T cells, mainly Th2 cells and partially Th1 cells, are necessary for chained responses of B cells including activation and differentiation of antigen-specific B cells to plasma cells and memory B cells to potent antibody-mediated immune responses and maintain long-lived memory. To express antibody-mediated elimination of *S. aureus*, infiltration and killing of neutrophils and macrophages are critical processes. It has been reported that both Th1 and Th17 cells play important roles in host defense against *S. aureus* infection. Our previous study showed that adaptive host defense is dependent on IFN-γ because IFN-γ gene-deficient mice and anti-IFN-γ mAb-treated mice fail to protect against intravenous re-challenge of *S. aureus* and the effect is independent of modification of *S. aureus*-specific antibody response [57]. Significance of IFN-γ in the adaptive immune phase of protection against systemic *S. aureus* infection has also been demonstrated using IFN-γ gene-deficient mice [59, 172]. In the peritonitis model, a prior intraperitoneal infection with *S. aureus* in mice, IFN-γ, which is derived from $CD4^+$ cells, is induced in peritoneal cavity soon after an intraperitoneal challenge with *S. aureus*. The protective effect is transferable to naïve mice by memory Th1 cells [173]. Macrophages activated by Th1 cell-derived IFN-γ are critically important in the protection [173].

The IL-17 family contains IL-17A, commonly known as IL-17, and IL-17B, IL-17C, IL-17D, IL-17E (IL-25) and IL-17E [174]. IL-17A produced by Th17 cells plays critical roles in adaptive host defense against *S. aureus* infection [172, 174, 175]. Main IL-17-producing cells are γδ T cells in innate immunity and Th17 cells in adaptive immunity, respectively [50, 172, 174, 176]. IL-17A induces activation of recruitment of neutrophils as well as innate immunity [174, 176]. Development of long-term memory is crucial for maintenance of adaptive immunity. Th1 and Th2 lineages are mutually exclusive and stable in their phenotypes and functions. However, plasticity of Th17 lineage has been argued. For instance, tuberculosis subunit vaccine-induced Th17 memory cells exhibit lineage stability by retaining both phenotypic and functional properties for nearly 2 years in mice [177]. It was also reported that Th17 cells are comparatively stable and retain the potential to produce IL-17 in the environment of allergic inflammation [178]. In contrast to these studies, it has been documented that Th17 cells exhibit a great degree of plasticity which allows a functional adaptation to various physiological situations acquiring capacity to produce IFN-γ or IL-10 as well as IL-17 or completely altering profiles of cytokine production [179–182]. Both in vivo-primed *C. albicans*- and *S. aureus*-specific memory Th17 cells down-regulate IL-17A production, whereas *S. aureus*-specific clones upregulate IL-10 production by the re-stimulation, and the increased IL-10 produced in memory T cells is significantly inhibited by IL-1β [2]. Moreover, Th17 cells induced systemic *S. aureus* infection can transdifferentiate into Treg cells in mice [181]. We investigated the functional properties of Th17 cells and the protective effect against *S. aureus* infection induced by vaccination with nontoxic mutant TSST-1 in the memory phase. The protective effect against systemic *S. aureus* infection was abrogated because of IL-10 production. The protective effect was recovered by blockade of IL-10. These results suggest that IL-10 produced by memory Th17 cells suppressed their ability to produce IL-17 through an autocrine mechanism, and that blockade of IL-10 restored IL-17 production in memory Th17 cells and IL-17-mediated protection against systemic *S. aureus* infection [183]. Therefore, it is not clear whether Th17 cells are stable and terminally differentiated or exhibit considerable plasticity.

5.5 Atopic Dermatitis

Atopic dermatitis (AD) is a chronic skin disease that is characterized by impaired epidermal barrier function, cutaneous inflammation and a high degree of *S. aureus* colonization in the skin lesions. Dysbiosis of skin microbiota is observed in both patients with AD and murine models of the disease. Decrease of beneficial commensals and increase in *S. aureus* occur in both humans and mice [184–186]. Increased colonization with *S. aureus* is associated with a lower diversity of skin microbiome. *S. aureus* colonizes skin in approximately 30–50% of healthy adults, but it is constantly detected in only 10–20% of such individuals [187]. In AD patients, more than 90% of the lesional skin exhibits an increase in *S. aureus* colonization as compared with the skin of healthy subjects [188]. Coagulase-negative Staphylococci including *S. epidermidis* produce AMPs such as LL-37, and are commonly found on healthy skin but rare on patients with AD [189]. *S. epidermidis* additionally produces a small lipopeptide that induces through host TLR2-dependent production of AMPs by keratinocytes [190]. These findings demonstrate that loss of beneficial commensals and AMPs causes *S. aureus* overgrowth, leading establishment or exacerbation of AD. Overgrowth of *S. aureus* sometimes precedes the development of AD [191]. Filaggrin is a filament aggregating protein and a natural moisturizing factor. It has been considered that mutations in filaggrin gene *FLG* are essential for patients with AD and impair the skin barrier functions [192, 193]. However, most of patients with AD do not harbor any mutation in the *FLG* gene and approximately 60% of mutation carriers do not present any clinical evidence of AD [194, 195]. This fact indicates that *FLG* mutations are neither necessary nor sufficient to cause AD and that other mechanisms affecting the skin barrier integrity may play a part in AD pathogenesis [196, 197]. The clinical hallmark of AD is the cutaneous inflammation, which is characterized by increased epidermal thickness, enhanced expression of inflammatory cytokines and the dense infiltration of activated CD4$^+$ T cell [198]. Based on cytokine response, AD can be divided into the acute phase and the chronic phase. Increase in the IL-1β level correlates with signs of clinical exacerbation and progression of AD [199]. The initiation of the acute inflammatory response is also triggered by an increased level of Th2 cytokines including of IL-4, IL-5, IL-13, and by a reduced production of IFN-γ. Th2 cytokines can downregulate the skin AMPs such as HBD-2, HBD-3, and LL-37, which can prevent from colonization/infection of the skin by pathogenic bacteria [200]. Recently, Th22 cell-derived IL-22 is involved in AD pathogenesis associating with Th2 cytokines. Th2 and Th22 cytokines can modulate the AD epidermal barrier, including suppression of keratinocyte differentiation and apoptosis, hyperplasia, and AMP production [201]. Furthermore, Th2 and Th22 cytokines inhibit the expression of skin barrier protein-encoding genes, such as filaggrin, loricrin and involcrin [201, 202]. Conversely, the chronic phase of inflammation associates with an increase of the Th1 response, with an overproduction of IL-12 as well as IL-8 and IFN-γ, which represent the typical markers of chronic skin inflammation [198]. Autosomal dominant hyper IgE syndrome (AD-HIES), also known as Job's syndrome, is a primary immune deficiency caused by dominant-negative loss-of-function mutations in the STAT3 gene. These patients reveal elevated serum IgE, atopic dermatitis, recurrent pulmonary infections, bone fragility, and vascular abnormalities, all of which may be related to impaired wound healing and tissue remodeling [203]. Recurrent *S. aureus* skin abscesses are one of characteristics in this syndrome. A defect in the peripheral blood $CD4^+$ T cell production of IL- 17 was discovered in patients with AD-HIES [204], and it was proposed that the patients' increased *S. aureus* susceptibility might be due to reduced IL-17 production. Alternatively, it has been reported that epithelial dysfunction by overproduction of TNF-α may increase *S. aureus* susceptibility in patients with AD-HIES whom cutaneous production of IL-17A, IL-17F, and most AMPs are normal [205].

It has been known that virulence factors of *S. aureus*-secreted toxins and superantigens play roles of AD pathogenesis. δ-Toxin induces the mast cell degranulation and IgE enhances its degranulation in the absence of antigen [206]. α-Toxin can lyse keratinocytes, especially in the presence of Th2 cytokines [207]. Roles of staphylococcal superantigens including SEB has been reported [208]. The severity of AD correlates with the number of SE-secreting *S. aureus* colonizing the skin [209]. Anti-SE IgE antibodies can be detected in on the skin of most AD patients and the presence of these IgE antibodies correlates with skin disease severity [210, 211]. SEB can induce eczematoid skin changes when applied to the atopic skin [212]. SEs are also a potential source of IL-22 in AD [213]. PSMα from *S. aureus* induces IL-1α and IL-36α production in keratinocytes and results in stimulation of IL-17-producing γδ T cells, ILC3and $CD4^+$ T cells as well as neutrophil recruitment [214, 215].

5.6 Conclusions

We had investigated interaction between bacteria and host immune system using *S. aureus* and *L. monocytogenes* in mice. We observed remarkable differences between these pathogens when intravenous infection was performed. Our study demonstrated that virulence of *S. aureus* is possibly low and often shifts acute infection to persistent infection [57]. Endogenous cytokine production is at low levels compared with *L. monocytogenes* infection [57, 216], although *S. aureus* produces miscellaneous virulence factors including adhesion molecules and exotoxins, and innate immune system recognizes virulence factors as well as *S. aureus* cell components including cell wall and nucleic acids and induces secretion of different molecules such as cytokines and AMPs to strengthen innate immunity. Alternatively, host defense against *L. monocytogenes* infection is strictly dependent on TNF-α and IFN-γ [54, 217]. However, the critical cytokine(s) has not verified in host defense against *S. aureus* infection, although different cytokines are reportedly involved [10, 35, 50–53, 57–59, 67, 68]. One of possible reasons is the existence of counteractive factors. For example, PSMs and PVL lyse neutrophils, critical cells for host defense against *S. aureus* infection [79, 82], and neutrophil recruitment and phagocytosis are inhibited by SSLs [88–90, 100], CHIPS and related factors [91–94], SpA [96–98], and adhesion molecules [96, 97]. Although superantigenic toxins such as SEs and TSST-1 can induce massive proinflammatory cytokines that play protective roles in host defense against *S. aureus* infection, activated T cells are deleted or fall into anergy [218]. Thus, host immune response is possibly balanced out by various virulence factors in *S. aureus* infection. In the same way, adaptive immunity to *S. aureus* infection is induced but rather ambiguous. It has been demonstrated that $CD4^+$ T cells are differentiated to Th17 cells and IL-17 produced by Th17 cells plays critical roles in adaptive host defense against *S. aureus* infection [172, 174–176]. However, plasticity of Th17 lineage has been argued. Our study indicated that IL-10 produced by memory Th17 cells suppressed their ability to produce IL-17 through an autocrine mechanism [183]. Moreover, anti-staphylococcal IgG and IgA antibodies are effective to eliminate *S. aureus* cells and their secreted exotoxins [164, 165], but SpA counteracts by blocking of Fcγ domain of IgG [108].

It is presumed that offense and defense responses are continuously executed between *S. aureus* and host to establish and maintain commensalism for *S. aureus*. Therefore, further elucidation of detailed immunological mechanism by *S. aureus* infection is expected to resolve prevention and therapy of MRSA infectious diseases as well as atopic dermatitis.

References

1. Williams RE (1963) Healthy carriage of *Staphylococcus aureus*: its prevalence and importance. Bacteriol Rev 27(1):56–71. https://doi.org/10.1128/br.27.1.56-71.1963
2. Wertheim HF, Melles DC, Vos MC, van Leewen W, van Belkum A, Verbrugh HA et al (2005) The role of nasal carriage in *Staphylococcus aureus* infections. Lancet Infect Dis 5(12):751–762. https://doi.org/10.1016/S1473-3099(05)70295-4
3. McCaig LF, McDonald LC, Mandal S, Jernigan DB (2006) *Staphylococcus aureus*-associated skin and soft tissue infections in ambulatory care. Emerg Infect Dis 12(11):1715–1723. https://doi.org/10.3201/eid1211.060190

4. David MZ, Daum RS (2010) Community-associated methicillin-resistant *Staphylococcus aureus*: epidemiology and clinical consequences of an emerging epidemic. Clin Microbiol Rev 23(3):616–687. https://doi.org/10.1128/CMR.00081-09
5. Garnier F, Tristan A, Francois B, Etienne J, Delage-Corre M, Martin C et al (2006) Pneumonia and new methicillin-resistant *Staphylococcus aureus* clone. Emerg Infect Dis 12(3):498–500. https://doi.org/10.3201/eid1203.051040
6. Joo HS, Cheung GY, Otto M (2011) Antimicrobial activity of community-associated methicillin-resistant *Staphylococcus aureus* is caused by phenol-soluble modulin derivatives. J Biol Chem 286(11):8933–8940. https://doi.org/10.1074/jbc.M111.221382
7. Prévost G, Mourey L, Colin DA, Menestrina G (2001) Staphylococcal pore-forming toxins. Curr Top Microbiol Immunol 257:53–83
8. Schreiner J, Kretschmer D, Klenk J, Otto M, Buhring HJ, Stevanovic S et al (2013) *Staphylococcus aureus* phenol-soluble modulin peptides modulate dendritic cell functions and increase in vitro priming of regulatory T cells. J Immunol 190(7):3417–3426. https://doi.org/10.4049/jimmunol.1202563
9. Grice EA, Segre JA (2011) The skin microbiome. Nat Rev Microbiol 9(3):244–253. https://doi.org/10.1038/nrmicro2537
10. Schauber J, Gallo RL (2009) Antimicrobial peptides and the skin immune defense system. J Allergy Clin Immunol 124(2):R13–R18. https://doi.org/10.1016/j.jaci.2009.07.014
11. Otto M (2010) *Staphylococcus* colonization of the skin and antimicrobial peptides. Expert Rev Dermatol 5(2):183–195. https://doi.org/10.1586/edm.10.6
12. Cho JS, Xuan C, Miller LS (2010) Lucky number seven: RNase 7 can prevent *Staphylococcus aureus* skin colonization. J Invest Dermatol 130(12):2703–2706. https://doi.org/10.1038/jid.2010.294
13. Foster TJ, Hook M (1998) Surface protein adhesins of *Staphylococcus aureus*. Trends Microbiol 6(12):484–488. https://doi.org/10.1016/s0966-842x(98)01400-0
14. Wilke GA, Bubeck WJ (2010) Role of a disintegrin and metalloprotease 10 in *Staphylococcus aureus* alpha-haemolysin-mediated cellular injury. Proc Natl Acad Sci U S A 107(30):13473–13478. https://doi.org/10.1073/pnas.1001815107
15. Ley K, Laudanna C, Cybulsky MI, Nourshargh S (2007) Getting to the site of inflammation: the leukocyte adhesion cascade updated. Nat Rev Immunol 7(9):678–689. https://doi.org/10.1038/nri2156
16. Kawai T, Akira S (2010) The role of pattern-recognition receptors in innate immunity: update on toll-like receptors. Nat Immunol 11(5):373–384. https://doi.org/10.1038/ni.1863
17. Nishiya T, DeFranco AL (2004) Ligand-regulated chimeric receptor approach reveals distinctive subcellular localization and signaling properties of the toll-like receptors. J Biol Chem 279(18):19008–19017
18. Ozinsky A, Underhill DM, Fontenot JD, Hajjar AM, Smith KD, Wilson CB et al (2000) The repertoire for pattern recognition of pathogens by the innate immune system is defined by cooperation between toll-like receptors. Proc Natl Acad Sci U S A 97(25):13766–13771. https://doi.org/10.1073/pnas.250476497
19. Travassos LH, Girardin SE, Philpott DJ, Blanot D, Nahori M-A, Werts C et al (2004) Toll-like receptor 2-dependent bacterial sensing does not occur via peptidoglycan recognition. EMBO Rep 5(10):1000–1006. https://doi.org/10.1038/sj.embor.7400248
20. Zivkovic A, Sharif O, Stich K, Doninger B, Biaggio M, Colinge J et al (2011) TLR 2 and CD14 mediate innate immunity and lung inflammation to staphylococcal Panton–Valentine leukocidin in vivo. J Immunol 186(3):1608–1617. https://doi.org/10.4049/jimmunol.1001665
21. Hajjar AM, O'Mahoney S, Ozinsky A, Underhill DM, Aderem A, Klebanoff SJ, Wilson CB (2001) Functional interactions between toll-like receptor (TLR) 2 and TLR1 or TLR6 in response to phenol-soluble modulin. J Immunol 166(1):15–19. https://doi.org/10.4049/jimmunol.166.1.15

22. Bardoel BW, Bouman VR, Aerts T, Bestebroer PC, Huizinga EG, Brondijk THC et al (2012) Evasion of toll-like receptor 2 activation by staphylococcal superantigen-like protein 3. J Mol Med 90(10):1109–1120. https://doi.org/10.1007/s00109-012-0926-8
23. Yokoyama R, Itoh S, Kamoshida G, Takii T, Fujii S, Tsuji T et al (2012) Staphylococcal superantigen-like protein 3 binds to the toll-like receptor 2 extracellular domain and inhibits cytokine production induced by *Staphylococcus aureus*, cell wall component, or lipopeptides in murine macrophages. Infect Immun 80(8):2816–2825. https://doi.org/10.1128/IAI.00399-12
24. Koymans KJ, Feitsma LJ, Brondijk THC, Aerts PC, Lukkien E, Lössl P et al (2015) Structural basis for inhibition of TLR2 by staphylococcal superantigen-like protein 3 (SSL3). Proc Natl Acad Sci U S A 112(35):11018–11023. https://doi.org/10.1073/pnas.1502026112
25. Bergstrøm B, Aune MH, Waugh JA, Kojen JF, Blix KJ, Ryan L et al (2015) TLR8 senses *Staphylococcus aureus* RNA in human primary monocytes and macrophages and induces IFN-β production via a TAK1-IKKβ-IRF5 signaling pathway. J Immunol 195(3):1100–1111. https://doi.org/10.4049/jimmunol.1403176
26. Ugolini M, Gerhard J, Burkert S, Jensen KJ, Georg P, Ebner F et al (2018) Recognition of microbial viability via TLR8 drives TFH cell differentiation and vaccine responses. Nat Immunol 19(3):386–396. https://doi.org/10.1038/s41590-018-0068-4
27. Parker D, Prince A (2012) *Staphylococcus aureus* induces type I IFN signaling in dendritic cells via TLR9. J Immunol 189(8):4040–4046. https://doi.org/10.4049/jimmunol.1201055
28. Deng GM, Nilsson IM, Verdrengh M, Collins LV, Tarkowski A (1999) Intra-articularly localized bacterial DNA containing CpG motifs induces arthritis. Nat Med 5(6):702–705. https://doi.org/10.1038/9554
29. Franchi L, Kanneganti TD, Dubyak GR, Nunez G (2007) Differential requirement of P2X7 receptor and intracellular K+ for caspase-1 activation induced by intracellular and extracellular bacteria. J Biol Chem 282(26):18810–18818. https://doi.org/10.1074/jbc.m610762200
30. Craven RR, Gao X, Allen IC, Gris D, Wardenburg JB, McElvania-TeKippe E et al (2009) *Staphylococcus aureus* α-hemolysin activates the NLRP3-inflammasome in human and mouse monocytic cells. PLoS One 4(10):e7446. https://doi.org/10.1371/journal.pone.0007446
31. Muñoz-Planillo R, Franchi L, Miller LS, Núñez G (2009) A critical role for hemolysis and bacterial lipoproteins in *Staphylococcus aureus*-induced activation of the Nlrp3 inflammasome. J Immunol 183(6):3942–3948. https://doi.org/10.4049/jimmunol.0900729
32. Miller LS, Pietras EM, Uricchio LH, Hirano K, Rao S, Lin H et al (2007) Inflammasome-mediated production of IL-1β is required for neutrophil recruitment against *Staphylococcus aureus* in vivo. J Immunol 179(10):6933–6942. https://doi.org/10.4049/jimmunol.179.10.6933
33. Hara H, Seregin SS, Yang D, Fukase K, Chamaillard M, Alnemri ES et al (2018) The NLRP6 inflammasome recognizes lipoteichoic acid and regulates gram-positive pathogen infection. Cell 175(6):1651–1664. https://doi.org/10.1016/j.cell.2018.09.047
34. Melenhani JH, James DBA, DuMont AL, Torres VJ, Duncan JA (2015) *Staphylococcus aureus* leucocidin a/B (LukAB) kills human monocytes via host NLRP3 and ASC when extracellular, but not intracellular. PLoS Pathog 11(6):e1004970. https://doi.org/10.1371/journal.ppat.1004970
35. Ishigame H, Kakuta S, Nagai T, Kadoki M, Nambu A, Komiyama Y et al (2009) Differential roles of interleukin-17A and -17F in host defense against mucoepithelial bacterial infection and allergic response. Immunity 30(1):108–119. https://doi.org/10.1016/j.immunl:2008.11.009
36. Fournier B, Philpott DJ (2005) Recognition of *Staphylococcus aureus* by the innate immune system. Clin Microbiol Rev 18(3):521–540. https://doi.org/10.1128/cmr.18.3.521-540.2005
37. Takeuchi O, Hoshino T, Akira S (2000) TLR2-deficient and MyD88-deficient mice are highly susceptible to *Staphylococcus aureus* infection. J Immunol 165(10):5392–5396. https://doi.org/10.4049/jimmunol.165.10.5392
38. Hoebe K, Georgel P, Rutschmann S, Du X, Mudd S, Crozat K et al (2005) CD36 is a sensor of diacylglycerides. Nature 433(7025):523–527. https://doi.org/10.1038/nature03253

39. Gonzalez-Zorn B, Senna JP, Fiette L, Shorte S, Testard A, Chignard M et al (2005) Bacterial and host factors implicated in nasal carriage of methicillin-resistant *Staphylococcus aureus* in mice. Infect Immun 73(3):1847–1851. https://doi.org/10.1128/iai.73.3.1847-1851.2005
40. Miller LS, O'Connell RM, Gutierrez MA, Pietras EM, Shahangian A, Gross CE et al (2006) MyD88 mediates neutrophil recruitment initiated by IL-1R but not TLR2 activation in immunity against *Staphylococcus aureus*. Immunity 24(1):79–91. https://doi.org/10.1016/j.immuni.2005.11.011
41. Stappers MHT, Thys Y, Oosting M, Plantinga TS, Ioana M, Reimnitz P et al (2014) TLR1, TLR2, and TLR6 gene polymorphisms are associated with increased susceptibility to complicated skin and skin structure infections. J Infect Dis 210(2):311–318. https://doi.org/10.1093/infdis/jiu080
42. Lentini G, Famá A, De Gaetano GV, Galbo R, Coppolino F, Venza M et al (2021) Role of endosomal TLRs in *Staphylococcus aureus* infection. J Immunol 207(5):1448–1455. https://doi.org/10.4049/jimmunol.2100389
43. Hruz P, Zinkernagel AS, Jenikova G, Botwin GJ, Hugot JP, Karin M et al (2009) NOD2 contributes to cutaneous defense against *Staphylococcus aureus* through alpha-toxin-dependent innate immune activation. Proc Natl Acad Sci U S A 106(31):12873–12878. https://doi.org/10.1073/pnas.0904958106
44. Deshmukh HS, Hamburger JB, Ahn SH, McCafferty DG, Yand SR, Fowler VG Jr (2009) Critical role of NOD2 in regulating the immune response to *Staphylococcus aureus*. Infect Immun 77(4):1376–1382. https://doi.org/10.1128/iai.00940-08
45. Kesavardhana S, Kanneganti T-D (2017) Mechanisms governing inflammasome activation, assembly and pyroptosis induction. Int Immunol 29(5):201–210. https://doi.org/10.1093/intimm/dxx018
46. Pastar I, Sawaya AP, Marjanovic J, Burgess JL, Strbo N, Rivas KE et al (2021) Intracellular *Staphylococcus aureus* triggers pyroptosis and contributes to inhibition of healing due to perforin-2 suppression. J Clin Invest 131(24):e133727. https://doi.org/10.1172/jci133727
47. Shimada T, Park BG, Wolf AJ, Brikos C, Goodridge HS, Becker CA et al (2010) *Staphylococcus aureus* evades lysozyme-based peptidoglycan digestion that links phagocytosis, inflammasome activation, and IL-1β secretion. Cell Host Microbe 7(1):38–49. https://doi.org/10.1016/j.chom.2009.12.008
48. Sokolovska A, Becker CE, Ip WKE, Rathinam VAK, Brudner M, Paquette N et al (2013) Activation of caspase-1 by the NLRP3 inflammasome regulates the NADPH oxidase NOX2 to control phagosome function. Nat Immunol 14(6):543–553. https://doi.org/10.1038/ni.2595
49. Cohen TS, Boland ML, Boland BB, Takahashi V, Tovchigrechko A, Lee Y et al (2018) *S. aureus* evades macrophage killing through NLRP3-dependent effects on mitochondrial trafficking. Cell Rep 22(9):2431–2441. https://doi.org/10.1016/j.celrep.2018.02.027
50. Cho JS, Pietras EM, Garcia NC, Ramos RI, Farzam DM, Monroe HR et al (2010) IL-17 is essential for host defense against cutaneous *Staphylococcus aureus* infection in mice. J Clin Invest 120(5):1762–1773. https://doi.org/10.1172/jci40891
51. Wanner S, Schade J, Keinhörster D, Weller N, George SE, Kull L et al (2017) Wall teichoic acids mediate increased virulence in *Staphylococcus aureus*. Nat Microbiol 2(1):16257. https://doi.org/10.1038/nmicrobiol.2016.257
52. Korn T, Betteli E, Oukka M, Kuchroo VK (2009) IL-17 and Th17 cells. Annu Rev Immunol 27:485–517. https://doi.org/10.1146/annurev.immunol.021908.132710
53. Cua DJ, Tato CM (2010) Innate IL-17-producing cells: the sentinels of the immune system. Nat Rev Immunol 10(6):479–489. https://doi.org/10.1038/nri2800
54. Nakane A, Minagawa T, Kato K (1988) Endogenous tumor necrosis factor (cachectin) is essential to host resistance against *Listeria monocytogenes* infection. Infect Immun 56(10):2563–2569. https://doi.org/10.1128/iai.56.10.2563-2569.1988
55. Flynn JL, Goldstein MM, Chan J, Triebold KJ, Pfeffersps K, Lowensteln CJ et al (1995) Tumor necrosis factor-α is required in the protective immune response against *Mycobacterium tuberculosis* in mice. Immunity 2(6):561–572. https://doi.org/10.1016/1074-7613(95)90001-2

56. Beutler B, Milsark IW, Cerami AC (1985) Passive immunization against cachectin/tumor necrosis factor protects mice from lethal effect of endotoxin. Science 229(4716):869–871. https://doi.org/10.1126/science.3895437
57. Nakane A, Okamoto M, Asano M, Kohanawa M, Minagawa T (1995) Endogenous gamma interferon, tumor necrosis factor, and interleukin-6 in *Staphylococcus aureus* infection in mice. Infect Immun 63(4):1165–1172. https://doi.org/10.1128/iai.63.4.1165-1172.1995
58. Giai C, Gonzalez C, Ledo C, Garofalo A, Di Genaro MS, Sordelli DO et al (2013) Shedding of tumor necrosis factor receptor 1 induced by protein a decreases tumor necrosis factor alpha availability and inflammation during systemic *Staphylococcus aureus* infection. Infect Immun 81(11):4200–4207. https://doi.org/10.1128/IAI.00593-13
59. Zhao Y-X, Tarkowski A (1995) Impact of interferon-γ receptor deficiency on experimental *Staphylococcus aureus* septicemia and arthritis. J Immunol 155(12):5736–5742. PMID:7499861
60. Martin FJ, Gomez MI, Wetzel DM, Memmi G, O'Seaghdha M, Soong G et al (2009) *Staphylococcus aureus* activates type I IFN signaling in mice and humans through the Xr repeated sequences of protein a. J Clin Invest 119(7):1931–1939. https://doi.org/10.1172/jci35879
61. Gries CM, Bruger EL, Moormeier DE, Scherr TD, Waters CM, Kielian T (2016) Cyclic di-AMP released from *Staphylococcus aureus* biofilm induces a macrophage type I interferon response. Infect Immun 84(12):3564–3576. https://doi.org/10.1128/iai.00447-16
62. Kaplan A, Ma J, Kyme P, Wolf AJ, Becker CA, Tseng CW et al (2012) Failure to induce IFN-β production during *Staphylococcus aureus* infection contributes to pathogenicity. J Immunol 189(9):4537–4545. https://doi.org/10.4049/jimmunol.1201111
63. Lizak M, Yarovinsky TO (2012) Phospholipid scramblase 1 mediates type I interferon-induced protection against staphylococcal α-toxin. Cell Host Microbe 11(1):70–80. https://doi.org/10.1016/j.chom.2011.12.004
64. Sabat RR, Grutz G, Warszawska K, Kirsch S, Witte E, Wolk K et al (2010) Biology of interleukin-10. Cytokine Growth Factor Rev 21(5):331–344. https://doi.org/10.1016/j.cytogfr.2010.09.002
65. Gérard C, Bruyns C, Marchant A, Abramowicz D, Vandenabeele P, Delvaux A et al (1993) Interleukin-10 reduces the release of tumor necrosis factor and prevents lethality in experimental endotoxemia. J Exp Med 177(2):547–550. https://doi.org/10.1084/jem.177.2.547
66. Berg DJ, Kühn R, Rajewsky K, Müller W, Menon S, Davidson N et al (1995) Interleukin-10 is a central regulator of the response to LPS in murine models of endotoxin shock and the Shwartzman reaction but not endotoxin tolerance. J Clin Invest 96(5):2339–2347. https://doi.org/10.1172/jci118290
67. Sasaki S, Nishikawa S, Miura T, Mizuki M, Yamada Y, Madarame H et al (2000) Interleukin-4 and interleukin-10 are involved in host resistance to *Staphylococcus aureus* infection through regulation of gamma interferon. Infect Immun 68(5):2424–2430. https://doi.org/10.1128/IAI.68.5.2424-2430.2000
68. Leech JM, Lacey KA, Mulcahy ME, Medina E, McLoughlin RM (2017) IL-10 plays opposite roles during *Staphylococcus aureus* systemic and local infections. J Immunol 198(6):2352–2365. https://doi.org/10.4049/jimmunol.1601018
69. Brinkmann V, Reichard U, Goosmann C, Fauler B, Uhlemann Y, Weiss D et al (2004) Neutrophil extracellular traps kill bacteria. Science 303(5663):1532–1535. https://doi.org/10.1126/science.1092385
70. Pilsczek FH, Salina D, Poon KKH, Fahey C, Yipp BG, Sibley CD et al (2010) A novel mechanism of rapid nuclear neutrophil extracellular trap formation in response to *Staphylococcus aureus*. J Immunol 185(12):7413–7425. https://doi.org/10.4049/jimmunol.1000675
71. Yipp BG, Petri B, Salina D, Jenne CN, Scott BNV, Zbytnuik LD et al (2012) Infection-induced NETosis is a dynamic process involving neutrophil multitasking in vivo. Nat Med 18(9):1386–1393. https://doi.org/10.1038/nm.2847

72. Standiford TJ, Arenberg DA, Danforth JM, Kunkel SL, VanOtteren GM, Strieter RM (1994) Lipoteichoic acid induces secretion of interleukin-8 from human blood monocytes: a cellular and molecular analysis. Infect Immun 62(1):119–125. https://doi.org/10.1128/iai.62.1.119-125.1994
73. Soell M, Diab M, Haan-Archipoff G, Beretz A, Herbelin C, Poutrel B et al (1995) Capsular polysaccharide types 5 and 8 of *Staphylococcus aureus* bind specifically to human epithelial (KB) cells, endothelial cells, and monocytes and induce release of cytokines. Infect Immun 63(4):1380–1386. https://doi.org/10.1128/iai.63.4.1380-1386.1995
74. Yao L, Lowy FD, Berman JW (1996) Interleukin-8 gene expression in *Staphylococcus aureus*-infected endothelial cells. Infect Immun 64(8):3407–3409. https://doi.org/10.1128/iai.64.8.3407-3409.1996
75. McLoughlin RM, Solinga RM, Rich J, Zaleski KJ, Cocchiaro JL, Risley A et al (2006) CD4^{+} T cells and CXC chemokines modulate the pathogenesis of *Staphylococcus aureus* wound infections. Proc Natl Acad Sci U S A 103(27):10408–10413. https://doi.org/10.1073/pnas.0508961103
76. Tzianabos AO, Wang JY, Lee JC (2001) Structural rationale for the modulation of abscess formation by *Staphylococcus aureus* capsular polysaccharides. Proc Natl Acad Sci U S A 98(16):9365–9370. https://doi.org/10.1073/pnas.161175598
77. Schmeling DJ, Peterson PK, Hammerschmidt DE, Kim Y, Verhoef J, Wilkinson BJ et al (1979) Chemotaxigenesis by cell surface components of *Staphylococcus aureus*. Infect Immun 26(1):57–63. https://doi.org/10.1128/iai.26.1.57-63.1979
78. Showell HJ, Freer RJ, Zigmond SH, Schiffmann E, Aswanikumar S, Corcoran B et al (1976) The structure–activity relations of synthetic peptides as chemotactic factors and inducers of lysosomal secretion for neutrophils. J Exp Med 143(5):1154–1169. https://doi.org/10.1084/jem.143.5.1154
79. Wang R, Braughton KR, Kretschmer D, Bach TH, Queck SY, Li M et al (2007) Identification of novel cytolytic peptides as key virulence determinants for community-associated MRSA. Nat Med 13(12):1510–1514. https://doi.org/10.1038/nm1656
80. Peschel A, Otto M (2013) Phenol-soluble modulins and staphylococcal infection. Nat Rev Microbiol 11(10):667–673. https://doi.org/10.1038/nrmicro3110
81. Kretschmer D, Gleske A-K, Rautenberg M, Wang R, Köberle M, Bohn E et al (2010) Human formyl peptide receptor 2 senses highly pathogenic *Staphylococcus aureus*. Cell Host Microbe 7(6):465–473. https://doi.org/10.1016/j.chom.2010.05.012
82. Aman MJ, Adhikari RP (2014) Staphylococcal bicomponent pore-forming toxins: targets for prophylaxis and immunotherapy. Toxins 6(3):950–972. https://doi.org/10.3390/toxins6030950
83. Lina G, Piemont Y, Godail-Gamot F, Bes M, Peter MO, Gauduchon V et al (1999) Involvement of Panton-valentine leukocidin—producing *Staphylococcus aureus* in primary skin infections and pneumonia. Clin Infect Dis 29(5):1128–1132. https://doi.org/10.1086/313461
84. Francis JS, Doherty MC, Lopatin U, Johnston CP, Sinha G, Ross T et al (2005) Severe community-onset pneumonia in healthy adults caused by methicillin-resistant *Staphylococcus aureus* carrying the Panton-valentine leukocidin genes. Clin Infect Dis 40(1):100–107. https://doi.org/10.1086/427148
85. Gillet Y, Issartel B, Vanhems P, Fournet J-C, Lina G, Bes M et al (2002) Association between *Staphylococcus aureus* strains carrying gene for Panton-valentine leukocidin and highly lethal necrotising pneumonia in young immunocompetent patients. Lancet 359(9308):753–759. https://doi.org/10.1016/s0140-6736(02)07877-7
86. Perret M, Badiou C, Lina G, Burbaud S, Benito Y, Bes M et al (2012) Cross-talk between *Staphylococcus aureus* leukocidins-intoxicated macrophages and lung epithelial cells triggers chemokine secretion in an inflammasome-dependent manner. Cell Microbiol 14(7):1019–1036. https://doi.org/10.1111/j.1462-5822.2012.01772.x
87. Fraser JD, Proft T (2008) The bacterial superantigen and superantigen-like proteins. Immunol Rev 225:226–243. https://doi.org/10.1111/j.1600-065x.2008.00681.x

88. Bestebroer J, Poppelier MJJG, Ulfman LH, Lenting PJ, Denis CV, van Kessel KPM et al (2007) Staphylococcal superantigen-like 5 binds PSGL-1 and inhibits P selectin-mediated neutrophil rolling. Blood 109(7):2936–2943. https://doi.org/10.1182/blood-2006-06-015461
89. Chung MC, Wines BD, Baker H, Langley RJ, Baker EN, Fraser JD (2007) The crystal structure of staphylococcal superantigen-like protein 11 in complex with sialyl Lewis X reveals the mechanism for cell binding and immune inhibition. Mol Microbiol 66(6):1342–1355. https://doi.org/10.1111/j.1365-2958.2007.05989.x
90. Chen C, Yang C, Barbieri JT (2019) Staphylococcal superantigen-like protein 11 mediates neutrophil adhesion and motility arrest, a unique bacterial toxin action. Sci Rep 9:4211. https://doi.org/10.1038/s41598-019-40817-x
91. de Haas CJC, Veldkamp KE, Peschel A, Weerkamp F, Van Wamel WJB, Heezius ECJM et al (2004) Chemotaxis inhibitory protein of *Staphylococcus aureus*, a bacterial antiinflammatory agent. J Exp Med 199(5):687–695. https://doi.org/10.1084/jem.20031636
92. Postma B, Poppelier MJ, van Galen JC, Prossnitz ER, van Strijp JAG, de Haas CJC et al (2004) Chemotaxis inhibitory protein of *Staphylococcus aureus* binds specifically to the C5a and formylated peptide receptor. J Immunol 172(11):6994–7001. https://doi.org/10.4049/jimmunol.172.11.6994
93. Prat C, Bestebroer J, de Haas CJ, van Strijp JAG, van Kessel KPM (2006) A new staphylococcal anti-inflammatory protein that antagonizes the formyl peptide receptor-like 1. J Immunol 177(11):8017–8026. https://doi.org/10.4049/jimmunol.177.11.8017
94. Prat C, Haas P-J, Bestebroer J, de Haas CJC, van Strijp JAG, van Kessel KPM (2009) A homolog of formyl peptide receptor-like 1 (FPRL1) inhibitor from *Staphylococcus aureus* (FPRL1 inhibitory protein) that inhibits FPRL1 and FPR1. J Immunol 183(10):6569–8026. https://doi.org/10.4049/jimmunol.0801523
95. Krishna S, Miller LS (2012) Innate and adaptive immune responses against *Staphylococcus aureus* skin infections. Semin Immunol 34(2):261–280. https://doi.org/10.1007/s00281-011-0292-6
96. Foster TJ (2005) Immune evasion by staphylococci. Nat Rev Microbiol 3(12):948–958. https://doi.org/10.1038/nrmicro1289
97. Rooijakkers SH, van Kesssel K, van Strijp JA (2005) Staphylococcal innate immune evasion. Trends Microbiol 13(12):596–601. https://doi.org/10.1016/j.tim.2005.10.002
98. Cruza AR, den Boer MA, Strasser J, Zwarthoff SA, Beurskens FJ, de Haas CJC et al (2012) Staphylococcal protein a inhibits complement activation by interfering with IgG hexamer formation. Proc Natl Acad Sci U S A 118(7):e2016772118. https://doi.org/10.1073/pnas.2016772118
99. Zhao Y, van Kessel KPM, de Haas CJC, Rogers MRC, van Strijp JAG, Haas P-J (2018) Staphylococcal superantigen-like protein 13 activates neutrophils via formyl peptide receptor 2. Cell Microbiol 20(11):e12941. https://doi.org/10.1111/cmi.12941
100. Bestebroer J, Aerts PC, Rooijakkers SHM, Pandey MK, Kohl J, van Strijp JAG et al (2010) Functional basis for complement evasion by staphylococcal superantigen-like 7. Cell Microbiol 12(10):1506–1516. https://doi.org/10.1111/j.1462-5822.2010.01486.x
101. Rigby KM, DeLeo FR (2012) Neutrophils in innate defense against *Staphylococcus aureus*. Semin Immunopathol 34(2):237–259. https://doi.org/10.1007/s00281-011-0295-3
102. Ley K, Hoffman HM, Kubes P, Cassatella MA, Zychlinsky A, Hedrick CC et al (2018) Neutrophils: new insights and open questions. Sci Immunol 3(30):eaat4579. https://doi.org/10.1126/sciimmunol.aat4579
103. Abuaita BH, Schultz TL, O'Riordan MX (2018) Mitochondria-derived vesicles deliver antimicrobial reactive oxygen species to control phagosome-localized *Staphylococcus aureus*. Cell Host Microbe 24(5):625–636. https://doi.org/10.1016/j.chom.2018.10.005
104. Femling JK, Nauseef WM, Weiss JP (2005) Synergy between extracellular group IIA phospholipase A2 and phagocyte NADPH oxidase in digestion of phospholipids of *Staphylococcus aureus* ingested by human neutrophils. J Immunol 175(7):4653–4661. https://doi.org/10.4049/jimmunol.175.7.4653

105. Papayannopoulos V (2018) Neutrophil extracellular traps in immunity and disease. Nat Rev Immunol 18(10):134–147. https://doi.org/10.1038/nri.2017.105
106. Mantovani A, Cassatella MA, Costantini C, Jaillon S (2011) Neutrophils in the activation and regulation of innate and adaptive immunity. Nat Rev Immunol 11(7):519–531. https://doi.org/10.1038/nri3024
107. Branzk N, Lubojemska A, Hardison SE, Wang Q, Gutierrrez MG, Brown GD et al (2014) Neutrophils sense microbe size and selectively release neutrophil extracellular traps in response to large pathogens. Nat Immunol 15(7):1017–1025. https://doi.org/10.1038/ni.2987
108. Thammavongsa V, Missiakas DM, Schneewind O (2013) *Staphylococcus aureus* degrades neutrophil extracellular traps to promote immune cell death. Science 342(6160):863–866
109. Thammavongsa V, Kim HK, Missiakas D, Schneewind O (2015) Staphylococcal manipulation of host immune responses. Nat Rev Microbiol 13(8):529–543. https://doi.org/10.1126/science.1242255
110. Tuchscherr L, Medina E, Hussain M, Völker W, Heitmann V, Niemann S et al (2011) *Staphylococcus aureus* phenotype switching: an effective bacterial strategy to escape host immune response and establish a chronic infection. EMBO Mol Med 3(3):129–141. https://doi.org/10.1002/emmm.201000115
111. Moldovan A, Fraunholz MJ (2018) In or out: phagosomal escape of *Staphylococcus aureus*. Cell Microbiol 21(3):e12997. https://doi.org/10.1111/cmi.12997
112. Flannagan RS, Heit B, Heinrichs DE (2016) Intracellular replication of *Staphylococcus aureus* in mature phagosomes in macrophages precedes host cell death, and bacterial escape and dissemination. Cell Microbiol 18(4):514–535. https://doi.org/10.1111/cmi.12527
113. Weinnick B, Dunman PM, McAleese F, Murphy E, Projan SJ, Fang Y et al (2004) Effect of mild acid on gene expression in *Staphylococcus aureus*. J Bacteriol 186(24):8407–8423. https://doi.org/10.1128/JB.186.24.8407-8423.2004
114. Tranchemontagne ZR, Camire RB, O'Donnell VJ, Baugh J, Burkholder KM (2015) *Staphylococcus aureus* strain USA300 perturbs acquisition of lysosomal enzymes and requires phagosomal acidification for survival inside macrophages. Infect Immun 84(1):241–253. https://doi.org/10.1128/iai.00704-15
115. Flannagan RS, Kulack RC, McGavin MJ, Heinrichs DE (2018) *Staphylococcus aureus* uses the GraXRS regulatory system to sense and adapt to the acidified phagosome in macrophages. MBio 9(4):e01143–e01118. https://doi.org/10.1128/mBio.01143-18
116. Kubica M, Guzik K, Koziel J, Zarebski M, Richter W, Gajkowaska B et al (2008) A potential new pathway for *Staphylococcus aureus* dissemination: the silent survival of *S. Aureus* phagocytosed by human monocyte-derived macrophages. PLoS One 3(1):e1409. https://doi.org/10.1371/journal.pone.0001409
117. Flannagen RS, Heit B, Heinrichs DE (2015) Antimicrobial mechanisms of macrophages and the immune evasion strategies of *Staphylococcus aureus*. Pathogens 4(4):826–868. https://doi.org/10.3390/pathogens4040826
118. Surewaard BGJ, Deniset JF, Zemp FJ, Amrein A, Otto M, Conly J et al (2016) Identification and treatment of the *Staphylococcus aureus* reservoir *in vivo*. J Exp Med 213(7):1141–1151. https://doi.org/10.1084/jem.20160334
119. Jorch SK, Surewaard BGJ, Hossain M, Peiseler M, Deppermann C, Deng J et al (2019) Peritoneal GATA6^{+} macrophages function as a portal for *Staphylococcus aureus* dissemination. J Clin Invest 129(11):4643–4656. https://doi.org/10.1172/jci127286
120. Thwaites GE, Gant V (2011) Are bloodstream leukocytes Trojan horses for the metastasis of *Staphylococcus aureus*? Nat Rev Microbiol 9(3):215–222. https://doi.org/10.1038/nrmicro2508
121. Hayworth JL, Mazzuca DM, Vareki SM, Welch I, McCormick JK, Haeryfar SM (2012) CD1d-independent activation of mouse and human iNKT cells by bacterial superantigens. Immunol Cell Biol 90(7):699–709. https://doi.org/10.1038/icb.2011.90

122. Kwiecinski J, Rhost S, Löfbom L, Blomqvist M, Månsson JE, Cardell SL et al (2013) Sulfatide attenuates experimental *Staphylococcus aureus* sepsis through a CD1d-dependent pathway. Infect Immun 81(4):1114–1120. https://doi.org/10.1128/IAI.01334-12
123. Salou M, Franciszkiewicz K, Lantz O (2017) MAIT cells in infectious diseases. Curr Opin Immunol 48:7–14. https://doi.org/10.1016/j.coi.2017.07.009
124. Shaler CR, Choi J, Rudak PT, Memarnejadian A, Szabo PA, Tun-Abraham ME et al (2017) MAIT cells launch a rapid, robust and distinct hyperinflammatory response to bacterial superantigens and quickly acquire an anergic phenotype that impedes their cognate antimicrobial function: defining a novel mechanism of superantigen-induced immunopathology and immunosuppression. PLoS Biol 15(6):e2001930. https://doi.org/10.1371/journal.pbio.2001930
125. Emgård J, Bergsten H, McCormick JK, Barrantes I, Skreda S, Sandberg JK et al (2019) MAIT cells are major contributors to the cytokine response in group a streptococcal toxic shock syndrome. Proc Natl Acad Sci U S A 116(51):25923–25931. https://doi.org/10.1073/pnas.1910883116
126. Steinman RM (1991) The dendritic cell system and its role in immunogenicity. Annu Rev Immunol 9:271–296. https://doi.org/10.1146/annurev.iy.09.040191.001415
127. Condamine T, Gabrilovich DI (2011) Molecular mechanisms regulating myeloid-derived suppressor cell differentiation and function. Trends Immunol 32(1):19–25. https://doi.org/10.1016/j.it.2010.10.002
128. Medina E, Hartl D (2018) Myeloid-derived suppressor cells in infection: a general overview. J Innate Immun 10(5–6):407–413. https://doi.org/10.1159/000489830
129. Heim CE, West SC, Ali H, Kieliana T (2018) Heterogeneity of $Ly6G^+$ $Ly6C^+$ myeloid-derived suppressor cell infiltrates during *Staphylococcus aureus* biofilm infection. Infect Immun 86(12):e00684–e00618. https://doi.org/10.1128/iai.00684-18
130. Heim CE, Vidlak D, Scherr TD, Kozel JA, Holzapfel M, Muirhead DE et al (2014) Myeloid-derived suppressor cells contribute to *Staphylococcus aureus* orthopedic biofilm infection. J Immunol 192(8):3778–3792. https://doi.org/10.4049/jimmunol.1303408
131. Heim CE, Vidlak D, Kielian T (2015) Interleukin-10 production by myeloid-derived suppressor cells contributes to bacterial persistence during *Staphylococcus aureus* orthopedic biofilm infection. J Leukoc Biol 98(6):1003–1013. https://doi.org/10.1189/jlb.4VMA0315-125RR
132. Heim CE, Vidlak D, Scherr TD, Hartman CW, Garvin KL, Kielian T (2015) IL-12 promotes myeloid-derived suppressor cell recruitment and bacterial persistence during *Staphylococcus aureus* orthopedic implant infection. J Immunol 194(8):3861–3872. https://doi.org/10.4049/jimmunol.1402689
133. Yamada KJ, Heim CE, Aldrich AL, Gries CM, Staudacher AG, Kieliana T (2018) Arginase-1 expression in myeloid cells regulates *Staphylococcus aureus* planktonic but not biofilm infection. Infect Immun 86(7):e00206–e00218. https://doi.org/10.1128/IAI.00206-18
134. Skabytska Y, Wolbing F, Gunther C, Koberle M, Kaesler S, Chen KM et al (2014) Cutaneous innate immune sensing of toll-like receptor 2–6 ligands suppreses T cell immunity by inducing myeloid-derived suppressor cells. Immunity 41(5):762–775. https://doi.org/10.1016/j.immuni.2014.10.009
135. Sakaguchi S, Miyara M, Costantino CM, Hafler DA (2010) $FOXP3^+$ regulatory T cells in the human immune system. Nat Rev Immunol 10(6):490–500. https://doi.org/10.1038/nri2785
136. Josefowicz SZ, Lu LF, Rudensky AY (2012) Regulatory T cells: mechanisms of differentiation and function. Annu Rev Immunol 30:531–564. https://doi.org/10.1146/annurev.immunol.25.022106.141623
137. Dominguez-Villar M, Hafler DA (2018) Regulatory T cells in autoimmune disease. Nat Immunol 19(6):665–673. https://doi.org/10.1038/s41590-018-0120-4
138. Kanamori M, Nakatsukasa H, Okada M, Lu Q, Yoshimura A (2016) Induced regulatory T cells: their development, stability, and applications. Trends Immunol 37(11):803–811. https://doi.org/10.1016/j.it.2016.08.012

139. Lee J, Park JY, Kapln BLF, Pruett SB, Park JW, Park YH et al (2018) Induction of immunosuppressive CD8⁺CD25⁺FOXP3⁺ regulatory T cells by suboptimal stimulation with staphylococcal enterotoxin C1. J Immunol 200(2):669–680. https://doi.org/10.4049/jimmunol.1602109
140. Rosser EC, Mauri C (2015) Regulatory B cells: origin, phenotype, and function. Immunity 42(4):607–612. https://doi.org/10.1016/j.immuni.2015.04.005
141. Parcina M, Miranda-Garcia MA, Durlanik S, Ziegler S, Over B, Georg P et al (2013) Pathogen-triggered activation of plasmacytoid dendritic cells induces IL-10-producing B cells in response to *Staphylococcus aureus*. J Immunol 190(4):1591–1602. https://doi.org/10.4049/jimmunol.1201222
142. Bowdish DM, Loffredo MS, Mukhopadhyay S, Mantovani A, Gordon S (2007) Macrophage receptors implicated in the "adaptive" form of innate immunity. Microbes Infect 9(14–15):1680–1687. https://doi.org/10.1016/j.micinf.2007.09.002
143. Netea MG, Schlitzer A, Placek K, Joosten LAB, Schultze JL (2019) Innate and adaptive immune memory: an evolutionary continuum in the host's response to pathogens. Cell Host Microbe 25(1):13–26. https://doi.org/10.1016/j.chom.2018.12.006
144. Quintin J, Cheng SC, van der Meer JW, Netea MG (2014) Innate immune memory: towards a better understanding of host defense mechanisms. Curr Opin Immunol 29:1–7. https://doi.org/10.1016/j.coi.2014.02.006
145. Barton CS, White DW, Cathelyn JS, Brett-McClellan KA, Engle M, Diamond MS et al (2007) Herpesvirus latency confers symbiotic protection from bacterial infection. Nature 447(7142):326–329. https://doi.org/10.1038/nature05762
146. van't Wount JW, Poell R, van Furth R (1992) The role of BCG/PPD-activated macrophages in resistance against systemic candidiasis in mice. Scand J Immunol 36(5):713–719. https://doi.org/10.1111/j.1365-3083.1992.tb03132.x
147. Spener JC, Ganguly R, Waldman RH (1977) Nonspecific protection of mice against influenza virus infection by local or systemic immunization with bacillus Calmette Guérin. J Infect Dis 136(2):171–175. https://doi.org/10.1093/infdis/136.2.171
148. Arts RJW, Moorlag S, Novakovic B, Li Y, Wang SY, Oosting M et al (2018) BCG vaccination protects against experimental viral infection in humans through the induction of cytokines associated with trained immunity. Cell Host Microbe 23(1):89–100.e5. https://doi.org/10.1016/j.chom.2017.12.010
149. Benn CS, Netea MG, Selin LK, Aaby P (2013) A small job—a big effect: nonspecific immunomodulation by vaccines. Trends Immunol 34(9):431–439. https://doi.org/10.1016/j.it.2013.04.004
150. Goodridge HS, Ahmed SS, Curtis N, Kollmann TR, Levy O, Neata MG et al (2016) Harnessing the beneficial heterologous effects of vaccination. Nat Rev Immunol 16(6):392–400. https://doi.org/10.1038/nri.2016.43
151. Quintin J, Saeed S, Martens JHA, Giamarellos-Bourboulis EJ, Ifrim DC, Logie C et al (2012) *Candida albicans* infection affords protection against reinfection via functional reprogramming of monocytes. Cell Host Microbe 12(2):223–232. https://doi.org/10.1016/j.chom.2012.06.006
152. Saeed S, Quintin J, Kerstens HHD, Rao NA, Aghajanirefah A, Matarese F et al (2014) Epigenetic programming of monocyte-to-macrophage differentiation and trained immunity. Science 345(6204):1251086. https://doi.org/10.1126/science.1251086
153. Di Luzio NR, Williams DL (1978) Protective effect of glucan against systemic *Staphylococcus aureus* septicemia in normal and leukemic mice. Infect Immun 20(3):804–810. https://doi.org/10.1128/iai.20.3.804-810.1978
154. Marakalala MJ, Williams DL, Hoving JC, Engstad R, Netea MG, Brown GD (2013) Dectin-1 plays a redundant role in the immunomodulatory activities of β-glucan-rich ligands in vivo. Microbes Infect 15(6–7):511–515. https://doi.org/10.1016/j.micinf.2013.03.002
155. Chan LC, Rossettic M, Miller LS, Fillera SG, Johnsona CW, Leea HK et al (2018) Protective immunity in recurrent *Staphylococcus aureus* infection reflects localized immune signatures

and macrophage-conferred memory. Proc Natl Acad Sci U S A 115(47):E11111–E11119. https://doi.org/10.1073/pnas.1808353115
156. Cerwenka A, Lanier LL (2016) Natural killer cell memory in infection, inflammation and cancer. Nat Rev Immunol 16(1):112–123. https://doi.org/10.1038/nri.2015.9
157. Murphy AG, O'Keeffe KM, Lalor SJ, Maher BM, Mills KHG, McLoughlin RM (2014) *Staphylococcus aureus* infection of mice expands a population of memory γδ T cells that are protective against subsequent infection. J Immunol 192(8):3697–3708. https://doi.org/10.4049/jimmunol.1303420
158. Dillen CA, Pinsker BL, Marusina AI, Merleev AA, Farber ON, Liu H et al (2018) Clonally expanded γδ T cells protect against *Staphylococcus aureus* skin reinfection. J Clin Invest 128(3):1026–1042. https://doi.org/10.1172/jci96481
159. Marchittoa MC, Dillena CA, Liua H, Miller RJ, Archer NK, Ortines RV et al (2019) Clonal $V\gamma6^{+}V\delta4^{+}$ T cells promote IL-17–mediated immunity against *Staphylococcus aureus* skin infection. Proc Natl Acad Sci U S A 116(22):10917–10926. https://doi.org/10.1073/pnas.1818256116
160. Bertrama T, Reimersa D, Lorya NC, Schmidta C, Schmida J, Heiniga LC et al (2023) Kidney-resident innate-like memory γδ T cells control chronic *Staphylococcus aureus* infection of mice. Proc Natl Acad Sci U S A 120(1):e2210490120. https://doi.org/10.1073/pnas.2210490120
161. Halim TY, Steer CA, Mathä L, Gold MJ, Martinez-Gonzalez I, McNagny KM et al (2014) Group 2 innate lymphoid cells are critical for the initiation of adaptive T helper 2 cell-mediated allergic lung inflammation. Immunity 40(3):425–435. https://doi.org/10.1016/j.immuni.2014.01.011
162. Martinez-Gonzalez I, Mathä L, Steer CA, Ghaedi M, Poon GF, Takei F (2016) Allergen-experienced group 2 innate lymphoid cells acquire memory-like properties and enhance allergic lung inflammation. Immunity 45(1):198–208. https://doi.org/10.1016/j.immuni.2016.06.017
163. Holtfreter S, Kolata J, Broker BM (2010) Towards the immune proteome of *Staphylococcus aureus*—the anti-*S. Aureus* antibody response. Int J Med Microbiol 300:176–192
164. Brocker BM, Holtfreter S, BekeredjiaDing I (2014) Immune control of *Staphylococcus aureus*-regulation and counter-regulation of the adaptive immune response. Int J Med Microbiol 304(2–3):204–214. https://doi.org/10.1016/j.ijmm.2009.10.002
165. Stentzela S, Sundaramoorthy N, Michalikb S, Nordengrüna M, Schulza S, Kolataa J et al (2015) Specific serum IgG at diagnosis of *Staphylococcus aureus* bloodstream invasion is correlated with disease progression. J Proteomics 128:1–7. https://doi.org/10.1016/j.jprot.2015.06.018
166. Keener AB, Thurlow LT, Kang S, Spidale NA, Clarke SH, Cunnion KJ et al (2017) *Staphylococcus aureus* protein a disrupts immunity mediated by long-lived plasma cells. J Immunol 198(3):1263–1273. https://doi.org/10.4049/jimmunol.1600093
167. Verkaik NJ, de Vogel CP, Boelens HA, Grumann D, Hoogenboezem T, Vink C et al (2009) Anti-staphylococcal humoral immune response in persistent nasal carriers and noncarriers of *Staphylococcus aureus*. J Infect Dis 199(5):625–632. https://doi.org/10.1086/596743
168. Kluytmans J, van Belkum A, Verbrugh H (1997) Nasal carriage of *Staphylococcus aureus*: epidemiology, underlying mechanisms, and associated risks. Clin Microbiol Rev 10(3):505–520. https://doi.org/10.1128/cmr.10.3.505
169. Ballow M (2002) Primary immunodeficiency disorders: antibody deficiency. J Allergy Clin Immunol 109(4):581–591. https://doi.org/10.1067/mai.2002.122466
170. Fowler VG Jr, Proctor RA (2014) Where does a *Staphylococcus aureus* vaccine stand? Clin Microbiol Infect 20(Suppl 5):S66–S75. https://doi.org/10.1111/1469-0691.12570
171. Spellberg B, Ibrahim AS, Yeaman MR, Lin L, Fu Y, Avanesian V et al (2008) The antifungal vaccine derived from the recombinant N terminus of Als3p protects mice against the bacterium *Staphylococcus aureus*. Infect Immun 76(10):4574–4580. https://doi.org/10.1128/IAI.00700-08

172. Lin L, Ibrahim AS, Xu X, Farber JM, Avanesian V, Baquir B et al (2009) Th1-Th17 cells mediate protective adaptive immunity against *Staphylococcus aureus* and *Candida albicans* infection in mice. PLoS Pathog 5(12):e1000703. https://doi.org/10.1371/journal.ppat.1000703
173. Brown AF, Murphy AG, Lalor SJ, Leech JM, O'Keeffe KM, Aogáin MM et al (2015) Memory Th1 cells are protective in invasive *Staphylococcus aureus* infection. PLoS Pathog 11(11):e1005226. https://doi.org/10.1371/journal.ppat.1005226
174. Narita K, Hu D-L, Mori F, Wakabayashi K, Iwakura Y, Nakane A (2010) Role of interleukin-17A in cell-mediated protection against *Staphylococcus aureus* infection in mice immunized with the fibronectin-binding domain of clumping factor. Infect Immun 78(10):4234–4242. https://doi.org/10.1128/IAI.00447-10
175. McGeachy MJ, Cua DJ, Gaffen SL (2019) The IL-17 family of cytokines in health and disease. Immunity 50(4):892–906. https://doi.org/10.1016/j.immuni.2019.03.021
176. Narita K, Hu D-L, Asano K, Nakane A (2015) Vaccination with non-toxic mutant toxic shock syndrome toxin-1 induces IL-17-depednent protection against *Staphylococcus aureus* infection. Pathog Dis 73(4):ffv023. https://doi.org/10.1093/femspd/ftv023
177. Lindenstrøm T, Woodworth J, Dietrich J, Aagaard C, Anderson P, Agger EM (2012) Vaccine-induced th17 cells are maintained long-term postvaccination as a distinct and phenotypically stable memory subset. Infect Immun 80(10):3533–3544. https://doi.org/10.1128/IAI.00550-12
178. Glosson-Byers NL, Sehra S, Stritesky GL, Yu Q, Awe O, Pham D et al (2014) Th17 cells demonstrate stable cytokine production in a proallergic environment. J Immunol 193(6):2631–2640. https://doi.org/10.4049/jimmunol
179. Annunziato F, Cosmi L, Santarlasci V, Maggi L, Liotta F, Mazzinghi B et al (2007) Phenotypic and functional features of human Th17 cells. J Exp Med 204(8):1849–1861. https://doi.org/10.1084/jem.20070663
180. Lee YK, Turner H, Maynard CL, Oliver JR, Chen D, Elson CO et al (2009) Late developmental plasticity in the T helper 17 lineage. Immunity 30(1):92–107. https://doi.org/10.1016/j.immuni.2008.11.005
181. Zielinski CE, Mele F, Aschenbrenner D, Aschenbrenner D, Jarrossay D, Ronchi F et al (2012) Pathogen-induced human TH17 cells produce IFN-γ or IL-10 and are regulated by IL-1β. Nature 484(7395):514–518. https://doi.org/10.1038/nature10957
182. Gagliani N, Vesely MCA, Iseppon A, Brockmann L, Xu H, Palm NW et al (2015) TH17 cells transdifferentiate into regulatory T cells during resolution of inflammation. Nature 523(7559):221–225. https://doi.org/10.1038/nature14452
183. Narita K, Hu DL, Asano K, Nakane A (2019) IL-10 produced by mutant toxic shock syndrome toxin-1 vaccine-induced memory T cells downregulates IL-17 production and abrogates the protective effect against *Staphylococcus aureus* infection. Infect Immun 87(10):e00494–e00419. https://doi.org/10.1128/IAI.00494-19
184. Kennedy EA, Connolly J, Hourihane JOB, Fallon PG, Mclean WHI, Murray D et al (2017) Skin microbiome before development of atopic dermatitis: early colonization with commensal staphylococci at 2 months is associated with a lower risk of atopic dermatitis at 1 year. J Allergy Clin Immunol 139(1):166–172. https://doi.org/10.1016/j.jaci.2016.07.029
185. Kong HH, Oh J, Deming C, Conlan S, Grice EA, Beatson MA et al (2012) Temporal shifts in the skin microbiome associated with disease flares and treatment in children with atopic dermatitis. Genome Res 22(5):850–859. https://doi.org/10.1101/gr.131029.111
186. Kobayashi T, Glatz M, Horiuchi K, Kawasaki H, Akiyama H, Kaplan DH et al (2015) Dysbiosis and *Staphylococcus aureus* colonization drives inflammation in atopic dermatitis. Immunity 42(4):756–766. https://doi.org/10.1016/j.immuni.2015.03.014
187. Lowy FD (1998) *Staphylococcus aureus* infections. N Engl J Med 339(8):520–532. https://doi.org/10.1056/NEJM199808203390806
188. Leung DY, Bieber T (2003) Atopic dermatitis. Lancet 361(9352):151–160. https://doi.org/10.1016/S0140-6736(03)12193-9

189. Nakatsuji T, Chen TH, Narala S, Chun KA, Two AM, Yun T et al (2017) Antimicrobials from human skin commensal bacteria protect against *Staphylococcus aureus* and are deficient in atopic dermatitis. Sci Transl Med 9(378):eaah4680. https://doi.org/10.1126/scitranslmed.aah4680
190. Lai Y, Cogen AL, Radek KA, Park HJ, MacLeod DT, Leichtle A et al (2010) Activation of TLR2 by a small molecule produced by *Staphylococcus epidermidis* increases antimicrobial defense against bacterial skin infections. J Invest Dermatol 130(9):2211–2221. https://doi.org/10.1038/jid.2010.123
191. Sandilands A, O'Regan GM, Liao H, Zhao Y, Terron-Kwiatkowski A, Watson RM et al (2006) Prevalent and rare mutations in the gene encoding filaggrin cause ichthyosis vulgaris and predispose individuals to atopic dermatitis. J Invest Dermatol 126(8):1770–1775. https://doi.org/10.1038/sj.jid.5700459
192. Cabanillas B, Novak N (2016) Atopic dermatitis and filaggrin. Curr Opin Immunol 42:1–8. https://doi.org/10.1016/j.coi.2016.05.002
193. Meylan P, Lang C, Mermoud S, Johannsen A, Norrenberg S, Hohl D et al (2017) Skin colonization by *Staphylococcus aureus* precedes the clinical diagnosis of atopic dermatitis in infancy. J Invest Dermatol 137(12):2497–2504. https://doi.org/10.1016/j.jid.2017.07.834
194. Marenholz I, Nickel R, Rüschendorf F, Schulz F, Esparza-Gordillo J, Kerscher T et al (2006) Filaggrin loss-of-function mutations predispose to phenotypes involved in the atopic march. J Allergy Clin Immunol 118(4):866–871. https://doi.org/10.1016/j.jaci.2006.07.026
195. Sandilands A, Sutherland C, Irvine AD, McLean WH (2009) Filaggrin in the frontline: role in skin barrier function and disease. J Cell Sci 122(Pt 9):1285–1294. https://doi.org/10.1242/jcs.033969
196. Irvine AD, McLean WHI, Leung DYM (2011) Filaggrin mutations associated with skin and allergic diseases. N Engl J Med 365(14):1315–1327. https://doi.org/10.1056/nejmra1011040
197. Kezic S, O'Regan GM, Yau N, Sandilands A, Chen H, Campbell LE et al (2011) Levels of filaggrin degradation products are influenced by both filaggrin genotype and atopic dermatitis severity. Allergy 66(7):934–940. https://doi.org/10.1111/j.1398-9995.2010.02540.x
198. Biedermann T, Skabytska Y, Kaesler S, Volz T (2015) Regulation of T cell immunity in atopic dermatitis by microbes: the yin and Yang of cutaneous inflammation. Front Immunol 6:353. https://doi.org/10.3389/fimmu.2015.00353
199. Krause K, Metz M, Makris M, Zuberbier T, Maurer M (2012) The role of interleukin-1 in allergy-related disorders. Curr Opin Allergy Clin Immunol 12(5):477–484. https://doi.org/10.1097/aci.0b013e3283574d0c
200. Gallo RL, Hooper LV (2012) Epithelial antimicrobial defence of the skin and intestine. Nat Rev Immunol 12:503–516. https://doi.org/10.1038/nri3228
201. Leung DY, Guttman-Yassky E (2014) Deciphering the complexities of atopic dermatitis: shifting paradigms in treatment approaches. J Allergy Clin Immunol 134(6):769–779. https://doi.org/10.1016/j.jaci.2014.08.008
202. Suárez-Fariñas M, Tintle SJ, Shemer A, Chiricozzi A, Nograles K, Cardinale I et al (2011) Nonlesional atopic dermatitis skin is characterized by broad terminal differentiation defects and variable immune abnormalities. J Allergy Clin Immunol 127(4):954–964. https://doi.org/10.1016/j.jaci.2010.12.1124
203. Farmand S, Sundin M (2015) Hyper-IgE syndromes: recent advances in pathogenesis, diagnostics and clinical care. Curr Opin Hematol 22:12–22. https://doi.org/10.1097/MOH.0000000000000104
204. Milner JD, Brenchley JM, Lawrence A, Fruman AF, Hill BJ, Elias KM et al (2008) Impaired TH17 cell differentiation in subjects with autosomal dominant hyperIgE syndrome. Nature 452(7188):773–776. https://doi.org/10.1038/nature06764
205. Myles IA, Anderson ED, Earland NJ, Zarember CA, Sastalla I, Williams KW et al (2018) TNF overproduction impairs epithelial staphylococcal response in hyper IgE syndrome. J Clin Invest 128(8):3595–3604. https://doi.org/10.1172/JCI121486

206. Nakamura Y, Oscherwitz J, Cease KB, Chan SM, Muñoz-Planillo R, Hasegawa M et al (2013) Staphylococcus δ-toxin induces allergic skin disease by activating mast cells. Nature 503(7476):397–401. https://doi.org/10.1038/natur12655
207. Brauweiler AM, Goleva E, Leung DYM (2014) Th2 cytokines increase *Staphylococcus aureus* α toxin induced keratinocyte death through the signal transducer and activator of transcription 6 (STAT6). J Invest Dermatol 134(8):2114–2121. https://doi.org/10.1038/jid.2014.43
208. Spaulding AR, Salgado-Pabón W, Kohler PL, Horswill R, Leung DYM, Schlievert PM (2013) Staphylococcal and streptococcal superantigen exotoxins. Clin Microbiol Rev 26(3):422–446. https://doi.org/10.1128/CMR.00104-12
209. Boguniewicz M, Leung DY (2010) Recent insights into atopic dermatitis and implications for management of infectious complications. J Allergy Clin Immunol 125(1):4–15. https://doi.org/10.1016/j.jaci.2009.11.027
210. Leung DY, Harbeck R, Bina P, Reiser RF, Yang E, Norris DA et al (1993) Presence of IgE antibodies to staphylococcal exotoxins on the skin of patients with atopic dermatitis. Evidence for a new group of allergens. J Clin Invest 92(3):1374–1380. https://doi.org/10.1172/JCI116711
211. Bunikowski R, Mielke M, Skarabis H, Herz U, Bergmann RL, Wahn U et al (1999) Prevalence and role of serum IgE antibodies to the *Staphylococcus aureus*-derived superantigens SEA and SEB in children with atopic dermatitis. J Allergy Clin Immunol 103(1 Pt 1):119–124. https://doi.org/10.1016/s0091-6749(99)70535-x
212. Skov L, Olsen JV, Giorno R, Schlievert PM, Baadsgaard O, Leung DY (2000) Application of staphylococcal enterotoxin B on normal and atopic skin induces up-regulation of T cells by a superantigen-mediated mechanism. J Allergy Clin Immunol 105(4):820–826. https://doi.org/10.1067/mai.2000.105524
213. Orfali RL, da Silva Oliveira LM, de Lima GJF, de Carvalho C, Ramos YAL et al (2018) *Staphylococcus aureus* enterotoxins modulate IL-22-secreting cells in adults with atopic dermatitis. Sci Rep 8:6655. https://doi.org/10.1038/s41598-018-25125-0
214. Nakagawa S, Matsumoto M, Katayama Y, Oguma R, Wakabayashi S, Nygaard T et al (2017) *Staphylococcus aureus* virulent PSMα peptides induce keratinocyte alarmin release to orchestrate IL-17-dependent skin inflammation. Cell Host Microbe 22(5):667–677. https://doi.org/10.1016/j.chom.2017.10.008
215. Liu H, Archer NK, Dillen CA, Wang Y, Ashbaugh AG, Ortines RV et al (2017) *Staphylococcus aureus* epicutaneous exposure drives skin inflammation via IL-36-mediated T cell responses. Cell Host Microbe 22(5):653–666. https://doi.org/10.1016/j.chom.2017.10.006
216. Nakane A, Numata A, Minagawa T (1992) Endogenous tumor necrosis factor, interleukin-6, and gamma interferon levels during *listeria monocytogenes* infection in mice. Infect Immun 60(2):523–528. https://doi.org/10.1128/iai.60.2.523-528.1992
217. Nakane A, Minagawa T, Kohanawa M, Chen Y, Sato H, Moriyama M et al (1989) Interactions between gamma interferon and tumor necrosis factor in host resistance against primary and secondary *listeria monocytogenes* infections. Infect Immun 57(11):3331–3337. https://doi.org/10.1128/iai.57.11.3331-3337.1989
218. Baschieri S, Lees KK, Lussow AR, MacDonald HR (1993) Clonal anergy to staphylococcal enterotoxin B in vivo: selective effects on T cell subsets and lymphokines. Eur J Immunol 23(10):2661–2666. https://doi.org/10.1002/eji.1830231041

Chapter 6
Immune Evasion and Interaction Between Autophagy and Intracellular *Staphylococcus aureus*

Krisana Asano

Abstract *Staphylococcus aureus* found on human skin is an opportunistic pathogen that causes a variety of infections. Although this bacterium has been considered as an extracellular pathogen, several studies have demonstrated that *S. aureus* is able to internalize and replicate in both professional and non-professional phagocytic cells. This intracellular niche is important for escape from humoral immunity and antibiotics, and is relevant to recurrent and chronic infections of *S. aureus*. For intracellular survival, *S. aureus* develops mechanisms to resist the innate immune response, particularly a cellular homeostatic autophagy. Autophagy is thought to be a cellular defense system that transfers invading pathogens to lysosomes for elimination. Therefore, understanding the interaction between *S. aureus* and autophagy is important to prevent recurrent and chronic *S. aureus* infections. This chapter focuses on the interaction between *S. aureus* and autophagy to understand how this bacterium evades this innate immune degradation or combats autophagy for intracellular survival.

Keywords Autophagy · Immune evasion

Abbreviations

agr	Accessary gene regulator
Atg5	Autophagy-related protein 5
BRCA1	Neighbor of breast cancer susceptibility gene 1
CALCOCO2	Calcium binding and coiled-coil domain 2
cAMP	Cyclic adenosine monophosphate
Hla	Alpha hemolysin
LC3	Microtubule-associated protein 1A/1B-light chain 3

K. Asano (✉)
Department of Microbiology and Immunology, Hirosaki University Graduate School of Medicine, Hirosaki, Aomori, Japan
e-mail: krisana@hirosaki-u.ac.jp

A. Nakane, K. Asano (eds.), *Staphylococcus aureus*,
https://doi.org/10.1007/978-981-99-9428-1_6

MAPK 14	Mitogen-activated protein kinase 14
PI3K	Phosphoinositide-3-kinase
PMN	Polymorphonuclear neutrophils
SCV	Small colony variant
TSST-1	Toxic shock syndrome toxin-1

6.1 Introduction

S. aureus had been considered to be an extracellular bacterium. However, over the past decades, it is becoming clear that *S. aureus* can invade into both professional and non-professional phagocytic cells. Moreover, this bacterium remains alive and persists in the host cells for an extended period of time after invading [1]. This evidence has changed the concept of *S. aureus* from being an extracellular pathogen to a facultative intracellular pathogen [2–5].

6.1.1 *Intracellular Survival of* S. aureus *in Professional Phagocytic Cells*

Professional phagocytes are known to actively phagocytose microorganisms and eliminate them in phagolysosomes. Only some pathogens such as *Mycobacterium tuberculosis* and *Streptococcus pyogenes* evolve the mechanisms to avoid this elimination and can survive after phagocytosis. *S. aureus* has also been shown to resist to this destructive activity and survive in neutrophils and macrophages [6–9]. Twenty-four hours after intraperitoneal infection of model mice, polymorphonuclear neutrophils (PMN) isolated from the peritoneal fluid harbor *S. aureus*. Furthermore, these infected PMN are sufficient to infect another naïve animal, indicating that this intracellular *S. aureus* is alive and active [8]. Moreover, *S. aureus* is shown to survive in human and animal macrophages [9–11]. Jakob and Green demonstrated that *S. aureus* remains in alveolar macrophages of mice infected with Sendai virus [11]. In particular, Kubota et al. showed that *S. aureus* survives in human monocyte-derived macrophages for 4 days and then causes host cell lysis [9]. Moreover, Hebert et al. showed that living *S. aureus* is present in both alveolar cells and macrophages in cow milk with chronically with bovine mastitis [10].

6.1.2 *Intracellular Persistence of* S. aureus *in Non-Professional Phagocytic Cells*

S. aureus infections have been shown to be associated with osteomyelitis [12]. Several surface molecules of *S. aureus* promote bone colonization [13–16] and internalize into osteoblasts [17–20]. Ellington and colleagues demonstrated that

S. aureus remains within human osteoblasts for 24 h and the bacterial cells that released from the dead osteoblasts can re-infect human osteoblasts [21]. These results suggest that *S. aureus* from the infected osteoblasts is still alive and active, as reported from PMN.

In addition to the osteoblasts, *S. aureus* is capable of internalization and survival in several non-professional phagocytic cells. Jarry and colleagues demonstrated that *S. aureus* invades and proliferates in airway epithelial cells derived from cystic fibrosis and these bacterial cells subsequently escape from the endocytic vesicles before apoptosis occurs [22]. By using an in vitro model, Garzoni et al. found that *S. aureus* persists in human epithelial cells for up to 2 weeks without host cell death or appearance of small colony variant (SCV) [23]. *S. aureus* cells can also adhere keratinocytes and are encapsulated in cytoplasmic vesicles where the bacterial cells can persist for 24–48 h [24]. Moreover, Menziesan and Kourteva have shown that the intracellular *S. aureus* remains viable up to 72 h after invading into human umbilical vein endothelial cells. Furthermore, the internalized *S. aureus* can exist in both vacuoles and cytoplasm, and only viable intracellular *S. aureus* induces apoptosis of endothelial cells [25]. This ability of *S. aureus* to internalize and survive in the endothelial cells is believed to relevant to the clinical characteristic of *S. aureus* infection that causes persistent endovascular infection with endothelial destruction.

It has been shown that *S. aureus* infections with severe peritoneal damage are often associated with a dramatic loss of mesothelial cells. Haslinger-Löffler et al. demonstrated that *S. aureus* can invade into cultured human peritoneal mesothelial cells. Although these cells are able to degrade *S. aureus*, a high percentage of alive *S. aureus* remains within these cells after prolonged infection and is released upon host cell death [26].

S. aureus also inhabits in the intestinal tract, a reservoir of antibiotics. Therefore, Hess et al. observed the ability of *S. aureus* to invade enterocytes for antibiotic avoidance. By using cultured enterocytes, Hess found that *S. aureus* survives within these cells for several days [27].

6.1.3 Intracellular Phenotype Switching of S. aureus

There is some evidence that intracellular *S. aureus* phenotype switches from the wildtype to a persistent form called SCV (small colony variant) [28–30]. The SCV subpopulation of *S. aureus* is characterized by slow growth due to decreased production of metabolic mediators as well as virulence factors [29, 31, 32]. This distinct phenotype is thought to protect SCVs from the host immune response and resist to lysosomal elimination [3, 4, 28, 32–35].

Intracellular persistence of *S. aureus* provides an ideal strategy for *S. aureus* to escape phagocytosis, humoral immunity and extracellular antibiotics, and consequently promotes recurrent and chronic infections. In 2020, Peyrusson and colleagues provided evidence that *S. aureus* can survive as intracellular persister in the antibiotic-treated cells. These cells include professional and non-professional phagocytic cells such as human macrophages, monocytes, epithelial cells and primary keratinocytes. The intracellular *S. aureus* is associated with multidrug resistance after single exposure to antibiotics such as oxacillin, clarithromycin, or moxifloxacin [36]. This accumulated data of *S. aureus* in the intracellular niches correlates with several clinical cases that are reported to be asymptomatic and apparently cured after antimicrobial treatment, but recurrent infections occur within several months or years [37–39]. In addition, the professional phagocytic cells carrying live *S. aureus* can contribute to the dissemination of infections. By these cells, *S. aureus* can spread to blood circulatory system and cause sepsis. Moreover, the intracellular persistence of *S. aureus* in the epithelial cells, endothelial cells and osteoblasts may be relevant to pneumonia, skin infections, wound infections and even osteomyelitis.

6.2 Interaction Between Autophagy and Intracellular Pathogens

Autophagy is a cellular homeostatic pathway that involves the formation of a double membrane layer covering the cytoplasmic matrix, forming organelle known as autophagosome [40]. Autophagosome is an essential organelle for cells to survive in nutrient-deficient conditions. To reduce cellular metabolism in the nutrient-deficient cells, autophagosome delivers cytoplasmic proteins, mitochondria and other organelles to the lysosome, and forms autophagolysosome [41]. Since the enzymes within autophagolysosome break down the intracellular substances and recycle them as metabolic precursors, autophagy pathway is shown to play a role in maintaining cellular metabolism [42, 43]. Autophagy is also important for cellular homeostasis because of its ability to regulate the quality of cellular components. Although autophagy is vital for cell survival, it is involved in the induction of host cell death under certain circumstances [44]. Therefore, autophagy is considered as a double-edges sword that can be both protective and injurious.

Autophagy is primarily described as nonselective process that eliminates cytoplasmic garbage to recycle metabolic precursors in low nutritional conditions. However, several studies have shown that intracellular organelles are selectively removed by autophagy [45]. These selective targets of autophagy are mediated by adaptor proteins including nuclear domain 10 protein 52 (NDP52), p62, neighbor of breast cancer susceptibility gene 1 (BRCA1) and optineurin [46]. The selective autophagy for organelles such as mitochondria, peroxisomes, lysosomes, ribosomes and endoplasmic reticulum has been elucidated and is called mitophagy pexophagy, lysophagy, ribophagy and ERphagy, respectively. In particular, the autophagy that is specifically selective for invading pathogens is called xenophagy.

Xenophagy, an autophagy that selectively acts on invading pathogens, has been suggested to be an innate immune response against harmful microorganisms. After internalization of pathogen, the formation of ubiquitinated protein aggregates is induced. The pathogen aggregated with ubiquitinated proteins is then recognized by adaptors proteins, phagophore, and ultimately destroyed in autophagolysosome [47,

48]. Although the xenophagy is an efficient host defense system, many successful intracellular bacteria have developed sophisticated mechanisms to overcome this degrading system and even proliferate [49]. The interaction between autophagy and successful intracellular pathogens can be classified into three groups; (1) pathogens that their growth is restricted in autophagosome, (2) pathogens that can evade autophagy recognition and (3) pathogens that can proliferate in either autophagosome or autophagolysosome. The pathogens that their growth is restricted in autophagosome include *Mycobacterium tuberculosis* [50], *Salmonella* Typhimurium [51] and *Streptococcus pyogenes* [52]. On the other hand, the bacteria such as *Shigella flexneri*, *Listeria monocytogenes* and *Burkholderia pseudomallei* induce actin polymerization to form actin comets in the cytoplasm after invasion [53–57]. These pathogens use the actin comets to migrate resulting in membrane protrusions, which allow them to spread from cell to cell and avoid autophagy targeting. The pathogens that have been reported to exploit autophagosome or autophagolysosome for proliferation included *Coxiella burnetiid, Legionella pneumophila, Yersinia pseudotuberculosis, Anaplasma phagocytophilum* and *Brucella abortus* [58–62]. They have evolved the sophisticated mechanisms to block or delay the fusion of autophagosome and lysosome and even to resist the acidification in autophagolysosome.

6.3 Interplay Between Autophagy and Intracellular *S. aureus*

To survive in the intracellular niche, *S. aureus* has to evolve some mechanisms to resist autophagy. Autophagy has been shown to be induced after *S. aureus* invasion. However, the mechanism of interaction between *S. aureus* and autophagy is still elusive. Mauthe et al. demonstrated that after invading into non-professional host cells, *S. aureus* is entrapped in autophagosome-like vesicle and targeted for lysosomal degradation [63]. On the other hand, some studies reported that the intracellular *S. aureus* have evolved mechanisms to evade autophagy. In the study of Mestre group, the replication of *S. aureus* in the cytoplasm was observed [64]. In addition, Bravo-Santano et al. found that intracellular *S. aureus* is able to disrupt autophagy and escape into the cytoplasm. In response to *S. aureus* invasion, low levels of glucose and several essential amino acids were observed, coupled with autophagy activation. However, most of the autophagosomes detected in the infected cells were not co-localized with bacterial cells. Thus, the data of Bravo-Santano group suggest that the internalized *S. aureus* induces autophagic flux for energy generation and nutrient scavenging [65]. Parallel to the finding of Mestre group, Neumann et al. showed that *S. aureus* proliferates in the cytoplasm. After entering the host cells, *S. aureus* is ubiquitinated and recognized by adapter proteins [such as p62, optineurin and calcium binding and coiled-coil domain 2 (CALCOCO2)]. However, Neumann et al. demonstrated that the invading *S. aureus* inhibits phagophore elongation by activating the phosphorylation of mitogen-activated protein kinase 14 (MAPK14) and autophagy-related protein 5 (Atg5) [66]. Asano group demonstrated that autophagosome formation is suppressed by toxic shock syndrome toxin-1 (TSST-1), a superantigen that is produced by the virulence strains of *S. aureus* [67]. Excluding the above-mentioned mechanisms, Neumann's group and Schnaith's group showed that autophagosome-lysosome fusion is inhibited after *S. aureus* infection. In particular, Schnaith's group reported that *S. aureus* utilizes autophagosomes for replication. The replicated *S. aureus* cells then eventually escape into the cytoplasm and induce host cell death [68]. These data suggest the complicated interaction between intracellular *S. aureus* and autophagy. After entering the host cells, post-invasion events depend on the strains of *S. aureus* and the produced virulence factors. In particular, the accessary gene regulator (*agr*), alpha-toxin and TSST-1 have been shown to play important role in autophagy escapable mechanism of *S. aureus*.

6.3.1 *Effect of* agr *on Autophagy*

The virulence strains of *S. aureus* produce a variety of virulence factors to enhance their ability of infection. The expression of these virulence factors is regulated by a series of regulators including *agr* [69, 70]. The *agr* that regulates a wide array expression of virulence factors of *S. aureus* is known to correlated with host cell death. Haslinger-Löffler et al. demonstrated that *agr* mutant of *S. aureus* retains the ability to invade human peritoneal mesothelial cells but fails to induce apoptosis [71]. *S. aureus* requires the expression of *agr* to escape from the endocytic vesicle into the cytosol of epithelial cells [72]. Furthermore, O'Keeffe et al. showed that *S. aureus* with high level of *agr* activity can block autophagic flux, resulting an accumulation of autophagosomes in the cytoplasm [73]. These studies correlate with the finding of Schnaith et al. which demonstrated that *agr*-deficient *S. aureus* is trapped in phagosomal pathway, whereas the wildtype *agr S. aureus* induces autophagy and localizes in the microtubule-associated protein 1A/1B-light chain 3 (LC3)-positive autophagosomes. Nevertheless, Schnaith et al. found that *agr*-positive *S. aureus* inhibits autophagosome-lysosome fusion, and the bacterial cells replicate in non-acidic LC3-positive vesicles, which then host cell death is induced after the replicated *S. aureus* is released into the cytoplasm [68].

6.3.2 *Effect of Alpha-Toxin on Autophagy*

Alpha-toxin, also known as alpha hemolysin (Hla), is one of cytotoxins that is produced by *S. aureus*. The expression of this toxin is regulated by the *agr* regulatory system. After expression and release outside the bacterial cells, the mature toxin folds into homoheptameric structure and forms pore in the phospholipid membrane of eukaryotic cells. As a result, it lyses host cells, especially erythrocytes [74]. Regarding pore-forming ability of Hla, host cell death is induced [75]. In addition,

Hla produced after *S. aureus* internalization is thought to lyse the membranes of phagosome, endosome and autophagosome. In the epithelial cells, the Hla activity is required for *S. aureus* to escape from endosome and be released to the cytoplasm [72]. Mestre et al. demonstrated that Hla promotes autophagy by decreasing intracellular cAMP (cyclic adenosine monophosphate) but does not activate autophagy in a phosphoinsitide-3-kinase (PI3K)/Beclin1-independent mechanism [64]. Hla-positive *S. aureus* is found to be localized in LC3-positive, non-acidic and non-degradative vesicles. In contrast, Hla-negative *S. aureus* is not able to escape from phagosome. These Hla-negative cells are localized in LC3-negative, acidic vesicles [64]. This data indicate that Hla activates autophagy and inhibits the fusion of autophagosome and lysosome.

6.3.3 Effect of TSST-1 on Autophagy

TSST-1 is a superantigen produced by *S. aureus* [74]. TSST-1 secreted from bacterial cells activates inflammatory response mediated by cross-linking between T cell and antigen-presenting cell via T cell receptor and major histocompatibility complex class II. In addition to the superantigenic activity of TSST-1, the toxin has been shown to internalize the epithelial cells [76] and contribute to infection of *S. aureus* [77]. Asano et al., reported that TSST-1 reduces autophagosome accumulation in the nutrient-starved cells. This effect is not involved in superantigenic activity [67]. Although the mechanism of TSST-1 for autophagy inhibition has not been yet elucidated, this activity is expected to inhibit autophagy-mediated bacterial elimination.

6.4 Perspectives

Several studies have confirmed that autophagy is induced after *S. aureus* invasion, and the complicated interaction between *S. aureus* and autophagy is suggested. The strategy of *S. aureus* to counteract autophagy appears to depend on strain specificity, virulence factors, cell types and intracellular environments. Accumulating data published today suggest that a part of intracellular *S. aureus* is eliminated by autophagy (Fig. 6.1a) [63]. *S. aureus* in the cytoplasm is ubiquitinated and aggregated with adapter proteins such as p62, optineurin and calcium binding and coiled-coil domain 2. Subsequently, the phagophore binds *S. aureus*-adapter protein complex and then elongates to form mature autophagosome. After lysosome fusion occurs, *S. aureus* is eliminated by enzymes in the acidic circumstance of autophagolysosome [66]. For the *agr*- and Hla-positive *S. aureus*, autophagy is induced, but autophagosome-lysosome fusion is inhibited [66, 68]. Within this non-acidic, LC3-positve autophagosome, *S. aureus* replicates [68] and escapes to the cytoplasm by the activity of Hla. Eventually, host cell death is induced [64, 68] (Fig. 6.1b). For TSST-1-producing *S. aureus*, the toxin suppresses autophagosome formation [67]

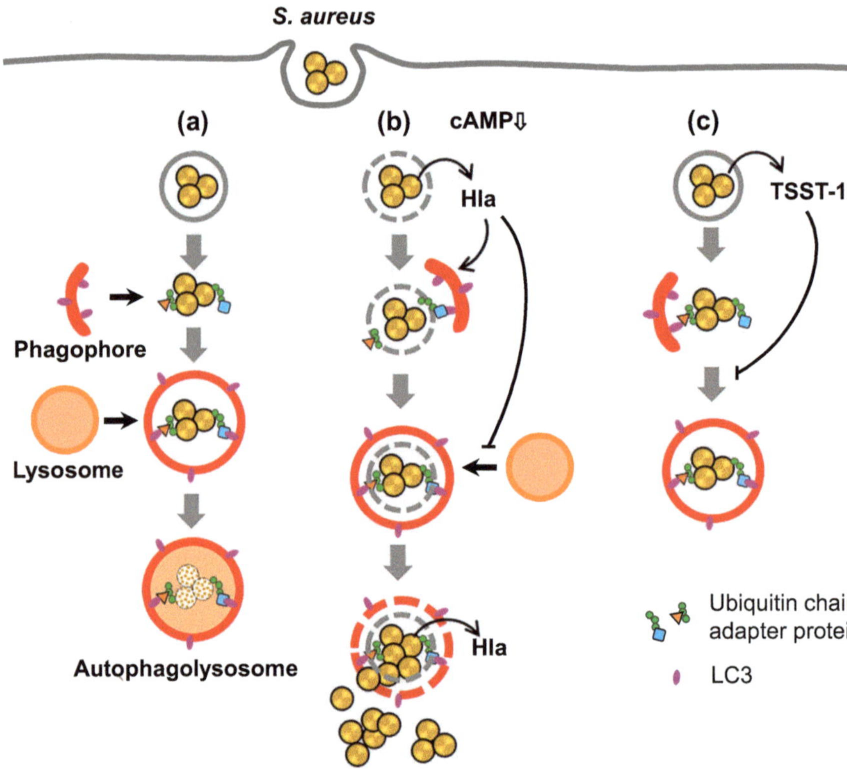

Fig. 6.1 Interplay between intracellular *S. aureus* and autophagy. (**a**) A part of intracellular *S. aureus* is eliminated by autophagolysosomes [63]. *S. aureus* in the cytoplasm becomes ubiquitinated and is recognized by adapter proteins [66]. *S. aureus* cells are entrapped in autophagosome-like vesicle and targeted for lysosomal degradation. (**b**) For the strains that are able to produce Hla under *agr* regulatory system, Hla destroys vesicle membrane. Subsequently, autophagy is promoted by decrease of intracellular cAMP [64] and lysosome fusion is inhibited by Hla [66, 68]. *S. aureus* replicates in autophagosomes [68]. Hla destroys autophagosome membrane, and then *S. aureus* is releases into the cytoplasm and induces host cell death [64, 68]. (**c**) For the strains that produce TSST-1, autophagosome formation is suppressed [67]. Degradation of intracellular *S. aureus* by autophagy system is partially reduced

(Fig. 6.1c). This may be a strategy that *S. aureus* utilizes not only to avoid bacterial elimination, but also to reduce host cell death.

6.5 Conclusions

The interaction between invading pathogens and autophagy is complicated. Functions of autophagy include regulation of metabolism, maintenance of cellular homeostasis and elimination of invading microorganisms. However, several intracellular pathogens have successfully evolved mechanisms to avoid autophagy and/

or to exploit autophagy for survival and proliferation. *S. aureus* is one of commensal bacteria that colonizes on human skin and persist in human nasal cavity. In the immunocompromised hosts, *S. aureus* becomes an opportunistic pathogen causing a variety of infections ranging from superficial skin infections to life-threatening sepsis. Several studies have shown that *S. aureus* can enter and persist in the host cells for long periods of time to evade antibody-mediated elimination and bactericidal or bacteriostatic effects of antibiotics. This intracellular persistence is implicated in recurrent and chronic infections of *S. aureus*. From the property and behavior of *S. aureus*, it is suggested that this bacterium establishes a dynamic relationship with its host and autophagy when it remains in the intracellular niche. The *agr* regulator and two toxins, Hla and TSST-1, are shown to be involved in autophagy. The *agr*-positive *S. aureus* induces autophagy and is entrapped in LC3-positive vesicles. However, these LC3-positive vesicles are non-acidic, suggesting that autophagosome is not fused with lysosome. Therefore, *S. aureus* can survive and proliferate. Similarly, Hla that is expressed under *agr* regulation activates autophagy and prevents the fusion of autophagosome and lysosome, subsequently prevents *S. aureus* degradation. On the other hand, TSST-1 has been reported to suppress autophagosome formation. Thus, the strategy of *S. aureus* to counteract autophagy may correlate with changes of Hla and TSST-1 expression. Future studies on other aspects of the detailed interaction between *S. aureus* and autophagy will be very important to understand its intracellular adaptation that associates with persistence stage or infection stage of *S. aureus*.

References

1. Kapral FA, Shayegani MG (1959) Intracellular survival of staphylococci. J Exp Med 110(1):123–138. https://doi.org/10.1084/jem.110.1.123
2. Fraunholz M, Sinha B (2012) Intracellular *Staphylococcus aureus*: live-in and let die. Front Cell Infect Microbiol 2:43. https://doi.org/10.3389/fcimb.2012.00043
3. Garzoni C, Kelley WL (2009) *Staphylococcus aureus*: new evidence for intracellular persistence. Trends Microbiol 17(2):59–65. https://doi.org/10.1016/j.tim.2008.11.005
4. Sendi P, Proctor RA (2009) *Staphylococcus aureus* as an intracellular pathogen: the role of small colony variants. Trends Microbiol 17(2):54–58. https://doi.org/10.1016/j.tim.2008.11.004
5. Lowy FD (2000) Is *Staphylococcus aureus* an intracellular pathogen? Trends Microbiol 8(8):341–343. https://doi.org/10.1016/s0966-842x(00)01803-5
6. Rogers DE, Tompsett R (1952) The survival of staphylococci within human leukocytes. J Exp Med 95(2):209–230. https://doi.org/10.1084/jem.95.2.209
7. Voyich JM, Braughton KR, Sturdevant DE, Whitney AR, Said-Salim B, Porcella SF et al (2005) Insights into mechanisms used by *Staphylococcus aureus* to avoid destruction by human neutrophils. J Immunol 175(6):3907–3919. https://doi.org/10.4049/jimmunol.175.6.3907
8. Gresham HD, Lowrance JH, Caver TE, Wilson BS, Cheung AL, Lindberg FP (2000) Survival of *Staphylococcus aureus* inside neutrophils contributes to infection. J Immunol 164(7):3713–3722. https://doi.org/10.4049/jimmunol.164.7.3713
9. Kubica M, Guzik K, Koziel J, Zarebski M, Richter W, Gajkowska B et al (2008) A potential new pathway for *Staphylococcus aureus* dissemination: the silent survival of *S. aureus* phagocytosed by human monocyte-derived macrophages. PLoS One 3(1):e1409. https://doi.org/10.1371/journal.pone.0001409

10. Hebert A, Sayasith K, Sénéchal S, Dubreuil P, Lagacé J (2000) Demonstration of intracellular *Staphylococcus aureus* in bovine mastitis alveolar cells and macrophages isolated from naturally infected cow milk. FEMS Microbiol Lett 193(1):57–62. https://doi.org/10.1111/j.1574-6968.2000.tb09402.x
11. Jakab GJ, Green GM (1976) Defect in intracellular killing of *Staphylococcus aureus* within alveolar macrophages in Sendai virus-infected murine lungs. J Clin Invest 57(6):1533–1539. https://doi.org/10.1172/JCI108423
12. Lew DP, Waldvogel FA (2004) Osteomyelitis. Lancet 364(9431):369–379. https://doi.org/10.1016/S0140-6736(04)16727-5
13. McDevitt D, Francois P, Vaudaux P, Foster TJ (1994) Molecular characterization of the clumping factor (fibrinogen receptor) of *Staphylococcus aureus*. Mol Microbiol 11(2):237–248. https://doi.org/10.1111/j.1365-2958.1994.tb00304.x
14. McGavin MH, Krajewska-Pietrasik D, Ryden C, Hook M (1993) Identification of a *Staphylococcus aureus* extracellular matrix-binding protein with broad specificity. Infect Immun 61(6):2479–2485. https://doi.org/10.1128/iai.61.6.2479-2485.1993
15. Park PW, Rosenbloom J, Abrams WR, Rosenbloom J, Mecham RP (1996) Molecular cloning and expression of the gene for elastin-binding protein (ebpS) in *Staphylococcus aureus*. J Biol Chem 271(26):15803–15809. https://doi.org/10.1074/jbc.271.26.15803
16. Patti JM, Bremell T, Krajewska-Pietrasik D, Abdelnour A, Tarkowski A, Rydén C et al (1994) The *Staphylococcus aureus* collagen adhesin is a virulence determinant in experimental septic arthritis. Infect Immun 62(1):152–161. https://doi.org/10.1128/iai.62.1.152-161.1994
17. Ellington JK, Reilly SS, Ramp WK, Smeltzer MS, Kellam JF, Hudson MC (1999) Mechanisms of *Staphylococcus aureus* invasion of cultured osteoblasts. Microb Pathog 26(6):317–323. https://doi.org/10.1006/mpat.1999.0272
18. Hudson MC, Ramp WK, Nicholson NC, Williams AS, Nousiainen MT (1995) Internalization of *Staphylococcus aureus* by cultured osteoblasts. Microb Pathog 19(6):409–419. https://doi.org/10.1006/mpat.1995.0075
19. Jevon M, Guo C, Ma B, Mordan N, Nair SP, Harris M et al (1999) Mechanisms of internalization of *Staphylococcus aureus* by cultured human osteoblasts. Infect Immun 67(5):2677–2681. https://doi.org/10.1128/IAI.67.5.2677-2681.1999
20. Reilly SS, Hudson MC, Kellam JF, Ramp WK (2000) In vivo internalization of *Staphylococcus aureus* by embryonic chick osteoblasts. Bone 26(1):63–70. https://doi.org/10.1016/s8756-3282(99)00239-2
21. Ellington JK, Harris M, Webb L, Smith B, Smith T, Tan K et al (2003) Intracellular *Staphylococcus aureus*. A mechanism for the indolence of osteomyelitis. J Bone Joint Surg Br 85(6):918–921. https://doi.org/10.1302/0301-620X.85B6.13509
22. Jarry TM, Cheung AL (2006) *Staphylococcus aureus* escapes more efficiently from the phagosome of a cystic fibrosis bronchial epithelial cell line than from its normal counterpart. Infect Immun 74(5):2568–2577. https://doi.org/10.1128/IAI.74.5.2568-2577.2006
23. Garzoni C, Francois P, Huyghe A, Couzinet S, Tapparel C, Charbonnier Y et al (2007) A global view of *Staphylococcus aureus* whole genome expression upon internalization in human epithelial cells. BMC Genomics 8:171. https://doi.org/10.1186/1471-2164-8-171
24. Mempel M, Schnopp C, Hojka M, Fesq H, Weidinger S, Schaller M et al (2002) Invasion of human keratinocytes by *Staphylococcus aureus* and intracellular bacterial persistence represent haemolysin-independent virulence mechanisms that are followed by features of necrotic and apoptotic keratinocyte cell death. Br J Dermatol 146(6):943–951. https://doi.org/10.1046/j.1365-2133.2002.04752.x
25. Menzies BE, Kourteva I (1998) Internalization of *Staphylococcus aureus* by endothelial cells induces apoptosis. Infect Immun 66(12):5994–5998. https://doi.org/10.1128/IAI.66.12.5994-5998.1998
26. Haslinger-Löffler B, Wagner B, Brück M, Strangfeld K, Grundmeier M, Fischer U et al (2006) *Staphylococcus aureus* induces caspase-independent cell death in human peritoneal mesothelial cells. Kidney Int 70(6):1089–1098. https://doi.org/10.1038/sj.ki.5001710

27. Hess DJ, Henry-Stanley MJ, Erickson EA, Wells CL (2003) Intracellular survival of *Staphylococcus aureus* within cultured enterocytes. J Surg Res 114(1):42–49. https://doi.org/10.1016/s0022-4804(03)00314-7
28. Proctor RA, van Langevelde P, Kristjansson M, Maslow JN, Arbeit RD (1995) Persistent and relapsing infections associated with small-colony variants of *Staphylococcus aureus*. Clin Infect Dis 20(1):95–102. https://doi.org/10.1093/clinids/20.1.95
29. Proctor RA, von Eiff C, Kahl BC, Becker K, McNamara P, Herrmann M et al (2006) Small colony variants: a pathogenic form of bacteria that facilitates persistent and recurrent infections. Nat Rev Microbiol 4(4):295–305. https://doi.org/10.1038/nrmicro1384
30. Tuchscherr L, Heitmann V, Hussain M, Viemann D, Roth J, von Eiff C et al (2010) *Staphylococcus aureus* small-colony variants are adapted phenotypes for intracellular persistence. J Infect Dis 202(7):1031–1040. https://doi.org/10.1086/656047
31. Balwit JM, van Langevelde P, Vann JM, Proctor RA (1994) Gentamicin-resistant, menadione and hemin auxotrophic *Staphylococcus aureus* persist within cultured endothelial cells. J Infect Dis 170(4):1033–1037. https://doi.org/10.1093/infdis/170.4.1033
32. von Eiff C, Heilmann C, Proctor RA, Woltz C, Peters G, Götz F (1997) A site-directed *Staphylococcus aureus* hemB mutant is a small-colony variant which persists intracellularly. J Bacteriol 179(15):4706–4712. https://doi.org/10.1128/jb.179.15.4706-4712.1997
33. Proctor RA, Kriegeskorte A, Kahl BC, Becker K, Löffler B, Peters G (2014) *Staphylococcus aureus* small colony variants (SCVs): a road map for the metabolic pathways involved in persistent infections. Front Cell Infect Microbiol 4:99. https://doi.org/10.3389/fcimb.2014.00099
34. von Eiff C, Becker K, Metze D, Lubritz G, Hockmann J, Schwarz T, Peters G (2001) Intracellular persistence of *Staphylococcus aureus* small-colony variants within keratinocytes: a cause for antibiotic treatment failure in a patient with Darier's disease. Clin Infect Dis 32(11):1643–1647. https://doi.org/10.1086/320519
35. Schröder A, Kland R, Peschel A, von Eiff C, Aepfelbacher M (2006) Live cell imaging of phagosome maturation in *Staphylococcus aureus* infected human endothelial cells: small colony variants are able to survive in lysosomes. Med Microbiol Immunol 195(4):185–194. https://doi.org/10.1007/s00430-006-0015-0
36. Peyrusson F, Varet H, Nguyen TK, Legendre R, Sismeiro O, Coppée JY et al (2020) Intracellular *Staphylococcus aureus* persisters upon antibiotic exposure. Nat Commun 11(1):2200. https://doi.org/10.1038/s41467-020-15966-7
37. Tuchscherr L, Medina E, Hussain M, Völker W, Heitmann V, Niemann S et al (2011) *Staphylococcus aureus* phenotype switching: an effective bacterial strategy to escape host immune response and establish a chronic infection. EMBO Mol Med 3(3):129–141. https://doi.org/10.1002/emmm.201000115
38. Chang FY, Peacock JE Jr, Musher DM, Triplett P, MacDonald BB, Mylotte JM et al (2003) *Staphylococcus aureus* bacteremia: recurrence and the impact of antibiotic treatment in a prospective multicenter study. Medicine (Baltimore) 82(5):333–339. https://doi.org/10.1097/01.md.0000091184.93122.09
39. Kreisel K, Boyd K, Langenberg P, Roghmann MC (2006) Risk factors for recurrence in patients with *Staphylococcus aureus* infections complicated by bacteremia. Diagn Microbiol Infect Dis 55(3):179–184. https://doi.org/10.1016/j.diagmicrobio.2006.01.021
40. Tanida I (2011) Autophagy basics. Microbiol Immunol 55(1):1–11. https://doi.org/10.1111/j.1348-0421.2010.00271.x
41. Tanida I (2011) Autophagosome formation and molecular mechanism of autophagy. Antioxid Redox Signal 14(11):2201–2214. https://doi.org/10.1089/ars.2010.3482
42. Wang CW, Klionsky DJ (2003) The molecular mechanism of autophagy. Mol Med 9(3–4):65–76. https://doi.org/10.1007/BF03402040
43. Glick D, Barth S, Macleod KF (2010) Autophagy: cellular and molecular mechanisms. J Pathol 221(1):3–12. https://doi.org/10.1002/path.2697
44. Cadwell K (2016) Crosstalk between autophagy and inflammatory signalling pathways: balancing defence and homeostasis. Nat Rev Immunol 16(11):661–675. https://doi.org/10.1038/nri.2016.100

45. Yang Z, Klionsky DJ (2009) An overview of the molecular mechanism of autophagy. Curr Top Microbiol Immunol 335:1–32. https://doi.org/10.1007/978-3-642-00302-8_1
46. Johansen T, Lamark T (2011) Selective autophagy mediated by autophagic adapter proteins. Autophagy 7(3):279–296. https://doi.org/10.4161/auto.7.3.14487
47. Collins CA, Brown EJ (2010) Cytosol as battleground: ubiquitin as a weapon for both host and pathogen. Trends Cell Biol 20(4):205–213. https://doi.org/10.1016/j.tcb.2010.01.002
48. Fujita N, Yoshimori T (2011) Ubiquitination-mediated autophagy against invading bacteria. Curr Opin Cell Biol 23(4):492–497. https://doi.org/10.1016/j.ceb.2011.03.003
49. Cemma M, Brumell JH (2012) Interactions of pathogenic bacteria with autophagy systems. Curr Biol 22(13):R540–R545. https://doi.org/10.1016/j.cub.2012.06.001
50. Gutierrez MG, Master SS, Singh SB, Taylor GA, Colombo MI, Deretic V (2004) Autophagy is a defense mechanism inhibiting BCG and *Mycobacterium tuberculosis* survival in infected macrophages. Cell 119(6):753–766. https://doi.org/10.1016/j.cell.2004.11.038
51. Kageyama S, Omori H, Saitoh T, Sone T, Guan JL, Akira S et al (2011) The LC3 recruitment mechanism is separate from Atg9L1-dependent membrane formation in the autophagic response against *Salmonella*. Mol Biol Cell 22(13):2290–2300. https://doi.org/10.1091/mbc.E10-11-0893
52. Nakagawa I, Amano A, Mizushima N, Yamamoto A, Yamaguchi H, Kamimoto T et al (2004) Autophagy defends cells against invading group a *Streptococcus*. Science 306(5698):1037–1040. https://doi.org/10.1126/science.1103966
53. Ogawa M, Yoshimori T, Suzuki T, Sagara H, Mizushima N, Sasakawa C (2005) Escape of intracellular *Shigella* from autophagy. Science 307(5710):727–731. https://doi.org/10.1126/science.1106036
54. Mostowy S, Bonazzi M, Hamon MA, Tham TN, Mallet A, Lelek M et al (2010) Entrapment of intracytosolic bacteria by septin cage-like structures. Cell Host Microbe 8(5):433–444. https://doi.org/10.1016/j.chom.2010.10.009
55. Yoshikawa Y, Ogawa M, Hain T, Yoshida M, Fukumatsu M, Kim M et al (2009) *Listeria monocytogenes* ActA-mediated escape from autophagic recognition. Nat Cell Biol 11(10):1233–1240. https://doi.org/10.1038/ncb1967
56. Dortet L, Mostowy S, Samba-Louaka A, Gouin E, Nahori MA, Wiemer EA et al (2011) Recruitment of the major vault protein by InlK: a *Listeria monocytogenes* strategy to avoid autophagy. PLoS Pathog 7(8):e1002168. https://doi.org/10.1371/journal.ppat.1002168
57. Gong L, Cullinane M, Treerat P, Ramm G, Prescott M, Adler B et al (2011) The *Burkholderia pseudomallei* type III secretion system and BopA are required for evasion of LC3-associated phagocytosis. PLoS One 6(3):e17852. https://doi.org/10.1371/journal.pone.0017852
58. Romano PS, Gutierrez MG, Berón W, Rabinovitch M, Colombo MI (2007) The autophagic pathway is actively modulated by phase II *Coxiella burnetii* to efficiently replicate in the host cell. Cell Microbiol 9(4):891–909. https://doi.org/10.1111/j.1462-5822.2006.00838.x
59. Amer AO, Swanson MS (2005) Autophagy is an immediate macrophage response to *Legionella pneumophila*. Cell Microbiol 7(6):765–778. https://doi.org/10.1111/j.1462-5822.2005.00509.x
60. Niu H, Yamaguchi M, Rikihisa Y (2008) Subversion of cellular autophagy by *Anaplasma phagocytophilum*. Cell Microbiol 10(3):593–605. https://doi.org/10.1111/j.1462-5822.2007.01068.x
61. Moreau K, Lacas-Gervais S, Fujita N, Sebbane F, Yoshimori T, Simonet M et al (2010) Autophagosomes can support *Yersinia pseudotuberculosis* replication in macrophages. Cell Microbiol 12(8):1108–1123. https://doi.org/10.1111/j.1462-5822.2010.01456.x
62. Starr T, Child R, Wehrly TD, Hansen B, Hwang S, López-Otin C et al (2012) Selective subversion of autophagy complexes facilitates completion of the *Brucella* intracellular cycle. Cell Host Microbe 11(1):33–45. https://doi.org/10.1016/j.chom.2011.12.002
63. Mauthe M, Yu W, Krut O, Krönke M, Götz F, Robenek H et al (2012) WIPI-1 positive autophagosome-like vesicles entrap pathogenic *Staphylococcus aureus* for lysosomal degradation. Int J Cell Biol 2012:179207. https://doi.org/10.1155/2012/179207
64. Mestre MB, Fader CM, Sola C, Colombo MI (2010) Alpha-hemolysin is required for the activation of the autophagic pathway in *Staphylococcus aureus*-infected cells. Autophagy 6(1):110–125. https://doi.org/10.4161/auto.6.1.10698

65. Bravo-Santano N, Ellis JK, Mateos LM, Calle Y, Keun HC, Behrends V et al (2018) Intracellular *Staphylococcus aureus* modulates host central carbon metabolism to activate autophagy. mSphere 3(4):e00374–e00318. https://doi.org/10.1128/mSphere.00374-18
66. Neumann Y, Bruns SA, Rohde M, Prajsnar TK, Foster SJ, Schmitz I (2016) Intracellular *Staphylococcus aureus* eludes selective autophagy by activating a host cell kinase. Autophagy 12(11):2069–2084. https://doi.org/10.1080/15548627.2016.1226732
67. Asano K, Asano Y, Ono HK, Nakane A (2014) Suppression of starvation-induced autophagy by recombinant toxic shock syndrome toxin-1 in epithelial cells. PLoS One 9(11):e113018. https://doi.org/10.1371/journal.pone.0113018
68. Schnaith A, Kashkar H, Leggio SA, Addicks K, Krönke M, Krut O (2007) *Staphylococcus aureus* subvert autophagy for induction of caspase-independent host cell death. J Biol Chem 282(4):2695–2706. https://doi.org/10.1074/jbc.M609784200
69. Novick RP (2003) Autoinduction and signal transduction in the regulation of staphylococcal virulence. Mol Microbiol 48(6):1429–1449. https://doi.org/10.1046/j.1365-2958.2003.03526.x
70. Ji G, Beavis R, Novick RP (1997) Bacterial interference caused by autoinducing peptide variants. Science 276(5321):2027–2030. https://doi.org/10.1126/science.276.5321.2027
71. Haslinger-Löffler B, Kahl BC, Grundmeier M, Strangfeld K, Wagner B, Fischer U et al (2005) Multiple virulence factors are required for *Staphylococcus aureus*-induced apoptosis in endothelial cells. Cell Microbiol 7(8):1087–1097. https://doi.org/10.1111/j.1462-5822.2005.00533.x
72. Jarry TM, Memmi G, Cheung AL (2008) The expression of alpha-haemolysin is required for *Staphylococcus aureus* phagosomal escape after internalization in CFT-1 cells. Cell Microbiol 10(9):1801–1814. https://doi.org/10.1111/j.1462-5822.2008.01166.x
73. O'Keeffe KM, Wilk MM, Leech JM, Murphy AG, Laabei M, Monk IR et al (2015) Manipulation of autophagy in phagocytes facilitates *Staphylococcus aureus* bloodstream infection. Infect Immun 83(9):3445–3457. https://doi.org/10.1128/IAI.00358-15
74. Dinges MM, Orwin PM, Schlievert PM (2000) Exotoxins of *Staphylococcus aureus*. Clin Microbiol Rev 13(1):16–34. https://doi.org/10.1128/cmr.13.1.16-34.2000
75. Bantel H, Sinha B, Domschke W, Peters G, Schulze-Osthoff K, Jänicke RU (2001) Alpha-toxin is a mediator of *Staphylococcus aureus*–induced cell death and activates caspases via the intrinsic death pathway independently of death receptor signaling. J Cell Biol 155(4):637–648. https://doi.org/10.1083/jcb.200105081
76. Kushnaryov VM, Bergdoll MS, MacDonald HS, Vellinga J, Reiser R (1984) Study of staphylococcal toxic shock syndrome toxin in human epithelial cell culture. J Infect Dis 150(4):535–545. https://doi.org/10.1093/infdis/150.4.535
77. Hu DL, Omoe K, Sasaki S, Sashinami H, Sakuraba H, Yokomizo Y, Shinagawa K, Nakane A (2003) Vaccination with nontoxic mutant toxic shock syndrome toxin 1 protects against *Staphylococcus aureus* infection. J Infect Dis 188(5):743–752. https://doi.org/10.1086/377308

Index

A. Nakane, K. Asano (eds.), *Staphylococcus aureus*,
https://doi.org/10.1007/978-981-99-9428-1

MIX
Papier aus verantwortungsvollen Quellen
Paper from responsible sources
FSC® C105338

If you have any concerns about our products,
you can contact us on
ProductSafety@springernature.com

In case Publisher is established outside the EU,
the EU authorized representative is:
Springer Nature Customer Service Center GmbH
Europaplatz 3, 69115 Heidelberg, Germany

Printed by Libri Plureos GmbH
in Hamburg, Germany